THE SCHUR COMPLEMENT AND ITS APPLICATIONS

Numerical Methods and Algorithms

VOLUME 4

THE SCHUR COMPLEMENT AND ITS APPLICATIONS

Edited by

FUZHEN ZHANG
Nova Southeastern University, Fort Lauderdale, U.S.A.
Shenyang Normal University, Shenyang, China

Springer

Library of Congress Cataloging-in-Publication Data

A C.I.P. record for this book is available from the Library of Congress.

ISBN 0-387-24271-6 e-ISBN 0-387-24273-2 Printed on acid-free paper.

Printed in the United States of America.

9 8 7 6 5 4 3 2 1 SPIN 11161356

springeronline.com

To our families, friends, and the matrix community

Issai Schur (1875-1941)

This portrait of Issai Schur was apparently made by the "Atelieir Hanni Schwarz, N. W. Dorotheenstraße 73" in Berlin, *c.* 1917, and appears in *Ausgewählte Arbeiten zu den Ursprüngen der Schur-Analysis: Gewidmet dem großen Mathematiker Issai Schur (1875–1941)* edited by Bernd Fritzsche & Bernd Kirstein, pub. B. G. Teubner Verlagsgesellschaft, Stuttgart, 1991.

Emilie Virginia Haynsworth (1916-1985)

This portrait of Emilie Virginia Haynsworth is on the Auburn University Web site www.auburn.edu/~fitzpjd/ben/images/Emilie.gif and in the book *The Education of a Mathematician* by Philip J. Davis, pub. A K Peters, Natick, Mass., 2000.

Contents

Preface

What's in a name? To paraphrase Shakespeare's Juliet, that which Emilie Haynsworth called the *Schur complement*, by any other name would be just as beautiful. Nevertheless, her 1968 naming decision in honor of Issai Schur (1875–1941) has gained lasting acceptance by the mathematical community. The Schur complement plays an important role in matrix analysis, statistics, numerical analysis, and many other areas of mathematics and its applications.

Our goal is to expose the Schur complement as a rich and basic tool in mathematical research and applications and to discuss many significant results that illustrate its power and fertility. Although our book was originally conceived as a research reference, it will also be useful for graduate and upper division undergraduate courses in mathematics, applied mathematics, and statistics. The contributing authors have developed an exposition that makes the material accessible to readers with a sound foundation in linear algebra.

The eight chapters of the book (Chapters 0–7) cover themes and variations on the Schur complement, including its historical development, basic properties, eigenvalue and singular value inequalities, matrix inequalities in both finite and infinite dimensional settings, closure properties, and applications in statistics, probability, and numerical analysis. The chapters need not be read in the order presented, and the reader should feel at leisure to browse freely through topics of interest.

It was a great pleasure for me, as editor, to work with a wonderful group of distinguished mathematicians who agreed to become chapter contributors: T. Ando (Hokkaido University, Japan), C. Brezinski (Université des Sciences et Technologies de Lille, France), R. A. Horn (University of Utah, Salt Lake City, USA), C. R. Johnson (College of William and Mary, Williamsburg, USA), J.-Z. Liu (Xiangtang University, China), S. Puntanen (University of Tampere, Finland), R. L. Smith (University of Tennessee, Chattanooga, USA), and G. P. H. Styan (McGill University, Canada).

I am particularly thankful to George Styan for his great enthusiasm in compiling the master bibliography for the book. We would also like to acknowledge the help we received from Gülhan Alpargu, Masoud Asgharian, M. I. Beg, Adi Ben-Israel, Abraham Berman, Torsten Bernhardt, Eva Brune, John S. Chipman, Ka Lok Chu, R. William Farebrother, Bernd Fritsche, Daniel Hershkowitz, Jarkko Isotalo, Bernd Kirstein, André Klein, Jarmo Niemelä, Geva Maimon Reid, Timo Mäkeläinen, Lindsey E. McQuade, Aliza K. Miller, Ingram Olkin, Emily E. Rochette, Vera Rosta,

Eugénie Roudaia, Burkhard Schaffrin, Hans Schneider, Shayle R. Searle, Daniel N. Selan, Samara F. Strauber, Evelyn M. Styan, J. C. Szamosi, Garry J. Tee, Götz Trenkler, Frank Uhlig, and Jürgen Weiß. We are also very grateful to the librarians in the McGill University Interlibrary Loan and Document Delivery Department for their help in obtaining the source materials for many of our references. The research of George P. H. Styan was supported in part by the Natural Sciences and Engineering Research Council of Canada.

Finally, I thank my wife Cheng, my children Sunny, Andrew, and Alan, and my mother-in-law Yun-Jiao for their understanding, support, and love.

Fuzhen Zhang
September 1, 2004
Fort Lauderdale, Florida

Chapter 0

Historical Introduction: Issai Schur and the Early Development of the Schur Complement

0.0 Introduction and mise-en-scène

In this introductory chapter we comment on the history of the Schur complement from 1812 through 1968 when it was so named and given a notation. As Chandler & Magnus [113, p. 192] point out, "The coining of new technical terms is an absolute necessity for the evolution of mathematics." And so we begin in 1968 when the mathematician Emilie Virginia Haynsworth (1916–1985) introduced a name and a notation for the Schur complement of a square nonsingular (or invertible) submatrix in a partitioned (two-way block) matrix [210, 211].

We then go back fifty-one years and examine the seminal lemma by the famous mathematician Issai Schur (1875–1941) published in 1917 [404, pp. 215–216], in which the *Schur determinant formula* (0.3.2) was introduced. We also comment on earlier implicit manifestations of the Schur complement due to Pierre Simon Laplace, later Marquis de Laplace (1749–1827), first published in 1812, and to James Joseph Sylvester (1814–1897), first published in 1851.

Following some biographical remarks about Issai Schur, we present the *Banachiewicz inversion formula* for the inverse of a nonsingular partitioned matrix which was introduced in 1937 [29] by the astronomer Tadeusz Banachiewicz (1882–1954). We note, however, that closely related results were obtained earlier in 1933 by Ralf Lohan [290], following results in the book [66] published in 1923 by the geodesist Hans Boltz (1883–1947).

We continue with comments on material in the book *Elementary Matrices and Some Applications to Dynamics and Differential Equations* [171], a classic by the three aeronautical engineers Robert Alexander Frazer (1891–1959), William Jolly Duncan (1894–1960), and Arthur Roderick Collar (1908–1986), first published in 1938, and in the book *Determinants and Matrices* [4] by the mathematician and statistician Alexander Craig Aitken (1895–1967), another classic, and first published in 1939.

We introduce the *Duncan inversion formula* (0.8.3) for the sum of two matrices, and the very useful *Aitken block-diagonalization formula* (0.9.1), from which easily follow the *Guttman rank additivity formula* (0.9.2) due to the social scientist Louis Guttman (1916–1987) and the *Haynsworth inertia additivity formula* (0.10.1) due to Emilie Haynsworth.

We conclude this chapter with some biographical remarks on Emilie Haynsworth and note that her thesis adviser was Alfred Theodor Brauer (1894–1985), who completed his Ph.D. degree under Schur in 1928.

This chapter builds on the extensive surveys of the Schur complement published (in English) by Brezinski [73], Carlson [105], Cottle [128, 129], Ouellette [345], and Styan [432], and (in Turkish) by Alpargu [8]. In addition, the role of the Schur complement in matrix inversion has been surveyed by Zielke [472] and by Henderson & Searle [219], with special emphasis on inverting the sum of two matrices, and by Hager [200], with emphasis on the inverse of a matrix after a small-rank perturbation.

0.1 The Schur complement: the name and the notation

The term *Schur complement* for the matrix

$$S - RP^{-1}Q, \tag{0.1.1}$$

where the nonsingular matrix P is the leading submatrix of the complex partitioned matrix

$$M = \begin{pmatrix} P & Q \\ R & S \end{pmatrix}, \tag{0.1.2}$$

was introduced in 1968 in two papers [210, 211] by Emilie Haynsworth published, respectively, in the *Basel Mathematical Notes* and in *Linear Algebra and its Applications.*

The notation

$$(M/P) = S - RP^{-1}Q \tag{0.1.3}$$

for the Schur complement of P in $M = \left(\begin{smallmatrix} P & Q \\ R & S \end{smallmatrix}\right)$ was apparently first used in 1968 by Haynsworth, in the *Basel Mathematical Notes* [210] but not in *Linear Algebra and its Applications* [211], where its first appearance seems to be in the 1970 paper by Haynsworth [212]. This notation does appear, however, in the 1969 paper [131] by Haynsworth with Douglas E. Crabtree in the *Proceedings of the American Mathematical Society* and is still in use today, see e.g., the papers by Brezinski & Redivo Zaglia [88] and N'Guessan [334] both published in 2003; the notation (0.1.3) is also used in the six surveys [8, 73, 128, 129, 345, 432].

The notation $(M|P)$, with a vertical line separator rather than a slash, was introduced in 1971 by Markham [295] and is used in the book by Prasolov [354, p. 17]; see also [296, 332, 343] published in 1972–1980. The notation $M|P$ without the parentheses was used in 1976 by Markham [297].

In this book we will use the original notation (0.1.3) but without the parentheses,

$$M/P = S - RP^{-1}Q, \tag{0.1.4}$$

for the Schur complement of the nonsingular matrix P in the partitioned matrix $M = \begin{pmatrix} P & Q \\ R & S \end{pmatrix}$. This notation (0.1.4) without the parentheses was introduced in 1974 by Carlson, Haynsworth & Markham [106] and seems to be very popular today, see, e.g., the recent books by Ben-Israel & Greville [45, p. 30], Berman & Shaked-Monderer [48, p. 24], and by C. R. Rao & M. B. Rao [378, p. 139], and the recent papers [160, 287, 471].

0.2 Some implicit manifestations in the 1800s

According to David Carlson in his 1986 survey article [105] entitled "What are Schur complements, anyway?":

> The idea of the Schur complement matrix goes back to the 1851 paper [436] by James Joseph Sylvester. It is well known that the entry a_{ij} of [the Schur complement matrix] A, $i = 1, \ldots, m-k$, $j = 1, \ldots, n-k$, is the minor of [the partitioned matrix] M determined by rows $1, \ldots, k, k+i$ and columns $1, \ldots, k, k+j$, a property which was used by Sylvester as his definition. For a discussion of this and other appearances of the Schur complement matrix in the 1800s, see the paper by Brualdi & Schneider [99].

Farebrother [162, pp. 116–117] discusses work by Pierre Simon Laplace, later Marquis de Laplace, and observes that Laplace [273, livre II, §21 (1812); *Œuvres*, vol. 7, p. 334 (1886)] obtained a ratio that we now recognize as the ratio of two successive leading principal minors of a symmetric positive definite matrix. Then the ratio $\det(M)/\det(M_1)$ is the determinant of what we now know as the Schur complement of M_1 in M, see the

Schur determinant formula (0.3.2) below. Laplace [273, §3 (1816); *Œuvres*, vol. 7, pp. 512–513 (1886)] evaluates the ratio $\det(M)/\det(M_1)$ with $n = 3$.

0.3 The lemma and the Schur determinant formula

The adjectival noun "Schur" in "Schur complement" was chosen by Haynsworth because of the lemma (Hilfssatz) in the paper [404] by Issai Schur published in 1917 in the *Journal für die reine und angewandte Mathematik*, founded in Berlin by August Leopold Crelle (1780–1855) in 1826 and edited by him until his death. Often called Crelle's *Journal* this is apparently the oldest mathematics periodical still in existence today [103]; Frei [174] summarizes the long history of the *Journal* in volume 500 (1998).

The picture of Issai Schur facing the opening page of this chapter appeared in the 1991 book *Ausgewählte Arbeiten zu den Ursprüngen der Schur-Analysis: Gewidmet dem großen Mathematiker Issai Schur (1875–1941)* [177, p. 20]; on the facing page [177, p. 21] is a copy of the title page of volume 147 (1917) of the *Journal für die reine und angewandte Mathematik* in which the Schur determinant lemma [404] was published.

This paper [404] is concerned with conditions for power series to be bounded inside the unit circle; indeed a polynomial with roots within the unit disk in the complex plane is now known as a *Schur polynomial*, see e.g., Lakshmikantham & Trigiante [271, p. 49].

The lemma appears in [404, pp. 215–216], see also [71, pp. 148–149], [177, pp. 33–34]. Our English translation, see also [183, pp. 33–34], follows. The Schur complement $S - RP^{-1}Q$ is used in the proof but the lemma holds even if the square matrix P is singular. We refer to this lemma as the *Schur determinant lemma.*

LEMMA. *Let P, Q, R, S denote four $n \times n$ matrices and suppose that P and R commute. Then the determinant* $\det(M)$ *of the $2n \times 2n$ matrix*

$$M = \begin{pmatrix} P & Q \\ R & S \end{pmatrix}$$

is equal to the determinant of the matrix $PS - RQ$.

Proof. We assume that the determinant of P is not zero. Then, with I denoting the $n \times n$ identity matrix,

$$\begin{pmatrix} P^{-1} & 0 \\ -RP^{-1} & I \end{pmatrix} \begin{pmatrix} P & Q \\ R & S \end{pmatrix} = \begin{pmatrix} I & P^{-1}Q \\ 0 & S - RP^{-1}Q \end{pmatrix}$$

Taking determinants yields $\det(P^{-1}) \cdot \det(M) = \det(S - RP^{-1}Q)$ and so

$$\det(M) = \det(P) \cdot \det(S - RP^{-1}Q) \tag{0.3.1}$$
$$= \det(PS - PRP^{-1}Q) = \det(PS - RQ).$$

If, however, $\det(P) = 0$, we replace matrix M with the matrix

$$M_1 = \begin{pmatrix} P + xI & Q \\ R & S \end{pmatrix}.$$

The matrices R and $P + xI$ commute. For the absolute value $|x|$ sufficiently small (but not zero), the determinant of $P + xI$ is not equal to 0 and so $\det(M_1) = \det\big((P + xI)S - RQ\big)$. Letting x converge to 0 yields the desired result. ∎

We may write (0.3.1) as the *Schur determinant formula*

$$\det(M) = \det(P) \cdot \det(M/P) = \det(P) \cdot \det(S - RP^{-1}Q) \tag{0.3.2}$$

and so determinant is multiplicative on the Schur complement, which suggests the notation M/P for the Schur complement of P in M.

Schur [404, pp. 215–216] used this lemma to show that the complex $2k \times 2k$ determinant

$$\delta_k = \det \begin{pmatrix} P_k & Q_k \\ Q_k^* & P_k^* \end{pmatrix} = \det(P_k P_k^* - Q_k^* Q_k), \quad k = 1, \ldots, n, \tag{0.3.3}$$

where

$$P_k = \begin{pmatrix} a_0 & 0 & \ldots & 0 \\ a_1 & a_0 & \ldots & 0 \\ \ldots & \ldots & \ldots & \ldots \\ a_{k-1} & a_{k-2} & \ldots & a_0 \end{pmatrix}, \quad Q_k = \begin{pmatrix} a_n & a_{n-1} & \ldots & a_{n-k+1} \\ 0 & a_n & \ldots & a_{n-k+2} \\ \ldots & \ldots & \ldots & \ldots \\ 0 & 0 & \ldots & a_n \end{pmatrix},$$

and so $P_k Q_k^* = Q_k^* P_k$, $k = 1, \ldots, n$. What are now known as *Schur conditions*,

$$\delta_k > 0, \ k = 1, \ldots, n,$$

are necessary and sufficient for the roots of the polynomial

$$f(x) = a_0 x^n + a_1 x^{n-1} + \cdots + a_{n-1} x + a_n = 0 \tag{0.3.4}$$

to lie within the unit circle of the complex plane, see e.g., Chipman [116, p. 371 (1950)].

Schur's paper [404] and its sequel [405] were selected by Fritzsche & Kirstein in the *Ausgewählte Arbeiten* [177] as two of the six influential papers considered as "fundamental for Schur analysis"; the book [177] is dedicated to the "great mathematician Issai Schur". The four other papers in [177] are by Gustav Herglotz (1881–1953), Rolf Nevanlinna (1895–1980), Georg Pick (1859–1942), and Hermann Weyl (1885–1955).

0.4 Issai Schur (1875–1941)

Issai Schur was born on 10 January 1875, the son of Golde Schur (née Landau) and the *Kaufmann* Moses Schur, according to Schur's *Biographische Mitteilungen* [406]. In a recent biography of Issai Schur, Vogt [449] notes that Schur used the first name "Schaia" rather than "Issai" until his mid-20s and that his father was a *Großkaufmann.*

Writing in German in [406], Schur gives his place of birth as Mohilew am Dnjepr (Russland)—in English: Mogilev on the Dnieper, Russia. Founded in the 13th century, Mogilev changed hands frequently among Lithuania, Poland, Sweden, and Russia, and was finally annexed to Russia in 1772 in the first partition of Poland [31, p. 155]. By the late 19th century, almost half of the population of Mogilev was Jewish [262]. About 200 km east of Minsk, Mogilev is in the eastern part of the country now known as Belarus (Belorussia, White Russia) and called Mahilyow in Belarusian [306].

In 1888 when he was 13, Schaia Schur, as he was then known [449], went to live with his older sister and brother-in-law in Libau (Kurland), about 640 km northwest of Mogilev. Also founded in the 13th century, Libau (Liepāja in Latvian or Lettish) is on the Baltic coast of what is now Latvia in the region of Courland (Kurland in German, Kurzeme in Latvian), which from 1562–1795 was a semi-independent duchy linked to Poland but with a prevailing German influence [60, 423]. Indeed the German way of life was dominant in Courland in 1888, with mostly German (not Yiddish) being the spoken language of the Jewish community until 1939 [39]. In the late 19th century there were many synagogues in Libau, the Great Synagogue in Babylonian style with three cupolas being a landmark [60].

Schur attended the German-language Nicolai Gymnasium in Libau from 1888–1894 and received the highest mark on his final examination and a gold medal [449]. It was here that he became fluent in German (we believe that his first language was probably Yiddish). In Germany the Gymnasium is a "state-maintained secondary school that prepares pupils for higher academic education" [158]. We do not know why the adjectival noun Nicolai is used here but in Leipzig the *Nikolaischule* was so named because of the adjacent *Nikolaikirche*, which was founded *c.* 1165 and named after Saint Nicholas of Bari [207, 224], the saint who is widely associated with Christmas and after whom Santa Claus in named [248, ch. 7].

In October 1894, Schur enrolled in the University of Berlin, studying mathematics and physics; on 27 November 1901 he passed his doctoral examination *summa cum laude* with the thesis entitled "Über eine Klasse von Matrizen, die sich einer gegebenen Matrix zuordnen lassen" [402]: his thesis adviser was Ferdinand Georg Frobenius (1849–1917). According to Vogt [449], in this thesis Schur used his first name "Issai" for the first time.

Feeling that he "had no chance whatsoever of sustaining himself as a mathematician in czarist Russia" [113, p. 197] and since he now wrote and spoke German so perfectly that one would guess that German was his native language, Schur stayed on in Germany. According to [406], he was *Privatdozent* at the University in Berlin from 1903 till 1913 and *außerordentlicher Professor* (associate professor) at the University of Bonn from 21 April 1913 till 1 April 1916 [425, p. 8], as successor to Felix Hausdorff (1868–1942); see also [276, 425]. In 1916 Schur returned to Berlin where in 1919 he was appointed full professor; in 1922 he was elected a member of the Prussian Academy of Sciences to fill the vacancy caused by the death of Frobenius in 1917. We believe that our portrait of Issai Schur in the front of this book was made in Berlin, *c.* 1917; for other photographs see [362].

Schur lived in Berlin as a highly respected member of the academic community and was a quiet unassuming scholar who took no part in the fierce struggles that preceded the downfall of the Weimar Republic. "A leading mathematician and an outstanding and highly successful teacher, [Schur] occupied for 16 years the very prestigious chair at the University of Berlin" [113, p. 197]. Until 1933 Schur's algebraic school at the University of Berlin was, without any doubt, the single most coherent and influential group of mathematicians in Berlin and among the most important in all of Germany. With Schur as its charismatic leader, the school centered around his research on group representations, which was extended by his students in various directions (soluble groups, combinatorics, matrix theory) [100, p. 25]. "Schur made fundamental contributions to algebra and group theory which, according to Hermann Weyl, were comparable in scope and depth to those of Emmy Amalie Noether (1882–1935)" [353, p. 178].

When Schur's lectures were canceled (in 1933) there was an outcry among the students and professors, for he was respected and very well liked [100, p. 27]. Thanks to his colleague Erhard Schmidt (1876–1959), Schur was able to continue his lectures till the end of September 1935 [353, p. 178], Schur being the last Jewish professor to lose his job at the Universität Berlin at that time [425, p. 8]. Schur's "lectures on number theory, algebra, group theory and the theory of invariants attracted large audiences. On 10 January 1935 some of the senior postgraduates congratulated [Schur] in the lecture theatre on his sixtieth birthday. Replying in mathematical language, Schur hoped that the good relationship between himself and his student audience would remain invariant under all the transformations to come" [353, p. 179].

Indeed Schur was a superb lecturer. His lectures were meticulously prepared and were exceedingly popular. Walter Ledermann (b. 1911) remembers attending Schur's algebra course which was held in a lecture theatre filled with about 400 students [276]: "Sometimes, when I had to be content with a seat at the back of the lecture theatre, I used a pair of opera glasses to get a glimpse of the speaker." In 1938 Schur was pressed to resign from the Prussian Academy of Sciences and on 7 April 1938 he resigned "voluntarily" from the Commissions of the Academy. Half a year later, he had to resign from the Academy altogether [100, p. 27].

The names of the 22 persons who completed their dissertations from 1917–1936 under Schur, together with the date in which the Ph.D. degree was awarded and the dissertation title, are listed in the *Issai Schur Gesammelte Abhandlungen* [71, *Band III*, pp. 479–480]; see also [100, p. 23], [249, p. xviii]. One of these 22 persons is Alfred Theodor Brauer (1894–1985), who completed his Ph.D. dissertation under Schur on 19 December 1928 and with Hans Rohrbach edited the *Issai Schur Gesammelte Abhandlungen* [71]. Alfred Brauer was a faculty member in the Dept. of Mathematics at The University of North Carolina at Chapel Hill for 24 years and directed 21 Ph.D. dissertations, including that of Emilie Haynsworth, who in 1968 introduced the term "Schur complement" (see §0.1 above).

A remark by Alfred Brauer [70, p. xiii], see also [100, p. 28], sheds light on Schur's situation after he finally left Germany in 1939: "When Schur could not sleep at night, he read the *Jahrbuch über die Fortschritte der Mathematik* (now *Zentralblatt MATH*). When he came to Tel Aviv (then British Mandate of Palestine, now Israel) and for financial reasons offered his library for sale to the Institute for Advanced Study in Princeton, he finally excluded the *Jahrbuch* in a telegram only weeks before his death."

Issai Schur died of a heart attack in Tel Aviv on his 66th birthday, 10 January 1941. Schur is buried in Tel Aviv in the Old Cemetery on Trumpeldor Street, which was "reserved for the Founders' families and persons of special note. Sadly this was the only tribute the struggling Jewish Home could bestow upon Schur" [249, p. clxxxvi]; see also [331, 362].

Schur was survived by his wife, medical doctor Regina (née Frumkin, 1881–1965), their son Georg (born 1907 and named after Frobenius), and daughter Hilde (born 1911, later Hilda Abelin-Schur), who in "A story about father" [1] in *Studies in Memory of Issai Schur* [249] writes

> One day when our family was having tea with some friends, [my father] was enthusiastically talking about his work. He said: "I feel like I am somehow moving through outer space. A particular idea leads me to a nearby star on which I decide to land. Upon my arrival I realize that somebody already lives there. Am I disappointed? Of course not. The inhabitant and I are cordially welcoming each other, and we are happy about our common discovery." This was typical of my father; he was never envious.

0.5 Schur's contributions in mathematics

Many of Issai Schur's contributions to linear algebra and matrix theory are reviewed in [152] by Dym & Katsnelson in *Studies in Memory of Issai Schur* [249]. Among the topics covered in [249] are estimates for matrix and integral operators and bilinear forms, the Schur (or Hadamard) product of matrices, Schur multipliers, Schur convexity, inequalities between eigenvalues and singular values of a linear operator, and triangular representations of matrices. Schur is considered as a "pioneer in representation theory" [136], and Haubrich [208] surveys Schur's contributions in linear substitutions, locations of roots of algebraic equations, pure group theory, integral equations, and number theory.

Soifer [425] discusses the origins of certain combinatorial problems nowadays seen as part of Ramsey theory, with special reference to a lemma, now known as Schur's theorem, embedded in a paper on number theory. Included in *Studies in Memory of Issai Schur* [249] are over 60 pages of biographical and related material (including letters and documents in German, with translations in English) on Issai Schur, as well as reminiscences by his former students Bernhard Hermann Neumann (1909–2002) and Walter Ledermann, and by his daughter Hilda Abelin-Schur [1] and his granddaughter Susan Abelin.

In the edited book [183] entitled *I. Schur Methods in Operator Theory and Signal Processing*, Thomas Kailath [252] briefly reviews some of the "many significant and technologically highly relevant applications in linear algebra and operator theory" arising from Schur's seminal papers [404, 405]. For some comments by Paul Erdős (1913–1996) on the occasion of the 120th anniversary of Schur's birthday in 1995, see [159].

0.6 Publication under J. Schur

Issai Schur published under "I. Schur" and under "J. Schur". As is pointed out by Ledermann in his biographical article [276] on Schur, this has caused some confusion: "For example I have a scholarly work on analysis which lists amongst the authors cited both J. Schur and I. Schur, and an author on number theory attributes one of the key results to I. J. Schur."

We have identified 81 publications by Issai Schur which were published before he died in 1941; several further publications by Schur were, however, published posthumously including the book [408] published in 1968. On the title page of the (original versions of the) articles [404, 405], the author is given as "J. Schur"; indeed for all but one of the other 11 papers by Issai Schur that we found published in the *Journal für die reine und angewandte Mathematik* the author is given as "J. Schur". For the lecture notes [407] published in Zürich in 1936, the author is given as J. Schur on the title page and so cited in the preface. For all other publications by Issai Schur that we have found, however, the author is given as "I. Schur", and posthumously as "Issai Schur"; moreover Schur edited the *Mathematische Zeitschrift* from 1917–1938 and he is listed there on the journal title pages as I. Schur.

The confusion here between "I" and "J" probably stems from there being two major styles of writing German: *Fraktur script*, also known as *black letter script* or *Gothic script*, in use since the ninth century and prevailing until 1941 [130, p. 26], and *Roman* or *Latin*, which is common today [237]. According to Mashey [302, p. 28], "it is a defect of most styles of German type that the same character $\mathfrak{J}$ is used for the capitals I (i) and J (j)"; when followed by a vowel it is the consonant "J" and when followed by a consonant, it is "I", see also [46, pp. 4–5], [220, pp. 166–167], [444, p. 397].

The way Schur wrote and signed his name, as in his *Biographische Mitteilungen* [406], his first name could easily be interpreted as "Jssai" rather than "Issai"; see also the signature at the bottom of the photograph in the front of this book and at the bottom of the photograph in the *Issai Schur Gesammelte Abhandlungen* [71, *Band I*, facing page v (1973)]. The official letter, reprinted in Soifer [425, p. 9], dated 28 September 1935 and signed by Kunisch [270], relieving Issai Schur of his duties at the University of Berlin, is addressed to "Jssai Schur"; the second paragraph starts with "Jch übersende Jhnen ... " which would now be written as "Ich übersende Ihnen ... "; see also [249, p. lxxiv (2003)]. Included in the article by Ledermann & Neumann [277, (2003)] are copies of many documents associated with Issai Schur. These are presented in chronological order, with a transcription first, followed by a translation. It is noted there [277, p. lx] that "Schur used Roman script" but "sometimes, particularly in typed official letters after 1933, initial letters I are rendered as J."

0.7 Boltz 1923, Lohan 1933, Aitken 1937, and the Banachiewicz inversion formula 1937

In 1937 the astronomer and mathematician Tadeusz Banachiewicz (1882–1954) established in [29, p. 50] the Schur determinant formula (0.3.2) with P nonsingular,

$$\det(M) = \det\begin{pmatrix} P & Q \\ R & S \end{pmatrix} = \det(P)\cdot\det(S - RP^{-1}Q)\,. \tag{0.7.1}$$

Also in 1937, the mathematician and statistician Alexander Craig Aitken (1895–1967) gave [3, p. 172] "a uniform working process for computing" the triple matrix product $RP^{-1}Q$, and noted explicitly that when the matrix

R is a row vector $-r'$, say, and Q is a column vector q, say, then

$$\det \begin{pmatrix} P & q \\ -r' & 0 \end{pmatrix} / \det(P) = r'P^{-1}q.$$

From (0.7.1), it follows at once that the square matrix M is nonsingular if and only if the Schur complement $M/P = S - RP^{-1}Q$ is nonsingular. We then obtain the *Banachiewicz inversion formula* for the inverse of a partitioned matrix

$$M^{-1} = \begin{pmatrix} P & Q \\ R & S \end{pmatrix}^{-1} = \begin{pmatrix} P^{-1} + P^{-1}Q(M/P)^{-1}RP^{-1} & -P^{-1}Q(M/P)^{-1} \\ -(M/P)^{-1}RP^{-1} & (M/P)^{-1} \end{pmatrix}$$

$$= \begin{pmatrix} P^{-1} & 0 \\ 0 & 0 \end{pmatrix} + \begin{pmatrix} -P^{-1}Q \\ I \end{pmatrix} (M/P)^{-1} \begin{pmatrix} -RP^{-1} & I \end{pmatrix}. \qquad (0.7.2)$$

Banachiewicz [29, p. 54] appears to have been the first to obtain (0.7.2); his proof used Cracovians, a special kind of matrix algebra in which columns multiply columns, and which is used, for example, in spherical astronomy (polygonometry), geodesy, celestial mechanics, and in the calculation of orbits; see e.g., Bujakiewicz-Korońska & Koroński [101], Ouellette [345, pp. 290–291],

Fourteen years earlier in 1923, the geodesist Hans Boltz (1883–1947) implicitly used partitioning to invert a matrix (in scalar notation), see [66, 181, 225, 240]. According to the review by Forsythe [170] of the book *Die Inversion geodätischer Matrizen* by Ewald Konrad Bodewig [63], Boltz's interest concerned the "inverse of a *geodetic matrix* G in which a large submatrix A is mostly zeros and depends only on the topology of the geodetic network of stations and observed directions. When the directions are given equal weights, A has 6 on the main diagonal and ± 2 in a few positions off the diagonal. Boltz proposed first obtaining A^{-1} (which can be done before the survey), and then using it to obtain G^{-1} by partitioning G; see also Wolf [460]. Bodewig [62] refers to the "method of Boltz and Banachiewicz". Nistor [335] used the "method of Boltz" applied to partitioning in the solution of *normal equations* in statistics; see also Householder [234].

The Banachiewicz inversion formula (0.7.2) appears in the original version of the book *Matrix Calculus* by Bodewig published in 1956 [64, Part IIIA, §2, pp. 188–192] entitled "Frobenius' Relation" and in the second edition, published in 1959 [64, Part IIIA, ch. 2, pp. 217–222] entitled "Frobenius–Schur's Relation". In [65, p. 20], Bodewig notes that it was Aitken who referred him to Frobenius. No specific reference to Frobenius is given in [64, 65]. Lokki [291, p. 22] refers to the "Frobenius–Schur–Boltz–Banachiewicz method for partitioned matrix inversion".

In 1933 Ralf Lohan, in a short note [290] "extending the results of Boltz [66]", solves the system of equations

$$\begin{pmatrix} P & Q \\ R & S \end{pmatrix} \begin{pmatrix} x \\ y \end{pmatrix} = \begin{pmatrix} v \\ w \end{pmatrix} \tag{0.7.3}$$

for the vectors x and y and explicitly gives the solution as

$$\begin{aligned} x &= (P^{-1} + P^{-1}Q(M/P)^{-1}RP^{-1})v - P^{-1}Q(M/P)^{-1}w, \\ y &= -(M/P)^{-1}RP^{-1}v + (M/P)^{-1}w. \end{aligned} \tag{0.7.4}$$

While Lohan [290] does not explicitly present the inversion formula (0.7.2), he does use it to compute the inverse (presented explicitly, correct to 4 decimal places) of a specific real symmetric indefinite 5×5 matrix A with positive and negative integer elements in the range $[-17, +36]$. Letting A_j denote the top left $j \times j$ principal leading submatrix of A with $j = 3, 4$, Lohan [290] first computes A_3^{-1}, and then using A_3^{-1} and the scalar Schur complement A_4/A_3 he obtains A_4^{-1}. His inversion of A is then completed using A_4^{-1} and the scalar Schur complement A/A_4. A similar method was given in 1940 by Jossa [250]; see also Forsythe [170].

Following up on the results of Banachiewicz (1937), the well-known mathematician and statistician Bartel Leendert van der Waerden (1903–1996) gives the formula

$$\begin{pmatrix} P & Q \\ R & S \end{pmatrix}^{-1} = \begin{pmatrix} I & -P^{-1}Q(M/P)^{-1} \\ 0 & (M/P)^{-1} \end{pmatrix} \begin{pmatrix} P^{-1} & 0 \\ -RP^{-1} & I \end{pmatrix} \tag{0.7.5}$$

in a short note [446] in the "Notizen" section of the *Jahresbericht der Deutschen Mathematiker Vereinigung* in 1938. The formula (0.7.5) follows at once from (0.7.2) and from the Schur determinant formula (0.3.2).

0.8 Frazer, Duncan & Collar 1938, Aitken 1939, and Duncan 1944

The three aeronautical engineers Robert Alexander Frazer (1891–1959), William Jolly Duncan (1894–1960) and Arthur Roderick Collar (1908–1986) established the Banachiewicz inversion formula (0.7.2) in their classic book entitled *Elementary Matrices and Some Applications to Dynamics and Differential Equations* [171, p. 113] first published in 1938, just one year after Banachiewicz (1937). The appearance in [171] of the Banachiewicz inversion formula is almost surely its first appearance in a book; the Schur determinant formula also appears here for the special case when the Schur complement is a scalar. We find no mention in [171], however, of Banachiewicz, Boltz or Schur.

Let us consider again the nonsingular partitioned matrix $M = \begin{pmatrix} P & Q \\ R & S \end{pmatrix}$ as above, but now with S nonsingular and where the Schur complement $M/S = P - QS^{-1}R$. Then, in parallel to the Banachiewicz inversion formula (0.7.2) above, we have

$$M^{-1} = \begin{pmatrix} (M/S)^{-1} & -(M/S)^{-1}QS^{-1} \\ -S^{-1}R(M/S)^{-1} & S^{-1} + S^{-1}R(M/S)^{-1}QS^{-1} \end{pmatrix} \tag{0.8.1}$$

with P not necessarily nonsingular (but square so that M is square). When, however, both S and P are nonsingular, then (0.7.2) also holds, i.e.,

$$M^{-1} = \begin{pmatrix} P^{-1} + P^{-1}Q(M/P)^{-1}RP^{-1} & -P^{-1}Q(M/P)^{-1} \\ -(M/P)^{-1}RP^{-1} & (M/P)^{-1} \end{pmatrix}. \tag{0.8.2}$$

Equating the top left-hand corners in (0.8.1) and (0.8.2) yields

$$(M/S)^{-1} = P^{-1} + P^{-1}Q(M/P)^{-1}RP^{-1},$$

or explicitly

$$(P - QS^{-1}R)^{-1} = P^{-1} + P^{-1}Q(S - RP^{-1}Q)^{-1}RP^{-1}, \tag{0.8.3}$$

which we refer to as the *Duncan inversion formula*. We believe that (0.8.3) was first explicitly established by William Jolly Duncan in 1944, see [151, equation (4.10), p. 666]. See also the 1946 paper by Guttman [197]. Piegorsch & Casella [351] call (0.8.3) the *Duncan–Guttman inverse* while Grewal & Andrews [189, p. 366] call (0.8.3) the *Hemes inversion formula* with reference to Bodewig [64, p. 218 (1959)], who notes that (0.8.3) "has, with another proof, been communicated to the author by H. Hemes."

The survey paper by Hager [200] focuses on the special case of (0.8.3) when $S = I$

$$(P - QR)^{-1} = P^{-1} + P^{-1}Q(I - RP^{-1}Q)^{-1}RP^{-1}, \tag{0.8.4}$$

which he calls the *inverse matrix modification formula* and observes that the matrix $I - RP^{-1}Q$ is often called the *capacitance matrix*, see also [356]. Hager [200] notes that (0.8.4) is frequently called the *Woodbury formula* and the special case of (0.8.4) when Q and R are vectors the *Sherman–Morrison formula*, following results by Sherman & Morrison [416, 417, 418] and Woodbury [325, 461] in 1949–1950; see also Bartlett [36] and our Chapter 6 on Schur complements in statistics and probability.

When P, Q, R and S are all $n \times n$ as in the Schur determinant lemma in §0.3 above, and if P, Q, R and S are all nonsingular, then Aitken [4, Example #27, p. 148] also obtained the additional formula involving four Schur complements:

$$M^{-1} = \begin{pmatrix} (M/S)^{-1} & (M/Q)^{-1} \\ (M/R)^{-1} & (M/P)^{-1} \end{pmatrix}, \tag{0.8.5}$$

where $M/Q = R - SQ^{-1}P$ and $M/R = Q - PR^{-1}S$. The formula (0.8.5) was obtained by Aitken in his classic book *Determinants and Matrices* [4] first published in 1939, just one year after Frazer, Duncan & Collar [171] was first published; the formula (0.8.5) appears in Example #27 in the section entitled "Additional Examples" in [4, p. 148].

Duncan [151, equation (3.3), p. 664] also gives the Banachiewicz inversion formula explicitly and notes there that it "has been given by A. C. Aitken in lectures to his students, together with some alternative equivalent forms which are now included in this paper", see also [65, p. 20].

0.9 The Aitken block-diagonalization formula 1939 and the Guttman rank additivity formula 1946

With P nonsingular, the useful *Aitken block-diagonalization formula*

$$\begin{pmatrix} I & 0 \\ -RP^{-1} & I \end{pmatrix} \begin{pmatrix} P & Q \\ R & S \end{pmatrix} \begin{pmatrix} I & -P^{-1}Q \\ 0 & I \end{pmatrix} = \begin{pmatrix} P & 0 \\ 0 & M/P \end{pmatrix} \tag{0.9.1}$$

was apparently first established explicitly by Aitken and first published in 1939, see [4, ch. III, §29]. In (0.9.1), neither M nor S need be square.

While the Aitken formula (0.9.1) holds even if neither M nor S is square, when both M and S are square, (0.9.1) immediately yields the Schur determinant formula (0.3.2), and when M is square and nonsingular, (0.9.1) immediately yields the Banachiewicz inversion formula (0.7.2).

From the Aitken formula (0.9.1) we obtain at once the *Guttman rank additivity formula*

$$\operatorname{rank}(M) = \operatorname{rank}(P) + \operatorname{rank}(M/P),$$

or equivalently

$$\operatorname{rank}\begin{pmatrix} P & Q \\ R & S \end{pmatrix} = \operatorname{rank}(P) + \operatorname{rank}(S - QP^{-1}R), \tag{0.9.2}$$

which we believe was first established in 1946 by the social scientist and statistician Louis Guttman (1916–1987) in [197, p. 339].

0.10 Emilie Virginia Haynsworth (1916–1985) and the Haynsworth inertia additivity formula

Emilie Haynsworth, in addition to introducing the term Schur complement in [210, 211], also showed there that inertia is "additive on the Schur complement". The *inertia* or *inertia triple* of the partitioned Hermitian matrix

$$H = \begin{pmatrix} H_{11} & H_{12} \\ H_{12}^* & H_{22} \end{pmatrix}$$

is defined to be the ordered integer triple

$$\operatorname{In}(H) = \{\pi,\, \nu,\, \delta\},$$

where the nonnegative integers $\pi = \pi(H)$, $\nu = \nu(H)$, and $\delta = \delta(H)$ give the numbers, respectively, of positive, negative and zero eigenvalues of H. Here H_{11} is nonsingular and H_{12}^* is the conjugate transpose of H_{12}. This leads to the *Haynsworth inertia additivity formula*

$$\operatorname{In}(H) = \operatorname{In}(H_{11}) + \operatorname{In}(H/H_{11}),$$

or equivalently

$$\operatorname{In}\begin{pmatrix} H_{11} & H_{12} \\ H_{12}^* & H_{22} \end{pmatrix} = \operatorname{In}(H_{11}) + \operatorname{In}(H_{22} - H_{12}^* H_{11}^{-1} H_{12}), \tag{0.10.1}$$

proved in 1968, apparently for the first time, by Haynsworth [210, 211]. From (0.10.1), it follows at once that rank is additive on the Schur complement in a Hermitian matrix. As Guttman showed, see (0.9.2) above, this rank additivity holds more generally: H need not even be square—we need only that H_{11} be square and nonsingular. As we will see in Chapter 6, however, such rank additivity also holds in a Hermitian matrix when H_{11} is rectangular or square and singular but with the generalized Schur complement $H_{22} - H_{12}^* H_{11}^- H_{12}$, where H_{11}^- is a generalized inverse of H_{11}; moreover inertia additivity then also holds provided H_{11} is square.

To prove the Haynsworth inertia additivity formula (0.10.1) we apply the Aitken factorization formula (0.9.1) to the Hermitian matrix H with H_{11} square and nonsingular, then we have

$$\begin{pmatrix} I & 0 \\ -H_{12}^* H_{11}^{-1} & I \end{pmatrix} \begin{pmatrix} H_{11} & H_{12} \\ H_{12}^* & H_{22} \end{pmatrix} \begin{pmatrix} I & -H_{11}^{-1} H_{12} \\ 0 & I \end{pmatrix} = \begin{pmatrix} H_{11} & 0 \\ 0 & H/H_{11} \end{pmatrix},$$

which immediately leads to (0.10.1) by Sylvester's Law of Inertia: The inertia $\operatorname{In}(H) = \operatorname{In}(THT^*)$ for any nonsingular matrix T, see also §1.3 of Chapter 1.

Emilie Virginia Haynsworth was born on 1 June 1916 and died on 4 May 1985, both at home in Sumter, South Carolina. As observed in the obituary article [108] by Carlson, Markham & Uhlig, "In her family there have been Virginia Emilies or Emilie Virginias for over 200 years. From childhood on, Emilie had a strong and independent mind, so that her intellectual pursuits soon gained her the respect and awe of all her relatives and friends".

Throughout her life Emilie Haynsworth was eager to discuss any issue whatsoever. From Carlson, Markham & Uhlig [108] we quote Philip J. Davis (b. 1923): "She was a strong mixture of the traditional and the unconventional and for years I could not tell beforehand on what side of the line she would locate a given action". In *The Education of a Mathematician* [144, p. 146], Davis observes that Emilie Haynsworth "had a fine sense of mathematical elegance—a quality not easily defined. Her research can be found in a number of books on advanced matrix theory under the topic: 'Schur complement'. Emilie taught me many things about matrix theory."

The portrait of Emilie Haynsworth reproduced on page ix in the frontal matter of this book is on the Auburn University Web site [214] and in the book *The Education of a Mathematician* by Philip J. Davis [144] We conjecture that the portrait was made *c.* 1968, the year in which the term Schur complement was introduced by Haynsworth [210, 211].

In 1952 Emilie Haynsworth received her Ph.D. degree in mathematics at The University of North Carolina at Chapel Hill with Alfred Brauer as her dissertation adviser. We note that Issai Schur was Alfred Brauer's Ph.D. dissertation adviser and that the topic of Haynsworth's dissertation was determinantal bounds for diagonally dominant matrices. From 1960 until retirement in 1983, Haynsworth taught at Auburn University (Auburn, Alabama) "with a dedication which honors the teaching profession" [108] and supervised 18 Ph.D. students.

The mathematician Alexander Markowich Ostrowski (1893–1986), with whom Haynsworth co-authored the paper [216] on the inertia formula for the apparently not-then-yet-publicly-named Schur complement, wrote the following upon her death:

> I lost a very good, life-long friend and mathematics [lost] an excellent scientist. I remember how on many occasions I had to admire the way in which she found a formulation of absolute originality.

Chapter 1

Basic Properties of the Schur Complement

1.0 Notation

Most of our notation is standard, and our matrices are complex or real (though greater algebraic generality is often possible). We designate the set of all $m \times n$ matrices over $\mathbb{C}$ (or $\mathbb{R}$) by $\mathbb{C}^{m\times n}$ (respectively $\mathbb{R}^{m\times n}$), and denote the conjugate transpose of a matrix A by $A^* = (\bar{A})^T$. A matrix A is *Hermitian* if $A^* = A$, and a Hermitian matrix is *positive semidefinite* (*positive definite*) if all its eigenvalues are nonnegative (positive). The *Löwner partial order* $A \geq B$ $(A > B)$ on Hermitian matrices means that $A - B$ is positive semidefinite (positive definite). For $A \in \mathbb{C}^{m\times n}$, we denote the *matrix absolute value* by $|A| = (A^*A)^{1/2}$. A nonsingular square matrix has *polar decompositions* $A = U\,|A| = |A^*|\,U$ in which the positive definite factors $|A|$ and $|A^*|$, and the unitary factor $U = A\,|A|^{-1} = |A^*|^{-1}\,A$ are uniquely determined; if A is singular then the respective positive semidefinite factors $|A|$ and $|A^*|$ are uniquely determined and the left and right unitary factor U may be chosen to be the same, but U is not uniquely determined. Two matrices A and B of the same size are said to be **-congruent* if there is a nonsingular matrix S of the same size such that $A = SAS^*$; *-congruence is an equivalence relation. We denote the (multi-) set of eigenvalues of A (its *spectrum*) by $s(A) = \{\lambda_i(A)\}$ (including multiplicities).

1.1 Gaussian elimination and the Schur complement

One way to solve an $n \times n$ system of linear equations is by row reduction—Gaussian elimination that transforms the coefficient matrix into upper triangular form. For example, consider a homogeneous system of linear equations $Mz = 0$, where M is an $n \times n$ coefficient matrix with a nonzero $(1,1)$ entry. Write $M = \begin{pmatrix} a & b^T \\ c & D \end{pmatrix}$, where b and c are column vectors of size $n-1$, D is a square matrix of size $n-1$, and $a \neq 0$. The equations

$$Mz = 0 \quad \text{and} \quad \begin{pmatrix} a & b^T \\ 0 & D - ca^{-1}b^T \end{pmatrix} \begin{pmatrix} x \\ y \end{pmatrix} = 0$$

are equivalent, so the original problem reduces to solving a linear equation system of size $n-1$: $(D - ca^{-1}b)y = 0$.

This idea extends to a linear system $Mz = 0$ with a nonsingular leading principal submatrix. Partition M as

$$M = \begin{pmatrix} A & B \\ C & D \end{pmatrix}, \tag{1.1.1}$$

suppose A is nonsingular, and partition $z = \begin{pmatrix} x \\ y \end{pmatrix}$ conformally with M. The linear system $Mz = 0$ is equivalent to the pair of linear systems

$$Ax + By = 0 \tag{1.1.2}$$
$$Cx + Dy = 0 \tag{1.1.3}$$

If we multiply (1.1.2) by $-CA^{-1}$ and add it to (1.1.3), the vector variable x is eliminated and we obtain the linear system of smaller size

$$(D - CA^{-1}B)y = 0.$$

We denote the matrix $D - CA^{-1}B$ by M/A and call it the *Schur complement of A in M*, or *the Schur complement of M relative to A*. In the same spirit, if D is nonsingular, the Schur complement of D in M is

$$M/D = A - BD^{-1}C.$$

For a non-homogeneous system of linear equations

$$\begin{pmatrix} A & B \\ C & D \end{pmatrix} \begin{pmatrix} x \\ y \end{pmatrix} = \begin{pmatrix} u \\ v \end{pmatrix},$$

we may use Schur complements to write the solution as (see Section 0.7)

$$x = (M/D)^{-1}(u - BD^{-1}v), \quad y = (M/A)^{-1}(v - CA^{-1}u).$$

The Schur complement is a basic tool in many areas of matrix analysis, and is a rich source of matrix inequalities. The idea of using the Schur complement technique to deal with linear systems and matrix problems is classical. It was certainly known to J. Sylvester in 1851 [436], and probably also to Gauss. A famous determinantal identity presented by I. Schur 1917 [404] was referred to as the *formula of Schur* by Gantmacher [180, p. 46]. The term *Schur complement*, which appeared in the sixties in a paper by Haynsworth [211] is therefore an apt appellation; see Chapter 0.

Theorem 1.1 (Schur's Formula) *Let M be a square matrix partitioned as in (1.1.1). If A is nonsingular, then*

$$\det(M/A) = \det M \,/\, \det A. \tag{1.1.4}$$

Proof. Block Gaussian elimination gives the factorization

$$\begin{pmatrix} A & B \\ C & D \end{pmatrix} = \begin{pmatrix} I & 0 \\ CA^{-1} & I \end{pmatrix} \begin{pmatrix} A & B \\ 0 & D - CA^{-1}B \end{pmatrix}.$$

The identity (1.1.4) follows by taking the determinant of both sides. ∎

It is an immediate consequence of the Schur formula (1.1.4) that if A is nonsingular, then M is nonsingular if and only if M/A is nonsingular.

Schur's formula may be used to compute characteristic polynomials of block matrices. Suppose A and C commute in (1.1.1). Then

$$\begin{aligned} \det(\lambda I - M) &= \det(\lambda I - A) \det\left[(\lambda I - M)/(\lambda I - A)\right] \\ &= \det\left[(\lambda I - A)(\lambda I - D) - CB\right]. \end{aligned}$$

The following useful formula, due to Babachiewicz (see Section 0.7), presents the inverse of a matrix in terms of Schur complements.

Theorem 1.2 *Let M be partitioned as in (1.1.1) and suppose both M and A are nonsingular. Then M/A is nonsingular and*

$$M^{-1} = \begin{pmatrix} A^{-1} + A^{-1}B\,(M/A)^{-1}\,CA^{-1} & -A^{-1}B\,(M/A)^{-1} \\ -\,(M/A)^{-1}\,CA^{-1} & (M/A)^{-1} \end{pmatrix}. \tag{1.1.5}$$

Thus, the $(2,2)$ block of M^{-1} is $(M/A)^{-1}$:

$$\left(M^{-1}\right)_{22} = (M/A)^{-1}. \tag{1.1.6}$$

Proof. Under the given hypotheses, one checks that

$$\begin{pmatrix} A & B \\ C & D \end{pmatrix} = \begin{pmatrix} I & 0 \\ CA^{-1} & I \end{pmatrix} \begin{pmatrix} A & 0 \\ 0 & M/A \end{pmatrix} \begin{pmatrix} I & A^{-1}B \\ 0 & I \end{pmatrix}.$$

Inverting both sides yields

$$\begin{aligned} M^{-1} &= \begin{pmatrix} I & A^{-1}B \\ 0 & I \end{pmatrix}^{-1} \begin{pmatrix} A & 0 \\ 0 & M/A \end{pmatrix}^{-1} \begin{pmatrix} I & 0 \\ CA^{-1} & I \end{pmatrix}^{-1} \\ &= \begin{pmatrix} I & -A^{-1}B \\ 0 & I \end{pmatrix} \begin{pmatrix} A^{-1} & 0 \\ 0 & (M/A)^{-1} \end{pmatrix} \begin{pmatrix} I & 0 \\ -CA^{-1} & I \end{pmatrix}. \end{aligned}$$

Multiplying out the block matrices on the right hand side gives the asserted presentation of M^{-1}, from which the identity (1.1.6) follows. ∎

In a similar fashion, one can verify each of the following alternative presentations of M^{-1} (see Sections 0.7 and 0.8):

$$M^{-1} = \begin{pmatrix} (M/D)^{-1} & -A^{-1}B(M/A)^{-1} \\ -D^{-1}C(M/D)^{-1} & (M/A)^{-1} \end{pmatrix};$$

$$M^{-1} = \begin{pmatrix} A^{-1} & 0 \\ 0 & 0 \end{pmatrix} + \begin{pmatrix} A^{-1}B \\ -I \end{pmatrix} (M/A)^{-1} \left(CA^{-1} \ \ -I \right);$$

and, if A, B, C, and D are all square and have the same size

$$M^{-1} = \begin{pmatrix} (M/D)^{-1} & (C - DB^{-1}A)^{-1} \\ (B - AC^{-1}D)^{-1} & (M/A)^{-1} \end{pmatrix}.$$

Comparing the $(1,1)$ blocks of M^{-1} in these gives the identities

$$\begin{aligned} (A - BD^{-1}C)^{-1} &= A^{-1} + A^{-1}B(D - CA^{-1}B)^{-1}CA^{-1} \\ &= -C^{-1}D(B - AC^{-1}D)^{-1} \\ &= -(C - DB^{-1}A)^{-1}DB^{-1} \\ &= C^{-1}D(D - CA^{-1}B)^{-1}CA^{-1} \\ &= A^{-1}B(D - CA^{-1}B)^{-1}DB^{-1} \end{aligned}$$

provided that each of the indicated inverses exists.

Of course, the Schur complement can be formed with respect to any nonsingular submatrix, not just a leading principal submatrix. Let α and β be given index sets, i.e., subsets of $\{1, 2, \ldots, n\}$. We denote the cardinality of an index set by $|\alpha|$ and its complement by $\alpha^c \equiv \{1, 2, \ldots, n\} \setminus \alpha$. Let $A[\alpha, \beta]$ denote the submatrix of A with rows indexed by α and columns indexed by β, both of which are thought of as increasingly ordered sequences, so the rows and columns of the submatrix appear in their natural order.

We often write $A[\alpha]$ for $A[\alpha, \alpha]$. If $|\alpha| = |\beta|$ and if $A[\alpha, \beta]$ is nonsingular, we denote by $A/A[\alpha, \beta]$ the *Schur complement* of $A[\alpha, \beta]$ in A:

$$A/A[\alpha, \beta] \equiv A[\alpha^c, \beta^c] - A[\alpha^c, \beta] \left(A[\alpha, \beta]\right)^{-1} A[\alpha, \beta^c]. \tag{1.1.7}$$

It is often convenient to write A/α for $A/A[\alpha]$.

Although it can be useful to have the Schur complement in the general form (1.1.7), it is equivalent to the simpler presentation (1.1.1): there are permutations of the rows and columns of A that put $A[\alpha, \beta]$ into the upper left corner of A, leaving the rows and columns of $A[\alpha, \beta^c]$ and $A[\alpha^c, \beta]$ in the same increasing order in A. If $\alpha = \beta$, the two permutations are the same, so there exists a permutation matrix P such that

$$P^T AP = \begin{pmatrix} A[\alpha] & A[\alpha, \alpha^c] \\ A[\alpha^c, \alpha] & A[\alpha^c] \end{pmatrix}.$$

Thus,

$$(P^T AP)/A[\alpha] = A/\alpha.$$

Schur's formula (1.1.4) may be extended to an arbitrary submatrix [18]. For an index set $\alpha = \{\alpha_1, \alpha_2, \ldots, \alpha_k\} \subseteq \{1, 2, \ldots, n\}$, we define

$$\operatorname{sgn}(\alpha) \equiv (-1)^{\sum_{i=1}^k \alpha_i - k(k+1)/2}.$$

The general form of Schur's formula is

$$\det A = \operatorname{sgn}(\alpha) \operatorname{sgn}(\beta) \det A[\alpha, \beta] \det (A/A[\alpha, \beta]) \tag{1.1.8}$$

whenever $A[\alpha, \beta]$ is nonsingular. The proof is similar to that for a leading principal submatrix. Similarly, the analog of (1.1.6) for an (α, β) block is

$$A^{-1}[\alpha, \beta] = (A/A[\beta^c, \alpha^c])^{-1}. \tag{1.1.9}$$

Although the Schur complement is a non-linear operation on matrices, we have $(kA)/\alpha = k(A/\alpha)$ for any scalar k, and $(A/\alpha)^* = A^*/\alpha$.

1.2 The quotient formula

In 1969, Crabtree and Haynsworth [131] gave a quotient formula for the Schur complement. Their formula was reproved by Ostrowski [342, 343]. Other approaches to this formula were found in [99, 106, 422] and [165, p. 22]. Applications of the quotient formula were given in [107, 279, 88].

We present a matrix identity [471] from which the quotient formula follows. Let M be partitioned as in (1.1.1) and suppose A is nonsingular. If $B = 0$ or $C = 0$, then $M/A = D$ and $M/D = A$; this is the case, for example, if M is upper or lower triangular.

Theorem 1.3 *Let*

$$M = \begin{pmatrix} A & B \\ C & D \end{pmatrix}, \; L = \begin{pmatrix} X & 0 \\ Y & Z \end{pmatrix}, \; \textit{and } R = \begin{pmatrix} U & V \\ 0 & W \end{pmatrix}$$

be conformally partitioned square matrices of the same size, suppose A, X, and U are nonsingular and $k \times k$, and let $\alpha = \{1, \cdots, k\}$. Then

$$(LMR)/\alpha = (L/\alpha)(M/\alpha)(R/\alpha) = L[\alpha^c](M/\alpha)R[\alpha^c],$$

that is,

$$(LMR)/(XAU) = (L/X)(M/A)(R/U) = Z(M/A)W.$$

Proof. First compute

$$LMR = \begin{pmatrix} XAU & XAV + XBW \\ YAU + ZCU & YAV + ZCV + YBW + ZDW \end{pmatrix}.$$

Then

$$\begin{aligned} (LMR)/(XAU) &= YAV + ZCV + YBW + ZDW \\ &\quad -(YAU + ZCU)(XAU)^{-1}(XAV + XBW) \\ &= YAV + ZCV + YBW + ZDW \\ &\quad -(YA + ZC)A^{-1}(AV + BW) \\ &= ZDW - ZCA^{-1}BW \\ &= Z(D - CA^{-1}B)W \\ &= Z(M/A)W. \blacksquare \end{aligned}$$

The following special case of the theorem ($R = I$) is often useful:

Corollary 1.1 *Let M and Q be square matrices of the same size, let α denote the index set of a nonsingular leading principal submatrix of Q, suppose $Q[\alpha, \alpha^c] = 0$, and suppose that $M[\alpha]$ is nonsingular. Then*

$$(QM)/\alpha = Q[\alpha^c](M/\alpha);$$

if also $Q[\alpha^c] = I$, then

$$(QM)/\alpha = M/\alpha.$$

In particular, if $Q = \Lambda$ is diagonal, then

$$(\Lambda M)/\alpha = (\Lambda[\alpha^c])(M/\alpha).$$

Here are some other special cases and applications of the theorem:

Case 1. Suppose $X = U = I$. Then

$$(LMR)/A = Z(M/A)W. \tag{1.1.10}$$

Now let J denote a square matrix whose entries are all 1. If $Z = W = J$, (1.1.10) shows that the Schur complement of A in the product

$$\begin{pmatrix} I & 0 \\ Y & J \end{pmatrix}\begin{pmatrix} A & B \\ C & D \end{pmatrix}\begin{pmatrix} I & V \\ 0 & J \end{pmatrix}$$

is sJ, where s denotes the sum of all entries of M/A. Of course, sJ is independent of Y and V and has rank 1.

If W is nonsingular and $Z = W^{-1}$, (1.1.10) shows that $(LMR)/A$ is similar to M/A. Thus the eigenvalues of $(LMR)/A$ can be obtained by computing those of M/A, and they do not depend on the choices of Y, V, and the nonsingular matrix W.

Finally, (1.1.10) shows that if a matrix N can be written as a product of a lower triangular matrix, a diagonal matrix, and an upper triangular matrix, say, $N = \mathcal{L}\Lambda\mathcal{U}$, then

$$N/\alpha = (\mathcal{L}/\alpha)(\Lambda/\alpha)(\mathcal{U}/\alpha)$$

is a factorization of N/α of the same form.

Case 2. Suppose $X = Z = U = W = I$. Then

$$(LMR)/A = M/A. \tag{1.1.11}$$

A closely related fact is the familiar identity

$$\begin{pmatrix} I & 0 \\ -CA^{-1} & I \end{pmatrix}\begin{pmatrix} A & B \\ C & D \end{pmatrix}\begin{pmatrix} I & -A^{-1}B \\ 0 & I \end{pmatrix} = \begin{pmatrix} A & 0 \\ 0 & M/A \end{pmatrix}.$$

If $V = 0$ (that is, $R = I$), then

$$(LM)/A = M/A. \tag{1.1.12}$$

The identities (1.1.11) and (1.1.12) show that block Gaussian elimination for rows (columns) applied to the complementary columns (rows) of A does not change the Schur complement of A; i.e., type three elementary row (column) operations on the columns (rows) complementary to A have no effect on the Schur complement of A. We will use this important fact to prove the quotient formula.

Case 3. Suppose $M = I$. Then $LMR = LR$ is the product of a block lower triangular matrix and a block upper triangular matrix, and

$$(LR)/\alpha = (L/\alpha)(R/\alpha) = L\left[\alpha^c\right] R\left[\alpha^c\right]. \tag{1.1.13}$$

A computation shows that for block lower triangular matrices L_1 and L_2

$$(L_1L_2)/\alpha = (L_1/\alpha)(L_2/\alpha),$$

and for block upper triangular matrices R_1 and R_2

$$(R_1R_2)/\alpha = (R_1/\alpha)(R_2/\alpha).$$

As a special case of (1.1.13), for any k and lower triangular matrix R

$$(LL^*)/\alpha = (L/\alpha)(L^*/\alpha) = (L[\alpha^c])(L[\alpha^c])^*. \tag{1.1.14}$$

Any positive definite matrix N can be written as $N = LL^*$ for some lower triangular matrix L. This is the *Cholesky factorization* of N, which is unique if we insist that L have positive diagonal entries. The identity (1.1.14) therefore provides the Cholesky factorization of the Schur complement N/α if we have the Cholesky factorization of N.

Although there does not seem to be a nice way to express $(RL)/\alpha$ in terms of R/α and L/α, one checks that

$$(L^*L)/\alpha \le (L^*/\alpha)(L/\alpha). \tag{1.1.15}$$

Suppose T is a square matrix that has an LU factorization (this would be the case, for example, if *every* leading principal submatrix of T were nonsingular), and consider any nonsingular leading principal submatrix indexed by α. Then (1.1.15) implies that

$$(T^*T)/\alpha \le (T^*/\alpha)(T/\alpha) \tag{1.1.16}$$

as follows:

$$\begin{aligned}(T^*T)/\alpha &= (U^*L^*LU)/\alpha\\ &= (U^*/\alpha)[(L^*L)/\alpha](U/\alpha)\\ &\le (U^*/\alpha)(L^*/\alpha)(L/\alpha)(U/\alpha) \quad \text{by (1.1.15)}\\ &= (T^*/\alpha)(T/\alpha) \quad \text{by (1.1.13).}\end{aligned}$$

Case 4. Suppose that

$$L = R^* = \begin{pmatrix} U^* & 0 \\ V^* & W^* \end{pmatrix}.$$

Theorem 1.3 tells us that

$$(R^*MR)/\alpha = (R^*/\alpha)(M/\alpha)(R/\alpha). \tag{1.1.17}$$

Although there does not seem to be any general analog of (1.1.17) for $(L^*ML)/\alpha$, if M is positive definite, then

$$(L^*ML)/\alpha \le (L^*ML)[\alpha^c] = (L^*/\alpha)M[\alpha^c](L/\alpha). \tag{1.1.18}$$

More generally, let N be positive semidefinite and let T be the same size as N. If $N[\alpha]$ and $T[\alpha]$ are nonsingular, then

$$(T^*NT)/\alpha \leq (T^*/\alpha)N[\alpha^c](T/\alpha). \tag{1.1.19}$$

This can be proved using (1.1.18), with T written in the form

$$T = \begin{pmatrix} I & 0 \\ \star & I \end{pmatrix} \begin{pmatrix} T(\alpha) & \star \\ 0 & T/\alpha \end{pmatrix},$$

in which blocks of entries irrelevant to the proof are indicated by $\star$.

Case 5. The fundamental identity

$$(A/\alpha)^{-1} = A^{-1}[\alpha^c] \tag{1.1.20}$$

in Theorem 1.2 is useful in many matrix problems. For example, it is the key to showing that the class of inverse M-matrices is closed under Schur complementation [244]. If A has an LU factorization, there is a nice proof using (1.1.13): Let $A = LU$ so that $A^{-1} = U^{-1}L^{-1}$. Then

$$\begin{aligned} (A/\alpha)^{-1} &= (L[\alpha^c]U[\alpha^c])^{-1} \\ &= \left(U[\alpha^c]^{-1}L[\alpha^c]\right)^{-1} \\ &= U^{-1}[\alpha^c]L^{-1}[\alpha^c] \\ &= A^{-1}[\alpha^c]. \end{aligned}$$

We now derive the Crabtree-Haynsworth quotient formula for the Schur complement.

Theorem 1.4 (Quotient Formula) *Let M, A, and E be given square nonsingular matrices such that*

$$M = \begin{pmatrix} A & B \\ C & D \end{pmatrix} \quad \textit{and} \quad A = \begin{pmatrix} E & F \\ G & H \end{pmatrix}.$$

Then A/E is a nonsingular principal submatrix of M/E and

$$M/A = (M/E)/(A/E).$$

Proof. Write

$$M = \begin{pmatrix} A & B \\ C & D \end{pmatrix} = \begin{pmatrix} E & F & B_1 \\ G & H & B_2 \\ C_1 & C_2 & D \end{pmatrix}$$

and compute

$$M/E = \begin{pmatrix} H & B_2 \\ C_2 & D \end{pmatrix} - \begin{pmatrix} G \\ C_1 \end{pmatrix} E^{-1} \begin{pmatrix} F & B_1 \end{pmatrix}$$
$$= \begin{pmatrix} H - GE^{-1}F & \star \\ \star & \star \end{pmatrix} = \begin{pmatrix} A/E & \star \\ \star & \star \end{pmatrix}.$$

Since A is nonsingular, so is A/E. Thus $(M/E)/(A/E)$ is well defined.

Now define

$$L = \begin{pmatrix} I & 0 \\ -CA^{-1} & I \end{pmatrix}$$

and compute

$$LM = \begin{pmatrix} A & B \\ 0 & D - CA^{-1}B \end{pmatrix} = \begin{pmatrix} E & F & B_1 \\ G & H & B_2 \\ 0 & 0 & M/A \end{pmatrix} \equiv \hat{M}.$$

The identity (1.1.12) ensures that $\hat{M}/E = M/E$. On the other hand,

$$\hat{M}/E = \begin{pmatrix} H & B_2 \\ 0 & M/A \end{pmatrix} - \begin{pmatrix} G \\ 0 \end{pmatrix} E^{-1} \begin{pmatrix} F & B_1 \end{pmatrix}$$
$$= \begin{pmatrix} A/E & B_2 - GE^{-1}B_1 \\ 0 & M/A \end{pmatrix},$$

so $(\hat{M}/E)/(A/E) = M/A$ and we have the desired formula. ∎

The quotient formula may also be derived from Theorem 1.3 directly by taking

$$X = \begin{pmatrix} I & 0 \\ -GE^{-1} & I \end{pmatrix}, \quad Y = \begin{pmatrix} -C_1E^{-1} & 0 \end{pmatrix}, \quad Z = I$$

and

$$U = \begin{pmatrix} I & -E^{-1}F \\ 0 & I \end{pmatrix}, \quad V = \begin{pmatrix} -E^{-1}B_1 \\ 0 \end{pmatrix}, \quad W = I.$$

Theorem 1.3 ensures that $(LMR)/E = M/E$. A computation shows that

$$LMR = \begin{pmatrix} E & 0 \\ 0 & (LMR)/E \end{pmatrix} = \begin{pmatrix} E & 0 \\ 0 & M/E \end{pmatrix}, \quad XAU = \begin{pmatrix} E & 0 \\ 0 & A/E \end{pmatrix}.$$

It follows that

$$(LMR)/(XAU) = \begin{pmatrix} E & 0 \\ 0 & M/E \end{pmatrix} \Big/ \begin{pmatrix} E & 0 \\ 0 & A/E \end{pmatrix} = (M/E)/(A/E).$$

On the other hand, $Z(M/A)W = M/A$, so we again have the formula.

1.3 Inertia of Hermitian matrices

The *inertia* of an $n \times n$ Hermitian matrix A is the ordered triple

$$\text{In}(A) \equiv (p(A),\ q(A),\ z(A))$$

in which $p(A)$, $q(A)$, and $z(A)$ (or π, ν, δ in Section 0.10) are the numbers of the positive, negative, and zero eigenvalues of A, respectively (including multiplicities). Of course, $\text{rank}\,(A) = p(A) + q(A)$.

By $\text{In}(A) \geq (a, b, c)$ we mean that $p(A) \geq a$, $q(A) \geq b$, and $z(A) \geq c$.

The inertia of a nonsingular Hermitian matrix and its inverse are the same since their (necessarily nonzero) eigenvalues are reciprocals of each other. The inertias of similar Hermitian matrices are the same because their eigenvalues are identical. The inertias of $*$-congruent matrices are also the same; this is *Sylvester's Law of Inertia.*

Theorem 1.5 (Sylvester's Law of Inertia) *Let A and B be $n \times n$ Hermitian matrices. Then there is a nonsingular $n \times n$ matrix G such that $B = G^*AG$ if and only if* $\text{In}\,(A) = \text{In}\,(B)$.

Proof. The spectral theorem ensures that there are positive diagonal matrices E and F with respective sizes $p(A)$ and $q\,(A)$ such that A is unitarily similar (*-congruent) to $E \oplus (-F) \oplus 0_{z(A)}$. With $G \equiv E^{-1/2} \oplus F^{-1/2} \oplus I_{z(A)}$, compute $G^*\,(E \oplus (-F) \oplus Z)\,G = I_{p(A)} \oplus (-I_{q(A)}) \oplus 0_{z(A)}$.The same argument shows that B is *-congruent to $I_{p(B)} \oplus (-I_{q(B)}) \oplus 0_{z(B)}$. If $\text{In}\,(A) = \text{In}\,(B)$, transitivity of *-congruence implies that A and B are *-congruent.

Conversely, suppose that A and B are *-congruent; for the moment, assume that A (and hence B) is nonsingular. Since A and B are *-congruent to $V \equiv I_{p(A)} \oplus (-I_{q(A)})$ and $W \equiv I_{p(B)} \oplus (-I_{q(B)})$, respectively, the unitary matrices V and W are also *-congruent. Let G be nonsingular and such that $V = G^*WG$. Let $G = PU$ be a (right) polar factorization, in which P is positive definite and U is unitary. Then $V = G^*WG = U^*PWPU$, so $P^{-1}\,(UVU^*) = WP$. This identity gives right and left polar factorizations of the same nonsingular matrix, whose (unique) right and left unitary polar factors UVU^* and W must therefore be the same [228, pp. 416-417]. Thus, $W = UVU^*$, so W and V are similar and hence have the same sets of eigenvalues. We conclude that $p(A) = p(B)$ and $q(A) = q(B)$, and hence that $\text{In}(A) = \text{In}(B)$.

If A and B are *-congruent and singular, they have the same rank, so $z(A) = z(B)$. Thus, if we set $A_1 \equiv I_{p(A)} \oplus (-I_{q(A)})$ and $B_1 \equiv I_{p(B)} \oplus (-I_{q(B)})$, the nonsingular matrices A_1 and B_1 are the same size and $A_1 \oplus 0_{z(A)}$ and $B_1 \oplus 0_{z(A)}$ are *-congruent: $A_1 \oplus 0_{z(A)} = G^*\left(B_1 \oplus 0_{z(A)}\right) G$ for some nonsingular G. Partition $G = [G_{ij}]_{i,j=1}^2$ conformally with $A_1 \oplus 0_{z(A)}$.

The (1,1) block of the congruence is $A_1 = G_{11}^* B_1 G_{11}$. This means that G_{11} is nonsingular and A_1 is *-congruent to B_1. The singular case therefore follows from the nonsingular case. ∎

The key point of the preceding argument is that two unitary matrices are *-congruent if and only if they are similar. This fact can be used to generalize Sylvester's Law of Inertia to normal matrices; see [236] or [246].

We can now state the addition theorem for Schur complements of Hermitian matrices, which, along with other results of this section, appeared in a sequel of E. Haynsworth's publications [211, 212, 213].

Theorem 1.6 *Let A be Hermitian and let A_{11} be a nonsingular principal submatrix of A. Then*

$$\operatorname{In}(A) = \operatorname{In}(A_{11}) + \operatorname{In}(A/A_{11}).$$

Proof. After a permutation similarity, if necessary, we may assume that

$$A = \begin{pmatrix} A_{11} & A_{12} \\ A_{21} & A_{22} \end{pmatrix} \quad \text{and we define} \quad G \equiv \begin{pmatrix} I & -A_{12}A_{11}^{-1} \\ 0 & I \end{pmatrix}.$$

Then

$$G^*AG = \begin{pmatrix} A_{11} & 0 \\ 0 & A/A_{11} \end{pmatrix}, \tag{1.1.21}$$

so $\sigma(G^*AG) = \sigma(A_{11}) \cup \sigma(A/A_{11})$ (with multiplicities). Since $\operatorname{In}(A) = \operatorname{In}(G^*AG)$, the conclusion follows from Sylvester's Law of Inertia. ∎

For any Hermitian matrix A and any index sets α and β it is clear that

$$\operatorname{In}(A) \geq \operatorname{In}(A[\alpha])$$

and

$$\operatorname{In}(A) \geq (\max_{\alpha} p(A[\alpha]),\ \max_{\beta} q(A[\beta]), 0). \tag{1.1.22}$$

Suppose A has a positive definite principal submatrix $A[\alpha]$ of order p. If it also has a negative definite principal submatrix of order q, then (1.1.22) ensures that $\operatorname{In}(A) \geq (p,\ q,\ 0)$. In particular, if $A[\alpha] > 0$ and $A[\alpha^c] < 0$, then $\operatorname{In}(A) = (p,\ n-p,\ 0)$. In order to prove a generalization of this observation, we introduce a lemma that is of interest in its own right. For a normal matrix A with spectral decomposition $A = U\Lambda U^*$, where U is unitary and $\Lambda = \operatorname{diag}(\lambda_1, \cdots, \lambda_n)$ is diagonal, $|A| = U|\Lambda|U^* = U \operatorname{diag}(|\lambda_1|, \cdots, |\lambda_n|) U^*$, which is always positive semidefinite. Of course, A is positive semidefinite if and only if $|A| = A$.

Lemma 1.1 *Let P be $m \times m$ and normal, let Q be $n \times m$, and let*

$$M = \begin{pmatrix} Q \\ P \end{pmatrix}.$$

Then

$$\operatorname{rank}(M) = \operatorname{rank}(|P| + Q^*Q).$$

As a consequence, $|P| + Q^*Q$ is positive definite if $\operatorname{rank}(M) = m$.

Proof. Let $A = U\Lambda U^*$ be a spectral decomposition of A and suppose $\Lambda = \Lambda_1 \oplus 0$, in which Λ_1 is nonsingular; partition $QU = ((QU)_1, (QU)_2)$ conformally with Λ. Then

$$\begin{aligned}
\operatorname{rank}(M) &= \operatorname{rank}\begin{pmatrix} Q \\ P \end{pmatrix} = \operatorname{rank}\left[\begin{pmatrix} I & 0 \\ 0 & U^* \end{pmatrix}\begin{pmatrix} Q \\ P \end{pmatrix} U\right] \\
&= \operatorname{rank}\begin{pmatrix} QU \\ \Lambda \end{pmatrix} = \operatorname{rank}\begin{pmatrix} (QU)_1 & (QU)_2 \\ \Lambda_1 & 0 \\ 0 & 0 \end{pmatrix} \\
&= \operatorname{rank}\begin{pmatrix} (QU)_1 & (QU)_2 \\ |\Lambda_1|^{1/2} & 0 \\ 0 & 0 \end{pmatrix} \\
&= \operatorname{rank}\begin{pmatrix} QU \\ |\Lambda|^{1/2} \end{pmatrix} = \operatorname{rank}\left[\begin{pmatrix} I & 0 \\ 0 & U \end{pmatrix}\begin{pmatrix} QU \\ |\Lambda|^{1/2} \end{pmatrix} U^*\right] \\
&= \operatorname{rank}\begin{pmatrix} Q \\ U|\Lambda|^{1/2}U^* \end{pmatrix} = \operatorname{rank}\begin{pmatrix} Q \\ |P|^{1/2} \end{pmatrix} \\
&= \operatorname{rank}\begin{pmatrix} Q \\ |P|^{1/2} \end{pmatrix}^*\begin{pmatrix} Q \\ |P|^{1/2} \end{pmatrix} \\
&= \operatorname{rank}(|P| + Q^*Q). \blacksquare
\end{aligned}$$

Theorem 1.7 *Let $A = [A_{ij}]_{i,j=1}^2$ be a partitioned $n \times n$ Hermitian matrix. Suppose that its leading principal submatrix A_{11} is $k \times k$ and positive definite, and that A_{22} is negative semidefinite. If the last $n - k$ columns of A are linearly independent, then A is nonsingular and*

$$\operatorname{In}(A) = (k,\ n-k,\ 0).$$

Proof. Let S be nonsingular and such that $S^* A_{11} S = I_k$; let

$$P = \begin{pmatrix} S & 0 \\ 0 & I \end{pmatrix}.$$

The last $n-k$ columns of

$$P^*AP = \begin{pmatrix} I_k & S^*A_{12} \\ A_{12}^*S & A_{22} \end{pmatrix}$$

are also linearly independent, and, by Sylvester's Law of Inertia, $\mathrm{In}(A) = \mathrm{In}(P^*AP) = \mathrm{In}(I_k) + \mathrm{In}\left((P^*AP)/I_k\right)$. Lemma 1.1 ensures that the Schur complement $-(P^*AP)/I_k = -A_{22} + (S^*A_{12})^*(S^*A_{12})$ is positive definite, so $\mathrm{In}(A) = (k,\ 0,\ 0) + (0,\ n-k,\ 0) = (k,\ n-k,\ 0)$. ∎

The next theorem gives information about the inertia of bordered Hermitian matrices.

Theorem 1.8 *Let A be an $n \times n$ Hermitian matrix partitioned as*

$$A = \begin{pmatrix} B & c \\ c^* & a \end{pmatrix}, \tag{1.1.23}$$

in which c is a row vector with $n-1$ complex entries and a is a real scalar. Suppose that $\mathrm{In}(B) = (p,\ q,\ z)$. *Then*

$$\mathrm{In}(A) \geq (p,\ q,\ z-1).$$

If, in addition, $z(A) = z-1$, then

$$\mathrm{In}(A) = (p+1,\ q+1,\ z-1).$$

Proof. Let the eigenvalues of B be $\beta_1 \geq \beta_2 \geq \cdots \geq \beta_{n-1}$ and let the eigenvalues of A be $\alpha_1 \geq \alpha_2 \geq \cdots \geq \alpha_n$. The Cauchy eigenvalue interlacing theorem ([230, Theorem 4.3.8] or [468, p. 222]) ensures that

$$\alpha_i \geq \beta_i \geq \alpha_{i+1},\ i = 1, 2, \ldots, n-1.$$

Since $p(B) = p$, we have

$$\alpha_i \geq \beta_i > 0,\ i = 1, 2, \ldots, p \text{ and } \alpha_{p+1} \geq \beta_{p+1} = 0$$

and since $q(B) = q$, we have

$$0 > \beta_i \geq \alpha_{i+1},\ i = p+z+1, \ldots, n-1 \text{ and } 0 = \beta_{p+z} \geq \alpha_{p+z+1}.$$

Thus $p(A) \geq p$ and $q(A) \geq q$. In addition,

$$0 = \beta_i \geq \alpha_{i+1} \geq \beta_{i+1} = 0,\ i = p+1, \ldots, p+z-1$$

so A has at least $z-1$ zero eigenvalues. If A has exactly $z-1$ zero eigenvalues, then we must have $\alpha_{p+1} > \beta_{p+1} = 0$ and $0 = \beta_{p+z} > \alpha_{p+z+1}$, so $p(A) = p+1$ and $q(A) = q+1$. ∎

Repeatedly applying Theorem 1.8 yields the following

Theorem 1.9 *Let A be an $n \times n$ Hermitian matrix and let B be a $k \times k$ principal submatrix of A. Then*

$$p(A) \geq p(B) \quad \text{and} \quad q(A) \geq q(B).$$

If $z(A) - z(B) = d > 0$, then

$$d \leq n-k,\ p(A) \geq p(B) + d \quad \text{and} \quad q(A) \geq q(B) + d.$$

If $d = n - k$, then

$$\text{In}(A) = (p(B) + n - k,\ q(B) + n - k,\ z(B) - n + k).$$

If A is nonsingular, then $z(B) = n - k$ and

$$\text{In}(A) = (p(B) + n - k,\ q(B) + n - k,\ 0).$$

Let A and B be square matrices of orders n and m, respectively, with $n \geq m$. If there is a solution X of rank m of the homogeneous linear matrix equation $AX - XB = 0$, it is known that the m eigenvalues of B are also eigenvalues of A. The following theorem exhibits a matrix (a Schur complement) whose eigenvalues are the remaining $n - m$ eigenvalues of A.

Theorem 1.10 *Suppose that $n \geq m$ and let $A \in \mathbb{C}^{n \times n}$ and $B \in \mathbb{C}^{m \times m}$. Let $X \in \mathbb{C}^{n \times m}$ be such that $AX = XB$, partition X and A conformally as*

$$X = \begin{pmatrix} X_1 \\ X_2 \end{pmatrix} \quad \text{and} \quad A = \begin{pmatrix} A_{11} & A_{12} \\ A_{21} & A_{22} \end{pmatrix},$$

and assume that X_1 is $m \times m$ and nonsingular. Let

$$C \equiv \begin{pmatrix} X_1 & A_{12} \\ X_2 & A_{22} \end{pmatrix}. \tag{1.1.24}$$

Then

$$\sigma(A) = \sigma(B) \cup \sigma(C/X_1).$$

Proof. Let

$$S = \begin{pmatrix} X_1 & 0 \\ X_2 & I \end{pmatrix} \text{ so that } S^{-1} = \begin{pmatrix} X_1^{-1} & 0 \\ -X_2X_1^{-1} & I \end{pmatrix}.$$

The equation $AX = XB$ ensures that $AS = A(X, \star) = (AX, \star) = (XB, \star)$, so

$$AS = \begin{pmatrix} X_1B & A_{12} \\ X_2B & A_{22} \end{pmatrix}$$

and

$$S^{-1}AS = \begin{pmatrix} B & \star \\ 0 & A_{22} - X_2 X_1^{-1} A_{12} \end{pmatrix}.$$

Since $A_{22} - X_2 X_1^{-1} A_{12} = C/X_1$, we have

$$\sigma(A) = \sigma\left(S^{-1}AS\right) = \sigma(B) \cup \sigma(C/X_1). \blacksquare$$

If $AX = XB$ and $\operatorname{rank}(X) = m$ but the first m rows of X are not independent, let P be a permutation matrix such that the first m rows of PX are independent. Then $(PAP^T)(PX) = (PX)B$ and we can apply the preceding theorem to PAP^T, B, and PX.

As an application of Theorem 1.10, suppose that A has m linearly independent (column) eigenvectors $x_1, \ldots, x_m$ corresponding, respectively, to the not-necessarily distinct eigenvalues $\lambda_1, \ldots, \lambda_m$. Let $X = (x_1, \ldots, x_m)$. Then

$$AX = X \operatorname{diag}(\lambda_1, \ldots, \lambda_m),$$

so $\operatorname{diag}(\lambda_1, \ldots, \lambda_m)$ plays the role of B in the preceding theorem. Partition X as in the theorem and suppose that X_1 is nonsingular. If C is defined by (1.1.24), then $\sigma(A) = \{\lambda_1, \ldots, \lambda_m\} \cup \sigma(C/X_1)$.

We now turn our attention to *skew block upper triangular matrices*.

Theorem 1.11 *Let $A \in \mathbb{C}^{m\times m}$ be Hermitian, $B \in \mathbb{C}^{m\times n}$ have rank r. Let*

$$M = \begin{pmatrix} A & B \\ B^* & 0 \end{pmatrix}.$$

Then $\operatorname{In}(M) \geq (r,\ r,\ 0)$. *If B is nonsingular, then* $\operatorname{In}(M) = (m,\ m,\ 0)$.

Proof. Let C and D be nonsingular matrices such that

$$CBD = \begin{pmatrix} I_r & 0 \\ 0 & 0 \end{pmatrix}.$$

Partition

$$CAC^* = \begin{pmatrix} E_1 & E_2 \\ E_2^* & E_3 \end{pmatrix}$$

and compute

$$\begin{pmatrix} C & 0 \\ 0 & D^* \end{pmatrix}\begin{pmatrix} A & B \\ B^* & 0 \end{pmatrix}\begin{pmatrix} C^* & 0 \\ 0 & D \end{pmatrix} = \begin{pmatrix} E_1 & E_2 & I_r & 0 \\ E_2^* & E_3 & 0 & 0 \\ I_r & 0 & 0 & 0 \\ 0 & 0 & 0 & 0 \end{pmatrix},$$

which is block permutationally $*$-congruent to

$$N \equiv \begin{pmatrix} E_1 & I_r & E_2 & 0 \\ I_r & 0 & 0 & 0 \\ E_2^* & 0 & E_3 & 0 \\ 0 & 0 & 0 & 0 \end{pmatrix}.$$

Let

$$K \equiv \begin{pmatrix} E_1 & I_r \\ I_r & 0 \end{pmatrix}$$

denote the upper left 2×2 block of N and compute

$$K^{-1} = \begin{pmatrix} 0 & I \\ I & -E_1 \end{pmatrix} \quad \text{and} \quad N/K = \begin{pmatrix} E_3 & 0 \\ 0 & 0 \end{pmatrix}.$$

Thus,

$$\text{In}(M) = \text{In}(N) = \text{In}(K) + \text{In}(N/K) = \text{In}\begin{pmatrix} E_1 & I_r \\ I_r & 0 \end{pmatrix} + \text{In}\begin{pmatrix} E_3 & 0 \\ 0 & 0 \end{pmatrix}.$$

Let $E_1 = U\Lambda U^*$ be a spectral decomposition, with $\Lambda = \text{diag}(\lambda_1, \cdots, \lambda_r)$ and all λ_i real. Then

$$\begin{pmatrix} U^* & 0 \\ 0 & I_r \end{pmatrix}\begin{pmatrix} E_1 & I_r \\ I_r & 0 \end{pmatrix}\begin{pmatrix} U & 0 \\ 0 & I_r \end{pmatrix} = \begin{pmatrix} \Lambda & I_r \\ I_r & 0 \end{pmatrix}$$

is permutation similar to

$$\begin{pmatrix} \lambda_1 & 1 \\ 1 & 0 \end{pmatrix} \oplus \cdots \oplus \begin{pmatrix} \lambda_r & 1 \\ 1 & 0 \end{pmatrix}. \tag{1.1.25}$$

The eigenvalues of the ith direct summand in (1.1.25) are $\left(\lambda_i \pm \sqrt{\lambda_i^2 + 4}\right)/2$, of which one is positive and one is negative. Thus,

$$\text{In}\begin{pmatrix} E_1 & I_r \\ I_r & 0 \end{pmatrix} = (r,\ r,\ 0)$$

and hence $\text{In}(M) \geq (r,\ r,\ 0)$.

If B is nonsingular, then $m = n = r$ and $\text{In}(M) = (m,\ m,\ 0)$. ∎

We note that the inertia of a general matrix is studied in [109, 344].

1.4 Positive semidefinite matrices

In this section we present some elementary matrix inequalities involving positive semidefinite matrices; more advanced results are in the later chapters. A fundamental and very useful fact is an immediate consequence of Theorem 1.6.

Theorem 1.12 *Let A be a Hermitian matrix partitioned as*

$$A = \begin{pmatrix} A_{11} & A_{12} \\ A_{12}^* & A_{22} \end{pmatrix}, \tag{1.1.26}$$

in which A_{11} is square and nonsingular. Then

(a) $A > 0$ *if and only if both* $A_{11} > 0$ *and* $A/A_{11} > 0$.

(b) $A \geq 0$ *if and only if* $A_{11} > 0$ *and* $A/A_{11} \geq 0$.

Thus, if $A \geq 0$ and $A_{11} > 0$, then $A/A_{11} = A_{22} - A_{12}^* A_{11}^{-1} A_{12} \geq 0$, so $A_{22} \geq A/A_{11}$. Consequently,

$$\det A_{22} \geq \det(A/A_{11}) = (\det A) \,/\, (\det A_{11}) \geq 0,$$

which (after a continuity argument) proves

Theorem 1.13 (Fischer's Inequality) *Let A be a positive semidefinite matrix partitioned as in (1.1.26). Then*

$$\det A \leq (\det A_{11})\,(\det A_{22}) \,.$$

Since $\det A = \det A_{11} \det A/A_{11}$ and $\det A_{11} \geq (\det A) \,/\, (\det A_{22})$, there is a reversed Fischer inequality if A_{22} is nonsingular (for example, if A is positive definite):

$$\det(A/A_{11}) \det(A/A_{22}) \leq \det A.$$

As an application of the Fischer inequality, we give a determinantal inequality. Let A, B, C, D be square matrices of the same size, so that

$$\begin{pmatrix} A & B \\ C & D \end{pmatrix} \begin{pmatrix} A^* & C^* \\ B^* & D^* \end{pmatrix} = \begin{pmatrix} AA^* + BB^* & AC^* + BD^* \\ CA^* + DB^* & CC^* + DD^* \end{pmatrix} \geq 0.$$

Then

$$\left| \det \begin{pmatrix} A & B \\ C & D \end{pmatrix} \right|^2 \leq \det(AA^* + BB^*) \det(CC^* + DD^*).$$

If A and C commute, then

$$|\det(AD - CB)|^2 \leq \det(AA^* + BB^*)\det(CC^* + DD^*).$$

The Fischer inequality and an induction gives the celebrated Hadamard inequality.

Corollary 1.2 (Hadamard's Inequality) *Let $A = (a_{ij})$ be an $n \times n$ positive semidefinite matrix. Then*

$$\det A \leq a_{11} \cdots a_{nn}$$

with equality if and only if either A is diagonal or has a zero main diagonal entry (and hence a zero row and column).

We next study the Schur complements of some elementary functions applied to positive semidefinite matrices. It is known that

$$A^2[\alpha] \geq (A[\alpha])^2, \quad A^{1/2}[\alpha] \leq (A[\alpha])^{1/2}, \quad A^{-1}[\alpha] \geq (A[\alpha])^{-1} \qquad (1.1.27)$$

if A is positive semidefinite; see [17] or [468, p. 177]. If we replace submatrices by Schur complements, the inequalities in (1.1.27) are reversed.

Theorem 1.14 *Let A be positive definite and α be a given index set. Then*

$$A^2/\alpha \leq (A/\alpha)^2, \quad (A/\alpha)^{1/2} \leq A^{1/2}/\alpha, \quad A^{-1}/\alpha \leq (A/\alpha)^{-1}.$$

Proof. The assertion for the inverse follows from the inverse part of (1.1.27) and two applications of (1.1.9):

$$\begin{aligned}(A/\alpha)^{-1} &= A^{-1}[\alpha^c] \geq (A[\alpha^c])^{-1} = (A/\alpha^c)^{-1} \\ &= \left((A^{-1}/\alpha^c)^{-1}\right)^{-1} = A^{-1}/\alpha.\end{aligned}$$

For the square, we follow the same steps and use the fact that the inverse function reverses the Löwner partial order:

$$\begin{aligned}A^2/\alpha &= \left((A^2)^{-1}[\alpha^c]\right)^{-1} = \left((A^{-1})^2[\alpha^c]\right)^{-1} \qquad (1.1.28)\\ &\leq \left(A^{-1}[\alpha^c]\right)^{-2} = \left((A^{-1}[\alpha^c])^{-1}\right)^2 = (A/\alpha)^2.\end{aligned}$$

Replacing A with $A^{1/2}$ in (1.1.28) gives $A/\alpha \leq (A^{1/2}/\alpha)^2$; using the fact that the square root preserves the Löwner partial order then gives the asserted inequality for the square root: $(A/\alpha)^{1/2} \leq A^{1/2}/\alpha$. ∎

Theorem 1.15 *Let A and B be $n \times n$ positive definite matrices, and let X and Y be $n \times m$. Then*

$$X^* A^{-1} X + Y^* B^{-1} Y \geq (X+Y)^* (A+B)^{-1} (X+Y).$$

Proof. Let

$$M = \begin{pmatrix} A & X \\ X^* & X^* A^{-1} X \end{pmatrix} \quad \text{and} \quad N = \begin{pmatrix} B & Y \\ Y^* & Y^* B^{-1} Y \end{pmatrix}.$$

Theorem 1.12 (b) ensures that M and N, and hence $M+N$, are positive semidefinite. The Schur complement of $A+B$ in $M+N$ is

$$(M+N)/(A+B) = X^* A^{-1} X + Y^* B^{-1} Y - (X+Y)^* (A+B)^{-1} (X+Y),$$

which Theorem 1.12 tells us is positive semidefinite. ∎

Theorem 1.15 can be found in [213]. It extends a result of M. Marcus [294] from vectors to matrices. The equality case was studied by Fiedler and Markham in [167] and the analogous inequality for the Hadamard product was proved in [297, 299, 453, 226]. The next theorem [468, p. 189] illustrates again how identities involving the inverse and Schur complement can be used to obtain matrix and determinantal inequalities.

Theorem 1.16 *Let A, B, and X be $n \times n$ matrices. Then*

$$\begin{aligned} AA^* + BB^* &= (B+AX)(I+X^*X)^{-1}(B+AX)^* \\ &\quad + (A-BX^*)(I+XX^*)^{-1}(A-BX^*)^*. \end{aligned} \tag{1.1.29}$$

Proof. Let

$$P = \begin{pmatrix} I & X^* \\ B & A \end{pmatrix} \begin{pmatrix} I & B^* \\ X & A^* \end{pmatrix} = \begin{pmatrix} I + X^*X & B^* + X^*A^* \\ B + AX & AA^* + BB^* \end{pmatrix}.$$

First assume that $A - BX^*$ is nonsingular, so P is nonsingular and

$$\begin{aligned} P^{-1} &= \begin{pmatrix} I & B^* \\ X & A^* \end{pmatrix}^{-1} \begin{pmatrix} I & X^* \\ B & A \end{pmatrix}^{-1} \\ &= \begin{pmatrix} I + B^*(A^* - XB^*)^{-1}X & -B^*(A^* - XB^*)^{-1} \\ -(A^* - XB^*)^{-1}X & (A^* - XB^*)^{-1} \end{pmatrix} \\ &\quad \cdot \begin{pmatrix} I + X^*(A - BX^*)^{-1}B & -X^*(A - BX^*)^{-1} \\ -(A - BX^*)^{-1}B & (A - BX^*)^{-1} \end{pmatrix}. \end{aligned}$$

Compute the (2,2) block of this product and use (1.1.6) to get the identity

$$\begin{aligned} (P/(I + X^*X))^{-1} &= (A^* - XB^*)^{-1} X X^* (A - BX^*)^{-1} \\ &\quad + (A^* - XB^*)^{-1} (A - BX^*)^{-1} \\ &= (A^* - XB^*)^{-1} (I + XX^*) (A - BX^*)^{-1}. \end{aligned}$$

Taking the inverse of both sides gives

$$P/(I + X^*X) = (A - BX^*)(I + XX^*)^{-1}(A^* - XB^*).$$

On the other hand, we can compute directly the Schur complement of $I + X^*X$ in P:

$$P/(I + X^*X) = AA^* + BB^* - (B + AX)(I + X^*X)^{-1}(B + AX)^*.$$

The asserted identity results from equating these two representations for $P/(I + X^*X)$.

If $A - BX^*$ is singular, the desired equality follows from a continuity argument, that is, replace A with $A + \varepsilon I$ and let $\varepsilon \to 0$. ∎

Since both summands on the right hand side of (1.1.29) are positive semidefinite, we obtain an inequality by omitting either of them, e.g.,

$$AA^* + BB^* \geq (B + AX)(I + X^*X)^{-1}(B + AX)^*,$$

which implies the determinant inequality

$$\det(AA^* + BB^*)\det(I + X^*X) \geq |\det(B + AX)|^2.$$

1.5 Hadamard products and the Schur complement

The *Hadamard product (Schur product)* of matrices $A = (a_{ij})$ and $B = (b_{ij})$ of the same size is the entrywise product

$$A \circ B = (a_{ij}b_{ij}).$$

Unlike the ordinary matrix product, $A \circ B = B \circ A$ always. The identity for the Schur product is the matrix J, all of whose entries are 1.

The Hadamard product of two Hermitian matrices is evidently Hermitian; it is positive semidefinite if both factors are positive semidefinite, and it is positive definite if one factor is positive definite and the other is positive semidefinite and has positive main diagonal entries (in particular, if both factors are positive definite). Proofs of these basic facts can be approached by writing each factor as a positive linear combination of rank one matrices and using bilinearity [228, Section 7.5], or by recognizing that the Hadamard product is a principal submatrix of the Kronecker product (tensor product) $A \otimes B$ [230, Section 5.1]. We shall focus on the Schur complement and the Hadamard product. For the Schur complement of the Kronecker product of matrices, see [286].

Schur complements can be useful in discovering and proving matrix inequalities involving Hadamard inequalities; after a preliminary lemma, we illustrate this principle with several examples.

Lemma 1.2 *If A and B are $n \times n$ positive semidefinite matrices, then*

$$\det(A+B) \geq \det A + \det B. \tag{1.1.30}$$

If A is positive definite, then equality holds if and only if $B = 0$.

Proof. To establish the inequality, suppose A is positive definite and let $C = A^{-1/2}BA^{-1/2}$. Then $C \geq 0$, $\det C = (\det B) / \det A$, and

$$\begin{aligned}
\det(A+B) &= \det\left(A^{1/2}\left(I + A^{-1/2}BA^{-1/2}\right)A^{1/2}\right) \\
&= (\det A)(\det(I+C)) \\
&= (\det A)\prod_{i=1}^{n}\lambda_i(I+C) \\
&= (\det A)\prod_{i=1}^{n}(1+\lambda_i(C)) \\
&\geq (\det A)(1+\operatorname{tr} C + \det C) \\
&\geq (\det A)(1+\det C) = \det A + \det B.
\end{aligned}$$

The last inequality is an equality if and only if $\operatorname{tr} C = 0$, that is, $C = 0$, since C is positive semidefinite, while $C = 0$ if and only if $B = 0$. The inequality (1.1.30) for a general $A \geq 0$ follows by a continuity argument. ∎

We now present an analog of (1.1.30) for Hadamard product [298, 341].

Theorem 1.17 (Oppenheim's Inequality) *Let $A = (a_{ij})$ and B be $n \times n$ positive definite matrices. Then*

$$\det(A \circ B) \geq (a_{11} \cdots a_{nn}) \det B$$

with equality if and only if B is diagonal.

Proof. We use induction on the order of A and B. The case $n = 1$ is obvious. Assume that $n > 1$ and that the assertion is true for all positive definite matrices of order less than n.

Partition A and B conformally as

$$A = \begin{pmatrix} a_{11} & \alpha \\ \alpha^* & A_{22} \end{pmatrix} \quad \text{and} \quad B = \begin{pmatrix} b_{11} & \beta \\ \beta^* & B_{22} \end{pmatrix},$$

in which A_{22} and B_{22} are of order $n-1$. Let $\tilde{A} = a_{11}^{-1}\alpha^*\alpha$ and $\tilde{B} = b_{11}^{-1}\beta^*\beta$. Then $A/a_{11} = A_{22} - \tilde{A} > 0$ and $B/b_{11} = B_{22} - \tilde{B} > 0$. A computation reveals that

$$\begin{aligned}
A_{22} \circ (B/b_{11}) + (A/a_{11}) \circ \tilde{B} &= A_{22} \circ \left(B_{22} - \tilde{B}\right) + \left(A_{22} - \tilde{A}\right) \circ \tilde{B} \\
&= A_{22} \circ B_{22} - A_{22} \circ \tilde{B} + A_{22} \circ \tilde{B} - \tilde{A} \circ \tilde{B} \\
&= A_{22} \circ B_{22} - \tilde{A} \circ \tilde{B} \\
&= (A \circ B)/(a_{11}b_{11}).
\end{aligned}$$

Using Lemma 1.2 and the induction hypothesis, it follows that

$$\begin{aligned}
\det(A \circ B) &= a_{11}b_{11} \det\left[(A \circ B)/(a_{11}b_{11})\right] \\
&= a_{11}b_{11} \det[A_{22} \circ (B/b_{11}) + (A/a_{11}) \circ \tilde{B}] \\
&\geq a_{11}b_{11} \det[A_{22} \circ (B/b_{11})] + a_{11}b_{11} \det[(A/a_{11}) \circ \tilde{B}] \\
&\geq a_{11}b_{11} \det[A_{22} \circ (B/b_{11})] \\
&\geq a_{11}b_{11}\,(a_{22} \cdots a_{nn}) \det(B/b_{11}) \\
&= (a_{11} \cdots a_{nn}) \det B.
\end{aligned}$$

If $\det(A \circ B) = (a_{11} \cdots a_{nn}) \det B$, then each of the preceding three inequalities is an equality. In particular,

$$\det[A_{22} \circ (B/b_{11}) + (A/a_{11}) \circ \tilde{B}] = \det[A_{22} \circ (B/b_{11})] + a_{11}b_{11} \det[(A/a_{11}) \circ \tilde{B}],$$

so the case of equality in Lemma 1.2 ensures that $(A/a_{11}) \circ \tilde{B} = 0$. But A/a_{11} is positive definite, so all its main diagonal entries are positive. We conclude that $\tilde{B} = 0$ and the induction is complete. ∎

Combining the Oppenheim inequality and the Hadamard determinantal inequality (and a continuity argument) shows that for any $n \times n$ positive semidefinite matrices A and B,

$$\det(A \circ B) \geq \det A \det B.$$

Theorem 1.18 *Let A and B be $n \times n$ positive definite matrices, and let $e \in \mathbb{C}^{n-1}$ denote the column vector all of whose entries are 1. Then*

$$A^{-1} \circ B^{-1} \geq (A \circ B)^{-1},$$

$$A \circ A^{-1} \geq I,$$

and

$$A \circ B \geq \left(e^T A^{-1} e\right)^{-1} B. \tag{1.1.31}$$

Proof. Define the Hermitian matrices

$$\mathcal{A} = \begin{pmatrix} A & I \\ I & A^{-1} \end{pmatrix} \quad \text{and} \quad \mathcal{B} = \begin{pmatrix} B & I \\ I & B^{-1} \end{pmatrix}.$$

Then $\mathcal{A}/A = A^{-1} - IA^{-1}I = 0$ and $\mathcal{B}/B = B^{-1} - IB^{-1}I = 0$, so Theorem 1.12 ensures that both $\mathcal{A}$ and $\mathcal{B}$ are positive semidefinite. Thus,

$$\mathcal{A} \circ \mathcal{B} = \begin{pmatrix} A \circ B & I \\ I & A^{-1} \circ B^{-1} \end{pmatrix}$$

is positive semidefinite and $A \circ B$ is positive definite. Theorem 1.12 now ensures that $(\mathcal{A} \circ \mathcal{B}) / (A \circ B) = A^{-1} \circ B^{-1} - (A \circ B)^{-1} \geq 0$, that is,

$$A^{-1} \circ B^{-1} \geq (A \circ B)^{-1}.$$

Now define

$$\mathcal{C} = \begin{pmatrix} A^{-1} & I \\ I & A \end{pmatrix}.$$

Then $\mathcal{C}$ is positive semidefinite, as is

$$\mathcal{A} \circ \mathcal{C} = \begin{pmatrix} A \circ A^{-1} & I \\ I & A \circ A^{-1} \end{pmatrix}. \tag{1.1.32}$$

Let $A \circ A^{-1} = U \Lambda U^*$ be a spectral decomposition of A so that $\Lambda = \operatorname{diag}(\lambda_1, \cdots, \lambda_n)$ is positive diagonal. Then

$$\begin{pmatrix} U & 0 \\ 0 & U \end{pmatrix}^* \begin{pmatrix} A \circ A^{-1} & I \\ I & A \circ A^{-1} \end{pmatrix} \begin{pmatrix} U & 0 \\ 0 & U \end{pmatrix} = \begin{pmatrix} \Lambda & I \\ I & \Lambda \end{pmatrix}$$

is positive semidefinite and permutation similar to

$$\begin{pmatrix} \lambda_1 & 1 \\ 1 & \lambda_1 \end{pmatrix} \oplus \cdots \oplus \begin{pmatrix} \lambda_n & 1 \\ 1 & \lambda_n \end{pmatrix}.$$

Since each direct summand is positive semidefinite if and only if each $\lambda_i \geq 1$, (1.1.32) implies that $\Lambda \geq I$ and hence that $A \circ A^{-1} = U \Lambda U^* \geq UIU^* = I$.

For the last assertion, consider

$$\mathcal{D} = \begin{pmatrix} A & ee^T \\ ee^T & (e^T A^{-1} e)\, ee^T \end{pmatrix}.$$

Then $\mathcal{D}/A = (e^T A^{-1} e)\, ee^T - ee^T A^{-1} ee^T = 0$, so Theorem 1.12 ensures that $\mathcal{D} \geq 0$ and hence that

$$\begin{aligned} \mathcal{D} \circ \mathcal{B} &= \begin{pmatrix} A & ee^T \\ ee^T & (e^T A^{-1} e)\, ee^T \end{pmatrix} \circ \begin{pmatrix} B & I \\ I & B^{-1} \end{pmatrix} \\ &= \begin{pmatrix} A \circ B & I \\ I & (e^T A^{-1} e)\, B^{-1} \end{pmatrix} \end{aligned}$$

is positive semidefinite. A final application of Theorem 1.12 tells us that

$$(\mathcal{D} \circ \mathcal{B}) / \left(\left(e^T A^{-1} e \right) B^{-1} \right) = A \circ B - \left(e^T A^{-1} e \right)^{-1} B \geq 0,$$

which is the desired inequality. ∎

The inverse inequalities on the Hadamard product in Theorem 1.18 are well known; see [243] and [30].

1.6 The generalized Schur complement

In the definition (1.1.7) of the Schur complement, we assumed that the submatrix $A[\alpha]$ is square and nonsingular. We now introduce generalized inverses and allow $A[\alpha]$ to be an arbitrary submatrix. A *generalized inverse* for a given $m \times n$ matrix M is an $n \times m$ matrix M^- (not necessarily unique) such that $MM^-M = M$. Of course, if M is square and nonsingular, its only generalized inverse is the ordinary inverse.

Two basic properties of a generalized inverse are:

$$M = M(M^*M)^-(M^*M) \tag{1.1.33}$$

and

$$M^* = (M^*M)(M^*M)^-M^*. \tag{1.1.34}$$

If an $m \times n$ matrix M has rank r, there are always nonsingular matrices P and Q such that

$$M = P \begin{pmatrix} A & 0 \\ 0 & 0 \end{pmatrix} Q$$

and A is $r \times r$ and nonsingular; in fact, we may even take $A = I_r$. The set of all generalized inverses of M is then

$$\left\{ Q^{-1} \begin{pmatrix} A^{-1} & X \\ Y & Z \end{pmatrix} P^{-1} : X,\ Y,\ Z \text{ arbitrary} \right\}.$$

The generalized inverses are closely related to column space inclusions. The matrix MM^- acts on a matrix N like an identity matrix, that is,

$$MM^-N = N,$$

if and only if the column space of N is contained in that of M, which we denote by $\mathcal{C}(N) \subseteq \mathcal{C}(M)$. It is known that two matrices M and N have the same sets of generalized inverses if and only if $M = N$. Also, it is known that for nonzero X and Y, XM^-Y is the same matrix for every choice of generalized inverse M^- if and only if

$$\mathcal{C}(Y) \subseteq \mathcal{C}(M) \quad \text{and} \quad \mathcal{C}(X^*) \subseteq \mathcal{C}(M^*).$$

The preceding criterion can be formulated as a pair of set inclusions for null spaces (denoted by $\mathcal{N}(\cdot)$) since $\mathcal{C}(B) \subseteq \mathcal{C}(A)$ if and only if $\mathcal{N}(A^*) \subseteq \mathcal{N}(B^*)$.

Because of its intimate connection with regression and least squares, perhaps the best known generalized inverse is the *Moore–Penrose generalized inverse* $A^\dagger$, which is the unique matrix X such that

$$AXA = A, \quad XAX = X, \quad (AX)^* = AX, \quad (XA)^* = XA. \tag{1.1.35}$$

If

$$A = V \begin{pmatrix} \Sigma & 0 \\ 0 & 0 \end{pmatrix} U^*$$

is a singular value decomposition of A, in which Σ is positive diagonal and U and V are unitary, then

$$A^\dagger = U \begin{pmatrix} \Sigma^{-1} & 0 \\ 0 & 0 \end{pmatrix} V^*.$$

We now use the Moore–Penrose inverse to define the Schur complement.

Let A be an $m \times n$ matrix, and let α and β be subsets of $\{1, 2, \ldots, m\}$ and $\{1, 2, \ldots, n\}$, respectively. The *Schur complement* of $A[\alpha, \beta]$ in A is

$$A/A[\alpha, \beta] = A[\alpha^c, \beta^c] - A[\alpha^c, \beta]\, A[\alpha, \beta]^\dagger\, A[\alpha, \beta^c]. \tag{1.1.36}$$

It is usually convenient to think of $A[\alpha, \beta]$ as being in the upper left corner of A (not necessarily square), a placement that can always be achieved with suitable row and column permutations, that is, with permutation matrices P and Q such that

$$PAQ = \begin{pmatrix} A[\alpha, \beta] & A[\alpha, \beta^c] \\ A[\alpha^c, \beta] & A[\alpha^c, \beta^c] \end{pmatrix}.$$

If $\alpha = \beta$ and $m = n$, $A[\alpha, \beta]$ is a principal submatrix of A and $P = Q^T$.

In order to consider replacing the Moore–Penrose generalized inverse in (1.1.36) with an unspecified generalized inverse, we would have to impose conditions sufficient to ensure that the generalized Schur complement obtained in this way did not depend on the choice of the generalized inverse. This would be the case if the row space of $A[\alpha^c, \beta]$ is contained in that of $A[\alpha, \beta]$ and the column space of $A[\alpha, \beta^c]$ is contained in that of $A[\alpha, \beta]$. For the standard presentation

$$M = \begin{pmatrix} A & B \\ C & D \end{pmatrix},$$

if we assume that $\mathcal{C}(B) \subseteq \mathcal{C}(A)$ and $\mathcal{C}(C^*) \subseteq \mathcal{C}(A^*)$, then $M/A = D - CA^-B$ is well-defined since the second term is independent of our choice of the generalized inverse. Therefore,

$$\begin{pmatrix} I & 0 \\ -CA^- & I \end{pmatrix} \begin{pmatrix} A & B \\ C & D \end{pmatrix} \begin{pmatrix} I & -A^-B \\ 0 & I \end{pmatrix} = \begin{pmatrix} A & 0 \\ 0 & M/A \end{pmatrix}$$

and hence

$$\operatorname{rank} M = \operatorname{rank} A + \operatorname{rank}(M/A).$$

If we impose no inclusion conditions on the row and column spaces, however, it is possible that $\operatorname{rank} M > \operatorname{rank} A + \operatorname{rank}(M/A)$.

Row and column space inclusions of the type relevant to the generalized Schur complement arise naturally in the context of positive definite block matrices.

Theorem 1.19 *Suppose M is positive semidefinite and partitioned as*

$$M = \begin{pmatrix} A & B \\ B^* & C \end{pmatrix},$$

in which A and C are square. Then i) there is a matrix R such that $B = AR$; ii) $\mathcal{C}(B) \subseteq \mathcal{C}(A)$, and iii) $B = AA^-B$. Also, iv) there is a matrix L such that $B = LC$; v) $\mathcal{R}(B) \subseteq \mathcal{R}(C)$; and vi) $B = BC^-C$.

Proof. The first three stated conditions are equivalent; we consider just i).

Since $M \geq 0$, it has a unique positive semidefinite square root, whose columns we partition conformally to those of M. Let $M^{1/2} = (S, T)$. Then

$$M = \left(M^{1/2}\right)^2 = \left(M^{1/2}\right)^* \left(M^{1/2}\right) = \begin{pmatrix} S^*S & S^*T \\ \star & \star \end{pmatrix}.$$

Let $S = XP$ be a polar decomposition, in which X has orthonormal columns and P is positive semidefinite. Then $A = S^*S = P^2$, so $P = A^{1/2}$ and $B = S^*T = PX^*T = P^2P^\dagger X^*T = A\left(P^\dagger X^*T\right)$. Thus, we may take $R = P^\dagger X^*T$ in i).

The second set of three conditions can be dealt with in a similar fashion by considering the second block row of M. ∎

For any positive semidefinite block matrix M partitioned as in the preceding theorem, the Schur complements $M/A = C - B^*A^-B$ and $M/C = A - BC^-B^*$ are well defined, so they may be computed using any generalized inverse.

We now rephrase Theorem 1.12 in terms of a singular principal submatrix as follows [6].

Theorem 1.20 *Suppose M is Hermitian and partitioned as*

$$M = \begin{pmatrix} A & B \\ B^* & C \end{pmatrix},$$

in which A and C are square (not necessarily of the same order). Then $M \geq 0$ if and only if $A \geq 0$, $\mathcal{C}(B) \subseteq \mathcal{C}(A)$, and $M/A \geq 0$.

Proof. The previous theorem ensures the necessity of the three stated conditions. To show that they are sufficient, observe that $M/A = C - B^*A^-B$ is well defined since the condition $\mathcal{C}(B) \subseteq \mathcal{C}(A)$ ensures the uniqueness of B^*A^-B. The matrix identity

$$\begin{pmatrix} I & 0 \\ -B^*A^- & I \end{pmatrix} \begin{pmatrix} A & B \\ B^* & C \end{pmatrix} \begin{pmatrix} I & -A^-B \\ 0 & I \end{pmatrix} = \begin{pmatrix} A & 0 \\ 0 & C - B^*A^-B \end{pmatrix}$$

now ensures that M is positive semidefinite. ∎

We now consider a generalized inverse analog of the representation (1.1.5) for the block form of the inverse, but only for positive semidefinite matrices that satisfy a special rank condition. When we do not have positive definiteness, the situation is more complicated; see Chapter 6.

Theorem 1.21 *Suppose M is positive semidefinite and partitioned as*

$$M = \begin{pmatrix} A & B \\ B^* & C \end{pmatrix},$$

in which A and C are square. Let

$$X = \begin{pmatrix} A^\dagger + A^\dagger B\,(M/A)^\dagger\, B^* A^\dagger & -A^\dagger B\,(M/A)^\dagger \\ -\,(M/A)^\dagger\, B^* A^\dagger & (M/A)^\dagger \end{pmatrix}. \tag{1.1.37}$$

Then $X = M^\dagger$ if and only if $\operatorname{rank} M = \operatorname{rank} A + \operatorname{rank} C$.

Proof. Denote the generalized Schur complement of A in M by $M/A = S = C - B^*A^\dagger B$. Use (1.1.33), (1.1.34), and (1.1.37) to compute

$$\begin{aligned} XMX &= \begin{pmatrix} A^\dagger A A^\dagger + A^\dagger B\left(S^\dagger S S^\dagger\right) B^* A^\dagger & -A^\dagger B\left(S^\dagger S S^\dagger\right) \\ -\left(S^\dagger S S^\dagger\right) B^* A^\dagger & \left(S^\dagger S S^\dagger\right) \end{pmatrix} \\ &= \begin{pmatrix} A^\dagger + A^\dagger B S^\dagger B^* A^\dagger & -A^\dagger B S^\dagger \\ -S^\dagger B^* A^\dagger & S^\dagger \end{pmatrix} = X, \end{aligned}$$

$$MXM = M,$$

$$MX = \begin{pmatrix} AA^\dagger & 0 \\ (I - SS^\dagger)B^*A^\dagger & SS^\dagger \end{pmatrix},$$

and

$$XM = \begin{pmatrix} A^\dagger A & A^\dagger B(I - S^\dagger S) \\ 0 & S^\dagger S \end{pmatrix}.$$

Thus, two of the four identities (1.1.35) that characterize the Moore-Penrose inverse of M are satisfied. The remaining two identities in (1.1.35) are satisfied if and only if MX and XM are Hermitian, that is, if and only if $A^\dagger B(I - S^\dagger S) = 0$. Use the spectral decomposition theorem to write

$$A = U \begin{pmatrix} 0 & 0 \\ 0 & \Lambda_1 \end{pmatrix} U^* \quad \text{and} \quad C = V \begin{pmatrix} \Lambda_2 & 0 \\ 0 & 0 \end{pmatrix} V^*,$$

in which U and V are unitary, and Λ_1 and Λ_2 are positive diagonal. Then

$$M = \begin{pmatrix} A & B \\ B^* & C \end{pmatrix} = \begin{pmatrix} U & 0 \\ 0 & V \end{pmatrix} \begin{pmatrix} \begin{matrix} 0 & 0 \\ 0 & \Lambda_1 \end{matrix} & U^*BV \\ V^*B^*U & \begin{matrix} \Lambda_2 & 0 \\ 0 & 0 \end{matrix} \end{pmatrix} \begin{pmatrix} U^* & 0 \\ 0 & V^* \end{pmatrix}.$$

Since a main diagonal entry in a positive semidefinite matrix is zero only if the entire row and column in which it lies is zero, U^*BV is a 2×2 block matrix in which three of the blocks must be zero, so we may write

$$M = \begin{pmatrix} U & 0 \\ 0 & V \end{pmatrix} \begin{pmatrix} 0 & 0 & 0 & 0 \\ 0 & \Lambda_1 & B_1 & 0 \\ 0 & B_1^* & \Lambda_2 & 0 \\ 0 & 0 & 0 & 0 \end{pmatrix} \begin{pmatrix} U^* & 0 \\ 0 & V^* \end{pmatrix}. \tag{1.1.38}$$

Let

$$N = \begin{pmatrix} \Lambda_1 & B_1 \\ B_1^* & \Lambda_2 \end{pmatrix}$$

denote the central 2×2 block matrix in (1.1.38). Then N is positive semidefinite, $\operatorname{rank} M = \operatorname{rank} N$, and the order of N is $\operatorname{rank} \Lambda_1 + \operatorname{rank} \Lambda_2 = \operatorname{rank} A + \operatorname{rank} C$. These two identities show that $\operatorname{rank} M = \operatorname{rank} A + \operatorname{rank} C$ if and only if N is nonsingular, that is, if and only if N is positive definite. Since Λ_1 is positive definite, we see that N is positive definite if and only if $\Gamma \equiv N/\Lambda_1 = \Lambda_2 - B_1^*\Lambda_1^{-1}B_1$ is positive definite.

Now compute

$$\begin{aligned} S &= C - B^*A^\dagger B \\ &= V\left(\begin{pmatrix} \Lambda_2 & 0 \\ 0 & 0 \end{pmatrix} - \begin{pmatrix} 0 & B_1^* \\ 0 & 0 \end{pmatrix} U^*U \begin{pmatrix} 0 & 0 \\ 0 & \Lambda_1^{-1} \end{pmatrix} U^*U \begin{pmatrix} 0 & 0 \\ B_1 & 0 \end{pmatrix}\right) V^* \\ &= V \begin{pmatrix} \Lambda_2 - B_1^*\Lambda_1^{-1}B_1 & 0 \\ 0 & 0 \end{pmatrix} V^* = V \begin{pmatrix} \Gamma & 0 \\ 0 & 0 \end{pmatrix} V^* \end{aligned}$$

and

$$S^\dagger = V \begin{pmatrix} \Gamma^\dagger & 0 \\ 0 & 0 \end{pmatrix} V^*.$$

Thus,

$$I - S^\dagger S = I - V \begin{pmatrix} \Gamma^\dagger \Gamma & 0 \\ 0 & 0 \end{pmatrix} V^* = V \begin{pmatrix} I - \Gamma^\dagger \Gamma & 0 \\ 0 & I \end{pmatrix} V^*$$

and

$$\begin{aligned} A^\dagger B(I - S^\dagger S) &= U \begin{pmatrix} 0 & 0 \\ 0 & \Lambda_1^{-1} \end{pmatrix} U^* U \begin{pmatrix} 0 & 0 \\ B_1 & 0 \end{pmatrix} V^* V \begin{pmatrix} I - \Gamma^\dagger \Gamma & 0 \\ 0 & I \end{pmatrix} V^* \\ &= U \begin{pmatrix} 0 & 0 \\ \Lambda_1^{-1} B_1 & 0 \end{pmatrix} \begin{pmatrix} I - \Gamma^\dagger \Gamma & 0 \\ 0 & I \end{pmatrix} V^* \\ &= U \begin{pmatrix} 0 & 0 \\ \Lambda_1^{-1} B_1 \left(I - \Gamma^\dagger \Gamma\right) & 0 \end{pmatrix} V^*. \end{aligned}$$

To conclude the proof, we must show that $B_1 \left(I - \Gamma^\dagger \Gamma\right) = 0$ if and only if $\Gamma > 0$. Of course, $\Gamma^\dagger \Gamma = I$ if $\Gamma > 0$, so one implication is clear. Now suppose that $B_1 \left(I - \Gamma^\dagger \Gamma\right) = 0$. Then the range of $I - \Gamma^\dagger \Gamma$ is contained in the null space of B_1. But the range of $I - \Gamma^\dagger \Gamma$ is the null space of Γ, so $\Gamma x = 0$ implies that $B_1 x = 0$, which implies that $0 = \Gamma x = \Lambda_2 x - B_1^* \Lambda_1^{-1} B_1 x = \Lambda_2 x$ and $x = 0$. We conclude that $B_1 \left(I - \Gamma^\dagger \Gamma\right) = 0$ only if $\Gamma > 0$. ∎

It is possible to obtain an explicit expression for $M^\dagger$ without assuming the rank condition that $\operatorname{rank} M = \operatorname{rank} A + \operatorname{rank} C$, but it is much more complicated than (1.1.37); see [192]. For more results on the generalized Schur complements and the discussions of generalized inverses of block matrices, see [385, 300, 56, 106, 102, 104, 192] and [204]. Comprehensive survey articles on the Schur complement include [73], [128], and [345] .

Chapter 2

Eigenvalue and Singular Value Inequalities of Schur Complements

2.0 Introduction

The purpose of this chapter is to study inequalities involving eigenvalues and singular values of products and sums of matrices.

In addition to denoting the $m \times n$ matrices with complex (real) entries by $\mathbb{C}^{m\times n}$ ($\mathbb{R}^{m\times n}$), we denote by $\mathbb{H}_n$ the set of $n \times n$ Hermitian matrices, and for an $A \in \mathbb{H}_n$, we arrange the eigenvalues of A in a decreasing order:

$$\lambda_1(A) \geq \lambda_2(A) \geq \cdots \geq \lambda_n(A).$$

The singular values of a matrix $A \in \mathbb{C}^{m\times n}$ are defined to be the square roots of the eigenvalues of the matrix A^*A, denoted and arranged as

$$\sigma_1(A) \geq \sigma_2(A) \geq \cdots \geq \sigma_n(A).$$

For a set of subscript indices $i_1, i_2, \ldots, i_k$, we always assume that $i_1 \leq i_2 \leq \cdots \leq i_k$. Furthermore, if $A \in \mathbb{H}_n$, then $\lambda_{i_t}(A)$ indicates $1 \leq i_t \leq n$.

One of the most important results in matrix analysis is the *Cauchy (eigenvalue) interlacing theorem* (see, e.g., [272, p. 294]). It asserts that the eigenvalues of any principal submatrix of a Hermitian matrix interlace those of the Hermitian matrix. To be precise, if $H \in \mathbb{H}_n$ is partitioned as

$$H = \begin{pmatrix} A & B \\ B^* & D \end{pmatrix}$$

in which A is an $r \times r$ principal submatrix, then for each $i = 1, 2, \ldots, r$,

$$\lambda_i(H) \geq \lambda_i(A) \geq \lambda_{i+n-r}(H).$$

Eigenvalue and singular value problems are a central topic of matrix analysis and have reached out to many other fields. A great number of inequalities on eigenvalues and singular values of matrices are seen in the literature (see, e.g., [228, 230, 272, 301, 438, 452]). Here, we single some of these out for later use.

Let A and B be $n \times n$ complex matrices. Let l be an integer such that $1 \le l \le n$. Then for any index sequence $1 \le i_1 \le \cdots \le i_l \le n$,

$$\prod_{t=1}^{l} \sigma_t(AB) \ge \prod_{t=1}^{l} \sigma_{i_t}(A)\sigma_{n-i_t+1}(B), \tag{2.0.1}$$

$$\prod_{t=1}^{l} \sigma_{i_t}(A)\sigma_t(B) \ge \prod_{t=1}^{l} \sigma_{i_t}(AB) \ge \prod_{t=1}^{l} \sigma_{i_t}(A)\sigma_{n-t+1}(B), \tag{2.0.2}$$

and

$$\min_{i+j=t+1}\{\sigma_i(A)\sigma_j(B)\} \ge \sigma_t(AB) \ge \max_{i+j=t+n}\{\sigma_i(A)\sigma_j(B)\}. \tag{2.0.3}$$

The inequalities on the product ($\prod$) yield the corresponding inequalities on the sum ($\sum$). This is done by majorization in the following sense.

Let $x_1, x_2, \ldots, x_n$ and $y_1, y_2, \ldots, y_n$ be two sequences of nonnegative numbers in the order $x_1 \ge x_2 \ge \cdots \ge x_n$ and $y_1 \ge y_2 \ge \cdots \ge y_n$. Then

$$\prod_{i=1}^{k} x_i \le \prod_{t=1}^{k} y_i, \quad k \le n \quad \Rightarrow \quad \sum_{i=1}^{k} x_i \le \sum_{t=1}^{k} y_i, \quad k \le n \tag{2.0.4}$$

and

$$\sum_{i=1}^{k} x_{(i)} \le \sum_{t=1}^{k} y_{(i)}, \quad k \le n \Rightarrow \prod_{i=1}^{k} x_{(i)} \le \prod_{t=1}^{k} y_{(i)}, \quad k \le n, \tag{2.0.5}$$

where $x_{(1)} \le x_{(2)} \le \cdots \le x_{(n)}$ and $y_{(1)} \le y_{(2)} \le \cdots \le y_{(n)}$ are rearrangements of $x_1, x_2, \ldots, x_n$ and $y_1, y_2, \ldots, y_n$, respectively.

Translations from product to sum or vice versa are often done through (2.0.4) and (2.0.5). For example, by (2.0.2) and (2.0.4), we can get

$$\sum_{t=1}^{l} \sigma_{i_t}(A)\sigma_t(B) \ge \sum_{t=1}^{l} \sigma_{i_t}(AB) \ge \sum_{t=1}^{l} \sigma_{i_t}(A)\sigma_{n-t+1}(B). \tag{2.0.6}$$

We point out that all the above singular value inequalities remain valid when AB is changed to BA; even though $\sigma_i(AB) \ne \sigma_i(BA)$ in general. Moreover they all hold with the replacement of the eigenvalues (λ) by the singular values (σ) when A and B are positive semidefinite. For instance,

$$\sum_{t=1}^{l} \lambda_{i_t}(AB) \ge \sum_{t=1}^{l} \lambda_{i_t}(A)\lambda_{n-t+1}(B). \tag{2.0.7}$$

For the sum of Hermitian matrices, two existing parallel results are

$$\sum_{t=1}^{l} \lambda_{i_t}(A) + \sum_{t=1}^{l} \lambda_t(B) \geq \sum_{t=1}^{l} \lambda_{i_t}(A+B) \geq \sum_{t=1}^{l} \lambda_{i_t}(A) + \sum_{t=1}^{l} \lambda_{n-t+1}(B) \quad (2.0.8)$$

and

$$\min_{i+j=t+1} \{\lambda_i(A) + \lambda_j(B)\} \geq \lambda_t(A+B) \geq \max_{i+j=t+n} \{\lambda_i(A) + \lambda_j(B)\}. \quad (2.0.9)$$

All the above inequalities appear explicitly in Chapter 2 of [451]. We note that the second inequality in (2.0.8) does not hold in general for singular values (σ) [451, p. 113].

2.1 The interlacing property

The Cauchy interlacing theorem states that the eigenvalues of any principal submatrix of a Hermitian matrix interlace those of the grand matrix. Does a Schur complement possess a similar property? That is, do the eigenvalues of a Schur complement in a Hermitian matrix interlace the eigenvalues of the original Hermitian matrix? The answer is negative in general: Take

$$H = \begin{pmatrix} 1 & 2 \\ 2 & 1 \end{pmatrix}, \qquad \alpha = \{1\}.$$

Then $H/\alpha = (-3)$, while the eigenvalues of H are -1 and 3.

In what follows, we show that with a slight modification of the Schur complement (augmented by 0s) the analogous interlacing property holds.

Theorem 2.1 *Let $H \in \mathbb{H}_n$ and let α be an index set with k elements, $1 \leq k < n$. If the principal submatrix $H[\alpha]$ is positive definite, then*

$$\lambda_i(H) \geq \lambda_i(H/\alpha \oplus 0) \geq \lambda_{i+k}(H), \quad i = 1, 2, \ldots, n-k, \quad (2.1.10)$$

and if $H[\alpha]$ is negative definite, i.e., $-H[\alpha]$ is positive definite, then

$$\lambda_i(H) \geq \lambda_{i+k}(H/\alpha \oplus 0) \geq \lambda_{i+k}(H), \quad i = 1, 2, \ldots, n-k. \quad (2.1.11)$$

Proof. Since permutation similarity preserves the eigenvalues, we may assume that $\alpha = \{n-k+1, \ldots, n\}$. With $\alpha^c = \{1, 2, \ldots, n-k\}$, we have

$$H = \begin{pmatrix} H/\alpha & 0 \\ 0 & 0 \end{pmatrix} + \begin{pmatrix} H[\alpha^c, \alpha](H[\alpha])^{-1}H[\alpha, \alpha^c] & H[\alpha^c, \alpha] \\ H[\alpha, \alpha^c] & H[\alpha] \end{pmatrix} \equiv E + F.$$

Let

$$P = \begin{pmatrix} I_{n-k} & -H[\alpha^c, \alpha](H[\alpha])^{-1} \\ 0 & I_k \end{pmatrix}.$$

Then $PFP^* = 0 \oplus H[\alpha]$, so F is positive semidefinite if $H[\alpha] \geq 0$. Moreover

$$\operatorname{rank}(F) = \operatorname{rank}(H[\alpha]) = k < n.$$

Now using (2.1.10) and by (2.0.9), we have

$$\lambda_{i+k}(H) = \lambda_{i+k}(E+F) \leq \lambda_i(E) + \lambda_{k+1}(F) = \lambda_i \left[\begin{pmatrix} H/\alpha & 0 \\ 0 & 0 \end{pmatrix} \right]$$

and

$$\lambda_i(H) = \lambda_i(E+F) \geq \lambda_i(E) + \lambda_n(F) = \lambda_i \left[\begin{pmatrix} H/\alpha & 0 \\ 0 & 0 \end{pmatrix} \right].$$

The inequalities (2.1.11) are proven in a similar manner. Note that if A is a Hermitian matrix, then $\lambda_i(-A) = -\lambda_{n-i+1}(A)$, $i = 1, 2, \ldots, n$. ∎

The theorem immediately yields the following results for positive semidefinite matrices; see [160, 288, 421].

Corollary 2.3 *Let H (or $-H$) be an $n \times n$ positive semidefinite matrix and let $H[\alpha]$ be a $k \times k$ nonsingular principal submatrix, $1 \leq k < n$. Then*

$$\lambda_i(H) \geq \lambda_i(H/\alpha) \geq \lambda_{i+k}(H), \quad i = 1, 2, \ldots, n-k. \tag{2.1.12}$$

Proof. When H is positive semidefinite, H/α is positive semidefinite. It is sufficient to notice that $\lambda_i(H/\alpha \oplus 0) = \lambda_i(H/\alpha)$ for $i = 1, 2, \ldots, n-k$. ∎

Corollary 2.4 *Let H be an $n \times n$ positive semidefinite matrix and let $H[\alpha]$ be a $k \times k$ nonsingular principal submatrix of H, $1 \leq k < n$. Then*

$$\lambda_i(H) \geq \lambda_i(H[\alpha^c]) \geq \lambda_i(H/\alpha) \geq \lambda_{i+k}(H), \quad i = 1, 2, \ldots, n-k. \tag{2.1.13}$$

Proof. Since H, $H[\alpha]$, and $H[\alpha^c]$ are all positive semidefinite, we obtain

$$H[\alpha^c] \geq H[\alpha^c] - H[\alpha^c, \alpha](H[\alpha])^{-1} H[\alpha, \alpha^c] = H/\alpha.$$

The second inequality in (2.1.13) follows at once, while the first inequality is the Cauchy interlacing theorem and the last one is (2.1.12). ∎

Corollary 2.5 *Let H be an $n \times n$ positive semidefinite matrix and let α and α' be nonempty index sets such that $\alpha' \subset \alpha \subset \{1, 2, \ldots, n\}$. If $H[\alpha]$ is nonsingular, then for every $i = 1, 2, \ldots, n - |\alpha|$,*

$$\lambda_i(H/\alpha') \geq \lambda_i(H[\alpha' \cup \alpha^c]/\alpha') \geq \lambda_i(H/\alpha) \geq \lambda_{i+|\alpha|-|\alpha'|}(H/\alpha'). \tag{2.1.14}$$

Proof. Note that $H[\alpha'] > 0$ since it is a principal submatrix of $H[\alpha] > 0$. By the quotient formula on the Schur complement (see Theorem 1.4),

$$H/\alpha = (H/\alpha')/(H[\alpha]/\alpha').$$

With H/α' and $H[\alpha]/\alpha'$ in place of H and $H[\alpha]$, respectively, in Corollary 2.4 and since $(H/\alpha')[\alpha^c] = H[\alpha' \cup \alpha^c]/\alpha'$, (2.1.14) follows. ∎

For the case where H is negative definite, we have the analogs:

$$\lambda_i(H) \geq \lambda_i(H/\alpha) \geq \lambda_i\left(H[\alpha^c]\right) \geq \lambda_{i+k}(H)$$

and

$$\lambda_i(H/\alpha') \geq \lambda_i(H/\alpha) \geq \lambda_i(H[\alpha' \cup \alpha^c]/\alpha') \geq \lambda_{i+|\alpha|-|\alpha'|}(H/\alpha').$$

As we saw, the Cauchy eigenvalue interlacing theorem does not hold for the Schur complement of a Hermitian matrix. We show, however, and interestingly, that it holds for the reciprocals of nonsingular Hermitian matrices. This is not surprising in view of the representation of a Schur complement in terms of a principal submatrix (see Theorem 1.2).

Lemma 2.3 *Let H be an $n \times n$ nonsingular Hermitian matrix and let A be a $k \times k$ nonsingular principal submatrix of H, where $1 \leq k < n$. Then*

$$\lambda_i(H^{-1}) \geq \lambda_i[(H/A)^{-1}] \geq \lambda_{i+k}(H^{-1}), \quad i = 1, 2, \ldots, n-k.$$

Proof. It is sufficient to notice, by Theorem 1.2, that $(H/A)^{-1}$ is a principal submatrix of the Hermitian matrix H^{-1}. ∎

We now extend this to a singular H. That is, we show that if H is any Hermitian matrix and A is a nonsingular principal submatrix of H, then the eigenvalues of $(H/A)^\dagger$ interlace the eigenvalues of $H^\dagger$.

Let $\mathrm{In}(H) = (p,\ q,\ z)$. The eigenvalues $H^\dagger$ are, in decreasing order,

$$\lambda_i(H^\dagger) = \begin{cases} \lambda^{-1}_{p+1-i}(H), & i = 1, \ldots, p, \\ 0, & i = p+1, \ldots, p+z, \\ \lambda^{-1}_{n+p+z+1-i}(H), & i = p+z+1, \ldots, n. \end{cases}$$

Since the eigenvalues of a matrix are continuous functions of the entries of the matrix, the eigenvalues of the Moore–Penrose inverse of a matrix are also continuous functions of the entries of the original matrix.

To establish the interlacing property for any Hermitian H, we need to use the usual trick – continuity argument. Let $H \in \mathbb{H}_n$ and $H_\varepsilon = H + \varepsilon I_n$, where ε is a positive number. Let A be a $k \times k$ nonzero principal submatrix of H and denote $A_\varepsilon = A + \varepsilon I_k$. Choose ε such that it is less than the absolute value of any nonzero eigenvalue of H and A. Thus H_ε, A_ε, and $H_\varepsilon/A_\varepsilon$ are all invertible. It follows that if $\lambda_s(H^\dagger) \neq 0$ and $\lambda_t\left[(H/A)^\dagger\right] \neq 0$,

$$\lim_{\varepsilon \to 0} \lambda_s(H_\varepsilon^{-1}) = \lambda_s(H^\dagger)$$

and

$$\lim_{\varepsilon \to 0} \lambda_t\left[(H_\varepsilon/A_\varepsilon)^{-1}\right] = \lambda_t\left[(H/A)^\dagger\right].$$

Now we are ready to present the following interlacing theorem [421].

Theorem 2.2 *Let H be an $n \times n$ Hermitian matrix and let A be a $k \times k$ nonsingular principal submatrix of H. Then for $i = 1, 2, \ldots, n-k$,*

$$\lambda_i(H^\dagger) \geq \lambda_i[(H/A)^\dagger] \geq \lambda_{i+k}(H^\dagger). \tag{2.1.15}$$

Proof. Let $\mathrm{In}(H) = (p,\ q,\ z)$ and $\mathrm{In}(A) = (p_1,\ q_1,\ 0)$. Consequently, $\mathrm{In}(H/A) = (p - p_1,\ q - q_1,\ z)$ by Theorem 1.6. Without loss of generality, we write $H = \begin{pmatrix} A & B \\ B^* & C \end{pmatrix}$. Let

$$H_\varepsilon = H + \varepsilon I_n = \begin{pmatrix} A + \varepsilon I_k & B \\ B^* & C + \varepsilon I_{n-k} \end{pmatrix} \equiv \begin{pmatrix} A_\varepsilon & B \\ B^* & C_\varepsilon \end{pmatrix},$$

in which ε is such a small positive number that both H_ε and A_ε are nonsingular. Note that $\mathrm{In}(H_\varepsilon) = (p + z,\ q,\ 0)$, $\mathrm{In}(A_\varepsilon) = \mathrm{In}(A)$, and also $\mathrm{In}(K) = \mathrm{In}(K^\dagger)$ for any Hermitian matrix K. Moreover, upon computation, we have $H_\varepsilon/A_\varepsilon = C_\varepsilon - B^* A_\varepsilon^{-1} B$, and thus $\lim_{\varepsilon \to 0} H_\varepsilon/A_\varepsilon = H/A$.

To show that $\lambda_i(H^\dagger) \geq \lambda_i[(H/A)^\dagger]$ for $i = 1, 2, \ldots, n-k$, we consider a set of exhaustive cases on the index i:

Case (1) If $i \leq p - p_1$, then $\lambda_i[(H/A]^\dagger] > 0$. By Lemma 2.3,

$$\lambda_i(H_\varepsilon^{-1}) \geq \lambda_i[(H_\varepsilon/A_\varepsilon)^{-1}] > 0.$$

The desired inequalities follow by taking the limits as $\varepsilon \to 0$.

Case (2) If $p - p_1 < i \leq p + z$, then $\lambda_i(H^\dagger) \geq 0 \geq \lambda_i\left[(H/A)^\dagger\right]$.

Case (3) If $p + z < i \leq n - k$, then, by Lemma 2.3,

$$0 > \lambda_i(H_\varepsilon^{-1}) \geq \lambda_i[(H_\varepsilon/A_\varepsilon)^{-1}].$$

By continuity, we arrive at $0 > \lambda_i(H^\dagger) \geq \lambda_i\left[(H/A)^\dagger\right]$.

To establish the second inequality in (2.1.15), we proceed by exhausting the cases of the index $i + k$:

Case (i) If $i + k \le p$, i.e., $i \le p - k \le n - k$, then, by Lemma 2.3, $\lambda_i[(H_\varepsilon/A_\varepsilon)^{-1}] \ge \lambda_{i+k}(H_\varepsilon^{-1}) > 0$. Letting $\varepsilon \to 0$ yields the inequalities.

Case (ii) If $p + 1 < i + k \le p + z$, then $i \le p - k + z \le p - p_1 + z$, so $\lambda_i[(H/A)^\dagger] \ge 0$ and $\lambda_{i+k}(H^\dagger) = 0$. The inequality then follow.

Case (iii) If $p + z < i + k \le p + k - p_1 + z = p + q_1 + z \le n$, then $i \le p - p_1 + z$, so $\lambda_i[(H/A)^\dagger] \ge 0$ and $\lambda_{i+k}(H^\dagger) < 0$ since $i + k > p + z$.

Case (iv) If $p+z \le p+k-p_1+z < i+k \le n$, then $p-p_1+z < i \le n-k$. By Lemma 2.3, $0 > \lambda_i[(H_\varepsilon/A_\varepsilon)^{-1}] \ge \lambda_{i+k}(H_\varepsilon^{-1})$. Letting $\varepsilon \to 0$ shows that $0 > \lambda_i[(H/A)^\dagger] \ge \lambda_{i+k}(H^\dagger)$. ∎

At the end of this section we note that the converse of the previous theorem is discussed by Hu and Smith in [235].

2.2 Extremal characterizations

The Courant–Fischer min-max principles, or the extremal characterizations, of eigenvalues for Hermitian matrices play an important role in deducing eigenvalue inequalities. For instance, the representation of the minimum eigenvalue $\lambda_{\min}(H)$ of a Hermitian matrix $H \in \mathbb{H}_n$

$$\lambda_{\min}(H) = \min_{x \in \mathbb{C}^n} \{x^*Hx : \ x^*x = 1\}$$

leads immediately to the eigenvalue inequalities: For $A,\ B \in \mathbb{H}_n$

$$\lambda_{\min}(A + B) \ge \lambda_{\min}(A) + \lambda_{\min}(B).$$

We now show extremal characterizations [280] for Schur complements.

Theorem 2.3 *Let H be an $n \times n$ positive semidefinite matrix partitioned as*

$$H = \begin{pmatrix} H_{11} & H_{12} \\ H_{21} & H_{22} \end{pmatrix},$$

where H_{11} is a $k \times k$ leading principal submatrix of H, $1 \le k < n$. Then

$$H/H_{11} = \max_{X \in \mathbb{C}^{(n-k)\times(n-k)}} \{X : \ H - (0_k \oplus X) \ge 0, \ X = X^*\}$$

and

$$H/H_{11} = \min_{Y \in \mathbb{C}^{(n-k)\times k}} \{Y : \ (Y, I_{n-k})H(Y, I_{n-k})^*\}. \tag{2.2.16}$$

Proof. Let X be an $(n-k) \times (n-k)$ Hermitian matrix and set

$$\hat{X} = \begin{pmatrix} 0_k & 0 \\ 0 & X \end{pmatrix}, \qquad T = \begin{pmatrix} I_k & 0 \\ -H_{21}H_{11}^\dagger & I_{n-k} \end{pmatrix}.$$

Since H_{11} is positive semidefinite, we have $(H_{11}^{1/2})^{\dagger} = (H_{11}^{\dagger})^{1/2}$. Since H is positive semidefinite, by Theorem 1.19, we have $H_{11}H_{11}^{\dagger}H_{12} = H_{12}$. Thus,

$$T(H - \hat{X})T^* = \begin{pmatrix} H_{11} & 0 \\ 0 & H_{22} - X - H_{21}H_{11}^{\dagger}H_{12} \end{pmatrix}.$$

So $H \geq \hat{X}$ if and only if the matrix on the right-hand side is positive semidefinite, and this occurs if and only if $H/H_{11} - X \geq 0$.

The maximum is attained when $X = H/H_{11}$ due to the fact that

$$H - \begin{pmatrix} 0 & 0 \\ 0 & H/H_{11} \end{pmatrix} = \begin{pmatrix} H_{11} & H_{12} \\ H_{21} & H_{21}H_{11}^{\dagger}H_{12} \end{pmatrix} \geq 0.$$

To show the minimum representation (2.2.16), observe that

$$(Y, I_{n-k})H(Y, I_{n-k})^* = H/H_{11} + (Y + H_{21}H_{11}^{\dagger})H_{11}(H_{11}^{\dagger}H_{12} + Y^*).$$

It follows that

$$(Y, I_{n-k})H(Y, I_{n-k})^* \geq H/H_{11},$$

and equality holds if and only if

$$(Y + H_{21}H_{11}^{\dagger})H_{11}(H_{11}^{\dagger}H_{12} + Y^*) = 0,$$

equivalently, $(Y + H_{21}H_{11}^{\dagger})H_{11} = 0$. One may take $Y = -H_{21}H_{11}^{\dagger}$. ∎

The following corollary will be used repeatedly in later sections.

Corollary 2.6 *Let H be $n \times n$ Hermitian. If $\alpha = \{1, 2, \ldots, k\}$, then*

$$H/\alpha = (Z, I)H(Z, I)^*$$

and if $\alpha = \{k+1, k+2, \ldots, n\}$, then

$$H/\alpha = (I, Z)H(I, Z)^*,$$

where, for both cases,

$$Z = -H[\alpha^c, \alpha]H[\alpha]^{\dagger}.$$

As consequences of the theorem, we have, for positive semidefinite A, B,

$$(A \star B)/\alpha \geq A/\alpha \star B/\alpha,$$

where α is an index set and $\star$ denotes sum $+$ or the Hadamard product $\circ$.

We now show a minimum representation for the product of the eigenvalues of a Schur complement [289]. Let integers l and k be such that $1 \leq l \leq k \leq n$. We consider the product of the eigenvalues of the Schur complement indexed by an increasing sequence $1 \leq i_1 \leq i_2 \leq \cdots \leq i_l \leq k$.

Theorem 2.4 *Let A be an $n \times n$ positive semidefinite matrix partitioned as*

$$A = \begin{pmatrix} A_{11} & A_{12} \\ A_{21} & A_{22} \end{pmatrix},$$

in which A_{22} is an $(n-k) \times (n-k)$ principal submatrix. Then

$$\prod_{t=1}^{l} \lambda_{i_t}(A/A_{22}) = \min_{Z \in \mathbb{C}^{k \times (n-k)}} \prod_{t=1}^{l} \lambda_{i_t}[(I_k, Z)A(I_k, Z)^*]. \tag{2.2.17}$$

Proof. For any $Z \in \mathbb{C}^{k \times (n-k)}$, by (2.2.16), we have

$$(I_k, Z)A(I_k, Z)^* \geq A/A_{22}$$

which yields

$$\lambda_{i_t}[(I_k, Z)A(I_k, Z)^*] \geq \lambda_{i_t}(A/A_{22})$$

for each i_t, $t = 1, 2, \ldots, l$, and equality holds by setting $Z = -A_{12}A_{22}^{\dagger}$. ∎

Putting $l = 1$ results in, for any $t = 1, 2, \ldots, k$,

$$\lambda_t(A/A_{22}) = \min_{Z \in \mathbb{C}^{k \times (n-k)}} \lambda_t[(I_k, Z)A(I_k, Z)^*]. \tag{2.2.18}$$

In a similar fashion, one proves that for positive $\theta_1, \theta_2, \ldots, \theta_l \in \mathbb{R}$

$$\begin{aligned} \sum_{t=1}^{l} \lambda_{i_t}(A/A_{22})\theta_t &= \min_{Z \in \mathbb{C}^{k \times (n-k)}} \sum_{t=1}^{l} \lambda_{i_t}[(I_k, Z)A(I_k, Z)^*]\theta_t \\ &= \sum_{t=1}^{l} \min_{Z \in \mathbb{C}^{k \times (n-k)}} \lambda_{i_t}[(I_k, Z)A(I_k, Z)^*]\theta_t. \end{aligned} \tag{2.2.19}$$

2.3 Eigenvalues of the Schur complement of a product

This section, based on [289], is focused on the eigenvalue inequalities of Schur complements concerning the product of positive semidefinite matrices that resemble those of Section 2.0.

Theorem 2.5 *Let A be $n \times n$ positive semidefinite. Let $\alpha \subset \{1, 2, \ldots, n\}$ denote an index set and $1 \leq i_1 < \cdots < i_l \leq k \equiv n - |\alpha|$, where l and k are positive integers such that $1 \leq l \leq k < n$. Then for any $B \in \mathbb{C}^{n \times n}$,*

$$\prod_{t=1}^{l} \lambda_{i_t}[(BAB^*)/\alpha] \geq \prod_{t=1}^{l} \lambda_{i_t}[(BB^*)/\alpha]\lambda_{n-t+1}(A), \tag{2.3.20}$$

$$\prod_{t=1}^{l} \lambda_t[(BAB^*)/\alpha] \geq \prod_{t=1}^{l} \lambda_{i_t}[(BB^*)/\alpha]\lambda_{n-i_t+1}(A), \tag{2.3.21}$$

and

$$\prod_{t=1}^{l} \lambda_{i_t}[(BAB^*)/\alpha] \le \prod_{t=1}^{l} \lambda_{i_t}(A)\lambda_t[(BB^*)/\alpha]. \tag{2.3.22}$$

Proof. There exists an $n \times n$ permutation matrix U such that

$$UAU^* = \begin{pmatrix} A[\alpha^c] & A[\alpha^c, \alpha] \\ A[\alpha, \alpha^c] & A[\alpha] \end{pmatrix}, \quad UBU^* = \begin{pmatrix} B[\alpha^c] & B[\alpha^c, \alpha] \\ B[\alpha, \alpha^c] & B[\alpha] \end{pmatrix}.$$

Let $\beta = \{k+1, \ldots, n\}$. Notice that for any $P \in \mathbb{C}^{k\times n}$, $Q \in \mathbb{C}^{n\times k}$, PQ and QP have the same nonzero eigenvalues. Using (2.2.17) and (2.0.2), we have

$$\begin{aligned}
&\prod_{t=1}^{l} \lambda_{i_t}[(BAB^*)/\alpha] \\
&= \prod_{t=1}^{l} \lambda_{i_t}[(UBAB^*U^*)/\beta] \\
&= \prod_{t=1}^{l} \lambda_{i_t}[(UBU^*UAU^*UB^*U^*)/\beta] \\
&= \min_{Z\in\mathbb{C}^{k\times(n-k)}} \prod_{t=1}^{l} \lambda_{i_t}[(I_k, Z)UBU^*UAU^*UB^*U^*(I_k, Z)^*] \\
&= \min_{Z\in\mathbb{C}^{k\times(n-k)}} \prod_{t=1}^{l} \lambda_{i_t}[(UAU^*)UB^*U^*(I_k, Z)^*(I_k, Z)UBU^*] \\
&\ge \min_{Z\in\mathbb{C}^{k\times(n-k)}} \prod_{t=1}^{l} \lambda_{n-t+1}(UAU^*)\lambda_{i_t}[UB^*U^*(I_k, Z)^*(I_k, Z)UBU^*] \\
&= \min_{Z\in\mathbb{C}^{k\times(n-k)}} \prod_{t=1}^{l} \lambda_{n-t+1}(A)\lambda_{i_t}[(I_k, Z)UBU^*UB^*U^*(I_k, Z)^*] \\
&= \prod_{t=1}^{l} \lambda_{n-t+1}(A) \min_{Z\in\mathbb{C}^{k\times(n-k)}} \lambda_{i_t}[(I_k, Z)UBB^*U^*(I_k, Z)^*] \\
&= \prod_{t=1}^{l} \lambda_{n-t+1}(A)\lambda_{i_t}[(UBB^*U^*)/\beta] \\
&= \prod_{t=1}^{l} \lambda_{i_t}[(BB^*)/\alpha]\lambda_{n-t+1}(A).
\end{aligned}$$

This proves (2.3.20). (2.3.21) and (2.3.22) can be proved similarly. ∎

An analogous result for (2.3.22) is

$$\prod_{t=1}^{l} \lambda_{i_t}[(BAB^*)/\alpha] \leq \prod_{t=1}^{l} \lambda_{i_t}[(BB^*)/\alpha]\lambda_t(A).$$

Setting $B = I$ in (2.3.20), (2.3.22), and (2.3.21), respectively, we obtain

$$\prod_{t=1}^{l} \lambda_{n-t+1}(A) \leq \prod_{t=1}^{l} \lambda_{i_t}(A/\alpha) \leq \prod_{t=1}^{l} \lambda_{i_t}(A)$$

and

$$\prod_{t=1}^{l} \lambda_t(A/\alpha) \geq \prod_{t=1}^{l} \lambda_{n-i_t+1}(A).$$

Putting $l = k$ in Theorem 2.5 reveals the inequalities

$$\prod_{t=1}^{k} \lambda_{n-t+1}(A) \det((BB^*)/\alpha) \leq \det((BAB^*)/\alpha) \leq \prod_{t=1}^{k} \lambda_t(A) \det((BB^*)/\alpha).$$

We point out that every matrix can be regarded as a Schur complement of some matrix. For instance, we may embed an $n \times n$ matrix A in

$$\tilde{A} \equiv \begin{pmatrix} 1 & 0 \\ 0 & A \end{pmatrix}.$$

If we take $\alpha = \{1\}$, then $\tilde{A}/\alpha = A$. With this observation, many of our inequalities on the Schur complements reduce to certain existing results on regular matrices (without involving the Schur complements).

Theorem 2.6 *Let A be $n \times n$ positive semidefinite. Let $\alpha \subset \{1, 2, \ldots, n\}$ denote an index set and $1 \leq i_1 < \cdots < i_l \leq k \equiv n - |\alpha|$, where l and k are positive integers such that $1 \leq l \leq k < n$. Then for any $B \in \mathbb{C}^{n \times n}$*

$$\sum_{t=1}^{l} \lambda_{i_t}[(BAB^*)/\alpha] \geq \sum_{t=1}^{l} \lambda_{i_t}[(BB^*)/\alpha]\lambda_{n-t+1}(A), \tag{2.3.23}$$

$$\sum_{t=1}^{l} \lambda_t[(BAB^*)/\alpha] \geq \sum_{t=1}^{l} \lambda_{i_t}[(BB^*)/\alpha]\lambda_{n-i_t+1}(A), \tag{2.3.24}$$

and

$$\sum_{t=1}^{l} \lambda_{i_t}[(BAB^*)/\alpha] \leq \sum_{t=1}^{l} \lambda_{i_t}(A)\lambda_t[(BB^*)/\alpha]. \tag{2.3.25}$$

Proof. This follows from (2.0.4) immediately. Following is a proof based upon (2.2.19) and (2.0.7). We may take $\alpha = \{k+1, \ldots, n\}$. Then

$$\begin{aligned}
&\sum_{t=1}^{l} \lambda_{i_t}[(BAB^*)/\alpha] \\
&\quad = \min_{Z\in\mathbb{C}^{k\times(n-k)}} \sum_{t=1}^{l} \lambda_{i_t}[(I_k, Z)BAB^*(I_k, Z)^*] \\
&\quad = \min_{Z\in\mathbb{C}^{k\times(n-k)}} \sum_{t=1}^{l} \lambda_{i_t}[AB^*(I_k, Z)^*(I_k, Z)B] \\
&\quad \geq \min_{Z\in\mathbb{C}^{k\times(n-k)}} \sum_{t=1}^{l} \lambda_{n-t+1}(A)\lambda_{i_t}[B^*(I_k, Z)^*(I_k, Z)B] \\
&\quad = \min_{Z\in\mathbb{C}^{k\times(n-k)}} \sum_{t=1}^{l} \lambda_{n-t+1}(A)\lambda_{i_t}[(I_k, Z)BB^*(I_k, Z)^*] \\
&\quad = \sum_{t=1}^{l} \lambda_{n-t+1}(A)\lambda_{i_t}[(BB^*)/\alpha]. \blacksquare
\end{aligned}$$

The following is a parallel result to the inequality (2.3.25):

$$\sum_{t=1}^{l} \lambda_{i_t}[(BAB^*)/\alpha] \leq \sum_{t=1}^{l} \lambda_{i_t}[(BB^*)/\alpha]\lambda_t(A).$$

Setting $B = I$ in (2.3.23), (2.3.25), and (2.3.24), respectively, we obtain

$$\sum_{t=1}^{l} \lambda_{n-t+1}(A) \leq \sum_{t=1}^{l} \lambda_{i_t}(A/\alpha) \leq \sum_{t=1}^{l} \lambda_{i_t}(A)$$

and

$$\sum_{t=1}^{l} \lambda_t(A/\alpha) \geq \sum_{t=1}^{l} \lambda_{n-i_t+1}(A).$$

Putting $l = k$ in Theorem 2.6, since $(BAB^*)/\alpha$ is $k \times k$, we have

$$\sum_{t=1}^{k} \lambda_t[(BB^*)/\alpha]\lambda_{n-t+1}(A) \leq \operatorname{tr}[(BAB^*)/\alpha)] \leq \sum_{t=1}^{k} \lambda_t[(BB^*)/\alpha]\lambda_t(A).$$

Theorem 2.7 *Let A be an $n \times n$ positive semidefinite matrix and let α be an index set of k elements. Then for any $B \in \mathbb{C}^{n\times n}$ and $t = 1, 2, \ldots, n-k$,*

$$\min_{i+j=t+1} \lambda_i(A)\lambda_j[(BB^*)/\alpha] \geq \lambda_t[(BAB^*)/\alpha] \geq \max_{i+j=t+n} \lambda_i(A)\lambda_j[(BB^*)/\alpha].$$

Proof. Taking $\alpha = \{n-k+1, \ldots, n\}$, by (2.2.18) and (2.0.3), we have

$$\begin{aligned}
&\lambda_t[(BAB^*)/\alpha] \\
&= \min_{Z \in \mathbb{C}^{(n-k)\times k}} \lambda_t[(I_{n-k}, Z)BAB^*(I_{n-k}, Z)^*] \\
&\geq \min_{Z \in \mathbb{C}^{(n-k)\times k}} \max_{i+j=t+n} \lambda_i(A)\lambda_j[(I_{n-k}, Z)BB^*(I_{n-k}, Z)^*] \\
&= \max_{i+j=t+n} \lambda_i(A) \min_{Z \in \mathbb{C}^{(n-k)\times k}} \lambda_j[(I_{n-k}, Z)BB^*(I_{n-k}, Z)^*] \\
&= \max_{i+j=t+n} \lambda_i(A)\lambda_j[(BB^*)/\alpha].
\end{aligned}$$

By (2.2.19), along with the first inequality in (2.0.3),

$$\begin{aligned}
&\lambda_t[(BAB^*)/\alpha] \\
&\leq \min_{Z \in \mathbb{C}^{(n-k)\times k}} \min_{i+j=t+1} \lambda_i(A)\lambda_j[(I_{n-k}, Z)BB^*(I_{n-k}, Z)^*] \\
&= \min_{i+j=t+1} \lambda_i(A) \min_{Z \in \mathbb{C}^{(n-k)\times k}} \lambda_j[(I_{n-k}, Z)BB^*(I_{n-k}, Z)^*] \\
&= \min_{i+j=t+1} \lambda_i(A)\lambda_j[(BB^*)/\alpha].
\end{aligned}$$

As we are interested in relating the eigenvalues of the matrix product AB to those of individual matrices A and B, our next result shows lower bounds for the eigenvalues of the Schur complement of the matrix product BAB^* in terms of the eigenvalues of the Schur complements of BB^* and A. The proof of the theorem is quite technical.

Theorem 2.8 *Let A be $n \times n$ positive semidefinite of rank r, $B \in \mathbb{C}^{m\times n}$, and $\alpha \subset \{1, 2, \ldots, m\}$. If* $\operatorname{rank}[(BAB^*)/\alpha] = s$, *then for each $l = 1, 2, \ldots, s$,*

$$\lambda_l[(BAB^*)/\alpha] \geq \max_{\substack{1\leq t\leq s-l+1 \\ 1\leq u\leq t}} [\lambda_{l+t+r-s-1}(A)\lambda_{r-u+1}(A)]^{\frac{1}{2}} \lambda_{s-t+u+n-r}[(BB^*)/\alpha].$$

Proof. Let $k = m - |\alpha|$. We may assume $\alpha = \{k+1, \ldots, m\}$. Then $\alpha^c = \{1, 2, \ldots, k\}$. Since $\operatorname{rank}(A) = r$, there exists unitary $U \in \mathbb{C}^{n\times n}$ such that

$$UAU^* = D \oplus 0 \equiv \operatorname{diag}(D, 0), \quad \text{where } D = \operatorname{diag}(\lambda_1(A), \ldots, \lambda_r(A)) \geq 0.$$

Let

$$X = -[(BAB^*)[\alpha^c, \alpha]]\,[(BAB^*)[\alpha]]^\dagger.$$

Then

$$\lambda_l[(BAB^*)/\alpha] = \lambda_l[(I_k, X)BAB^*(I_k, X)^*] = \lambda_l[AB^*(I_k, X)^*(I_k, X)B].$$

Thus

$$\operatorname{rank}[AB^*(I_k, X)^*(I_k, X)B] = \operatorname{rank}[(BAB^*)/\alpha] = s.$$

Let

$$\tilde{B} = B^*(I_k, X)^*(I_k, X)B, \quad U = \begin{pmatrix} U_1 \\ U_2 \end{pmatrix}, \; U_1 \in \mathbb{C}^{r\times m}.$$

Then

$$\begin{aligned} \operatorname{rank}(A\tilde{B}) &= \operatorname{rank}(UAU^*U\tilde{B}U^*) \\ &= \operatorname{rank}[\operatorname{diag}(D,0)(U\tilde{B}U^*)] \\ &= \operatorname{rank}[\operatorname{diag}(D^{\frac{1}{2}},0)(U\tilde{B}U^*)\operatorname{diag}(D^{\frac{1}{2}},0)] \\ &= \operatorname{rank}(D^{\frac{1}{2}}U_1\tilde{B}U_1^*D^{\frac{1}{2}}) \\ &= \operatorname{rank}(U_1\tilde{B}U_1^*). \end{aligned}$$

Since $U_1\tilde{B}U_1^*$ is $r \times r$ positive semidefinite and $\operatorname{rank}[(BAB^*)/\alpha] = s$, there exists an $r \times r$ unitary matrix V_1 such that

$$V_1U_1\tilde{B}U_1^*V_1^* = \operatorname{diag}(G, 0),$$

where

$$G = \operatorname{diag}(\lambda_1(U_1\tilde{B}U_1^*), \ldots, \lambda_s(U_1\tilde{B}U_1^*)).$$

Set $\tilde{D} = V_1DV_1^*$ and partition it as $\begin{pmatrix} D_1 & D_2 \\ D_2^* & D_3 \end{pmatrix}$ with D_1 of order $s \times s$. Let

$$L = \begin{pmatrix} I & 0 \\ -D_2^*D_1^\dagger & I \end{pmatrix}.$$

Then

$$LV_1DV_1^*L^* = \operatorname{diag}(D_1, D_3 - D_2^*D_1^\dagger D_2).$$

Let $\tilde{B}_1 = D^{\frac{1}{2}}U_1\tilde{B}U_1^*D^{\frac{1}{2}}$. Then

$$\begin{aligned} &(L^{*-1}V_1D^{-\frac{1}{2}})\tilde{B}_1(L^{*-1}V_1D^{-\frac{1}{2}})^{-1} \\ &= L^{*-1}V_1D^{-\frac{1}{2}}D^{\frac{1}{2}}U_1\tilde{B}U_1^*D^{\frac{1}{2}}D^{\frac{1}{2}}V_1^*L^* \\ &= L^{*-1}V_1U_1\tilde{B}U_1^*DV_1^*L^* \\ &= L^{*-1}V_1U_1\tilde{B}U_1^*V_1^*L^{-1}(LV_1DV_1^*L^*) \\ &= (L^{-1})^*\operatorname{diag}(G,0)L^{-1}\operatorname{diag}(D_1, D_3 - D_2^*D_1^\dagger D_2) \\ &= \operatorname{diag}(GD_1, 0). \end{aligned}$$

So $\tilde{B}_1$ and GD_1 have the same nonzero eigenvalues. On the other hand,

$$\begin{aligned}
\lambda_l[(BAB^*)/\alpha] &= \lambda_l[AB^*(I_k, X)^*(I_k, X)B] \\
&= \lambda_l(A\tilde{B}) \\
&= \lambda_l[(UAU^*)(U\tilde{B}U^*)] \\
&= \lambda_l[\operatorname{diag}(D, 0)(U\tilde{B}U^*)] \\
&= \lambda_l[\operatorname{diag}(D^{\frac{1}{2}}, 0)\operatorname{diag}(D^{\frac{1}{2}}, 0)(U\tilde{B}U^*)] \\
&= \lambda_l[\operatorname{diag}(D^{\frac{1}{2}}, 0)(U\tilde{B}U^*)\operatorname{diag}(D^{\frac{1}{2}}, 0)] \\
&= \lambda_l(D^{\frac{1}{2}}U_1\tilde{B}U_1^*D^{\frac{1}{2}}) \\
&= \lambda_l(\tilde{B}_1).
\end{aligned}$$

Noticing that

$$D_1^{\frac{1}{2}}(GD_1)D_1^{-\frac{1}{2}} = D_1^{\frac{1}{2}}GD_1^{\frac{1}{2}}$$

and

$$G^{-\frac{1}{2}}(GD_1)G^{\frac{1}{2}} = G^{\frac{1}{2}}D_1G^{\frac{1}{2}},$$

we see that $\tilde{B}_1$, $D_1^{\frac{1}{2}}GD_1^{\frac{1}{2}}$, and $G^{\frac{1}{2}}D_1G^{\frac{1}{2}}$ have the same nonzero eigenvalues, including multiplicities. It follows that, for $l = 1, 2, \ldots, s$,

$$\lambda_l[(BAB^*)/\alpha] = \lambda_l(\tilde{B}_1) = \lambda_l(D_1^{\frac{1}{2}}GD_1^{\frac{1}{2}}) = \lambda_l(G^{\frac{1}{2}}D_1G^{\frac{1}{2}}).$$

For $l = s + 1, \ldots, k$, since $\operatorname{rank}[((BAB^*)/\alpha] = s$, we have

$$\lambda_l[(BAB^*)/\alpha] = 0.$$

By the Cauchy interlacing theorem, we have, for $i = 1, 2, \ldots, s$,

$$\lambda_i(D_1) \geq \lambda_{i+r-s}(V_1DV_1^*) = \lambda_{i+r-s}(A) \tag{2.3.26}$$

and for $i = 1, 2, \ldots, r$,

$$\lambda_i(U_1\tilde{B}U_1^*) = \lambda_i(U_1\tilde{B}U_1^*) \geq \lambda_{i+n-r}(\tilde{B}). \tag{2.3.27}$$

By (2.0.3) and (2.2.17), we have, for $t = 1, \ldots, s-l+1$, $u = 1, \ldots, t$,

$$\begin{aligned}
&\lambda_l[(BAB^*)/\alpha] \\
&= \lambda_l(D_1^{\frac{1}{2}} G D_1^{\frac{1}{2}}) \\
&\geq \lambda_{l+t-1}(D_1^{\frac{1}{2}})\sigma_{s-t+1}(G D_1^{\frac{1}{2}}) \quad \text{[by (2.0.3)]} \\
&\geq \lambda_{l+t-1}(D_1^{\frac{1}{2}})\lambda_{s-t+1+u-1}(G)\lambda_{s-u+1}(D_1^{\frac{1}{2}}) \quad \text{[by (2.0.3)]} \\
&\geq [\lambda_{l+t-1+r-s}(A)\lambda_{r-u+1}(A)]^{\frac{1}{2}} \\
&\quad \cdot \lambda_{s-t+u+n-r}(\tilde{B}) \quad \text{[by (2.3.26) and (2.3.27)]} \\
&= [\lambda_{l+t+r-s-1}(A)\lambda_{r-u+1}(A)]^{\frac{1}{2}} \\
&\quad \cdot \lambda_{s-t+u+n-r}[B^*(I_k \ X)^*(I_k \ X)B] \\
&= [\lambda_{l+t+r-s-1}(A)\lambda_{r-u+1}(A)]^{\frac{1}{2}} \\
&\quad \cdot \lambda_{s-t+u+n-r}[(I_k \ X)BB^*(I_k \ X)^*] \\
&\geq [\lambda_{l+t+r-s-1}(A)\lambda_{r-u+1}(A)]^{\frac{1}{2}} \\
&\quad \cdot \min_{Z \in C^{k\times(m-k)}} \lambda_{s-t+u+n-r}[(I_k \ Z)BB^*(I_k \ Z)^*] \\
&= [\lambda_{l+t+r-s-1}(A)\lambda_{r-u+1}(A)]^{\frac{1}{2}} \\
&\quad \cdot \lambda_{s-t+u+n-r}[(BB^*)/\alpha] \quad \text{[by (2.2.17)].} \ \blacksquare
\end{aligned}$$

In a similar manner, one can obtain the following additional inequalities

$$\lambda_l[(BAB^*)/\alpha] \geq$$

$$\max_{\substack{t=1,\ldots,s-l+1 \\ u=1,\ldots,t}} \begin{cases}
\{\lambda_{l+t-1+n-r}[(BB^*)/\alpha]\lambda_{s-u+1+n-r}[(BB^*)/\alpha]\}^{\frac{1}{2}}\lambda_{r-t+u}(A), \\
[\lambda_{r-t+u}(A)\lambda_{l+t+r-s-1}(A)]^{\frac{1}{2}}\lambda_{s-u+1+n-r}[(BB^*)/\alpha], \\
[\lambda_{r-u+1}(A)\lambda_{r-t+1}(A)]^{\frac{1}{2}}\lambda_{l+t+u-2+n-r}[(BB^*)/\alpha], \\
[\lambda_{l+t+u-2+r-s}(A)\lambda_{r-t+1}(A)]^{\frac{1}{2}}\lambda_{s-u+1+n-r}[(BB^*)/\alpha], \\
\{\lambda_{l+t-1+n-r}[(BB^*)/\alpha]\lambda_{s-t+u+n-r}[(BB^*)/\alpha]\}^{\frac{1}{2}}\lambda_{r-u+1}(A), \\
\{\lambda_{s-t+1+n-r}[(BB^*)/\alpha]\lambda_{s-u+1+n-r}[(BB^*)/\alpha]\}^{\frac{1}{2}}\lambda_{r+t+u-2}(A), \\
\{\lambda_{s-t+1+n-r}[(BB^*)/\alpha]\lambda_{s+t+u-2+n-r}[(BB^*)/\alpha]\}^{\frac{1}{2}}\lambda_{r-u+1}(A).
\end{cases}$$

Setting $r = n$ in the first inequality above, we arrive at

$$\lambda_l[(BAB^*)/\alpha] \geq \max_{\substack{t=1,\ldots,s-l+1 \\ u=1,\ldots,t}} \{\lambda_{l+t-1}[(BB^*)/\alpha]\lambda_{s-u+1}[(BB^*)/\alpha]\}^{\frac{1}{2}}\lambda_{n-t+u}(A).$$

In particular, letting $t = u = 1$ reveals that

$$\lambda_l[(BAB^*)/\alpha] \geq [\lambda_l[(BB^*)/\alpha]\lambda_s[(BB^*)/\alpha]]^{\frac{1}{2}}\lambda_n(A).$$

If we take $\alpha = \{1\}$ and set $\tilde{X} = \left(\begin{smallmatrix} 1 & 0 \\ 0 & X \end{smallmatrix}\right)$ for the matrix X, then for any $n \times n$ matrices A and B, we obtain

$$\lambda_l(BAB^*) \geq [\lambda_l(BB^*)\lambda_s(BB^*)]^{\frac{1}{2}}\,\lambda_{n+1}(\tilde{A}).$$

The result below presents a lower bound for the product of eigenvalues.

Theorem 2.9 *Let all assumptions of Theorem 2.8 be satisfied, let u be a positive integer with $1 \leq u \leq k$, and let $1 \leq i_1 < \cdots < i_u \leq k$. Then*

$$\prod_{t=1}^{u} \lambda_t[(BAB^*)/\alpha] \geq \prod_{t=1}^{u} [\lambda_{r-i_t+1}(A)\lambda_{r-t+1}(A)]^{\frac{1}{2}}\lambda_{n-r+i_t}[(BB^*)/\alpha].$$

Proof. Following the line of the proof of the previous theorem, we have

$$\begin{aligned}
&\prod_{t=1}^{u} \lambda_t[(BAB^*)/\alpha] \\
&\quad= \prod_{t=1}^{u} \lambda_t(D_1^{\frac{1}{2}} G D_1^{\frac{1}{2}}) \\
&\quad\geq \prod_{t=1}^{u} \lambda_{s-i_t+1}(D_1^{\frac{1}{2}})\lambda_{i_t}(G D_1^{\frac{1}{2}}) \quad \text{[by (2.0.1)]} \\
&\quad\geq \prod_{t=1}^{u} \lambda_{s-i_t+1}(D_1^{\frac{1}{2}})\lambda_{s-t+1}(D_1^{\frac{1}{2}})\lambda_{i_t}(G) \quad \text{[by (2.0.2)]} \\
&\quad\geq \prod_{t=1}^{u} [\lambda_{r-i_t+1}(A)\lambda_{r-t+1}(A)]^{\frac{1}{2}} \\
&\qquad\cdot \lambda_{n-r+i_t}(\tilde{B}) \quad \text{[by (2.3.26) and (2.3.27)]} \\
&\quad= \prod_{t=1}^{u} [\lambda_{r-i_t+1}(A)\lambda_{r-t+1}(A)]^{\frac{1}{2}} \\
&\qquad\cdot \lambda_{n-r+i_t}[(I_k\ X)BB^*(I_k\ X)^*] \\
&\quad\geq \prod_{t=1}^{u} [\lambda_{r-i_t+1}(A)\lambda_{r-t+1}(A)]^{\frac{1}{2}} \\
&\qquad\cdot \min_{Z \in C^{k\times(m-k)}} \lambda_{n-r+i_t}[(I_k\ Z)BB^*(I_k\ Z)^*] \\
&\quad= \prod_{t=1}^{u} [\lambda_{r-i_t+1}(A)\lambda_{r-t+1}(A)]^{\frac{1}{2}} \\
&\qquad\cdot \lambda_{n-r+i_t}[(BB^*)/\alpha] \quad \text{[by (2.2.17)]}. \ \blacksquare
\end{aligned}$$

Similar results are

$$\prod_{t=1}^{u} \lambda_t[(BAB^*)/\alpha] \geq$$

$$\begin{cases} \prod_{t=1}^{u} [\lambda_{r-i_t+1}(A)\lambda_{r-s+i_t}(A)]^{\frac{1}{2}} \lambda_{s-t+1+n-r}[(BB^*)/\alpha], \\ \prod_{t=1}^{u} \{\lambda_{s-i_t+1+n-r}[(BB^*)/\alpha]\lambda_{n-r+i_t}[(BB^*)/\alpha]\}^{\frac{1}{2}} \lambda_{r-t+1}(A), \\ \prod_{t=1}^{u} \{\lambda_{s-i_t+1+n-r}[(BB^*)/\alpha]\lambda_{s-t+1+n-r}[(BB^*)/\alpha]\}^{\frac{1}{2}} \lambda_{r-s+i_t}(A). \end{cases}$$

2.4 Eigenvalues of the Schur complement of a sum

This section is concerned with inequalities involving the eigenvalues of Schur complements of sums of positive semidefinite matrices [289].

Theorem 2.10 *Let A, B be $n\times n$ positive semidefinite. Let $\alpha \subset \{1, 2, \ldots, n\}$ and $k = n - |\alpha|$. If $1 \leq i_1 < \cdots < i_l \leq n$, where $1 \leq l \leq k$, then*

$$\sum_{t=1}^{l} \lambda_{i_t}[(A+B)/\alpha] \geq \sum_{t=1}^{l} \lambda_{i_t}(A/\alpha) + \sum_{t=1}^{l} \lambda_{k-t+1}(B/\alpha).$$

Proof. This actually follows immediately from (2.0.8) and the fact that $(A+B)/\alpha \geq A/\alpha + B/\alpha$. It can also be proven by (2.2.19) as follows. As in the proof of Theorem 2.5, we may take $\alpha = \{k+1, \ldots, n\}$ and have

$$\begin{aligned} &\sum_{t=1}^{l} \lambda_{i_t}[(A+B)/\alpha] \\ &\quad = \min_{Z\in\mathbb{C}^{k\times(n-k)}} \sum_{t=1}^{l} \lambda_{i_t}[(I_k, Z)(A+B)(I_k, Z)^*] \\ &\quad = \min_{Z\in\mathbb{C}^{k\times(n-k)}} \sum_{t=1}^{l} \lambda_{i_t}[(I_k, Z)A(I_k, Z)^* + (I_k, Z)B(I_k, Z)^*] \\ &\quad \geq \min_{Z\in\mathbb{C}^{k\times(n-k)}} \sum_{t=1}^{l} \lambda_{i_t}[(I_k, Z)A(I_k, Z)^*] \\ &\qquad + \min_{Z\in\mathbb{C}^{k\times(n-k)}} \sum_{t=1}^{l} \lambda_{k-t+1}[(I_k, Z)B(I_k, Z)^*] \\ &\quad = \sum_{t=1}^{l} \lambda_{i_t}(A/\alpha) + \sum_{t=1}^{l} \lambda_{k-t+1}(B/\alpha). \blacksquare \end{aligned}$$

Note that $\prod_{i=1}^{l}(x_i+y_i)^{1/l} \geq (\prod_{i=1}^{l} x_i)^{1/l}+(\prod_{i=1}^{l} y_i)^{1/l}$ for nonnegative x, y's and $(a+b)^p \geq a^p + b^p$ for $a, b \geq 0$, $p \geq 1$. Setting $i_t = k - t + 1$, $x_t = \lambda_t[(A+B)/\alpha]$ and $y_t = \lambda_t(A/\alpha) + \lambda_t(B/\alpha)$ and by (2.0.5), we have

Corollary 2.7 *Let A, B be $n \times n$ positive semidefinite. Let $\alpha \subset \{1, 2, \ldots, n\}$ and $k = n - |\alpha|$. Then for any integer l, $1 \leq l \leq k$, and real number $p > 1$,*

$$\prod_{t=1}^{l} \lambda_{k-t+1}^{p/l}[(A+B)/\alpha] \geq \prod_{t=1}^{l} \lambda_{k-t+1}^{p/l}(A/\alpha) + \prod_{t=1}^{l} \lambda_{k-t+1}^{p/l}(B/\alpha).$$

Putting $l = k$ and $p = 1$ in the corollary reveals the known result:

$$\left(\frac{\det(A+B)}{\det(A+B)[\alpha]}\right)^{1/k} \geq \left(\frac{\det A}{\det A[\alpha]}\right)^{1/k} + \left(\frac{\det B}{\det B[\alpha]}\right)^{1/k}.$$

By mathematical induction, we may extend our results to multiple copies of positive semidefinite matrices.

Corollary 2.8 *Let $A_1, \ldots, A_m$ be $n \times n$ positive semidefinite matrices. Let $\alpha \subset \{1, 2, \ldots, n\}$ and $k = n - |\alpha|$. Then for any integer l, $1 \leq l \leq k$, and real number $p > 1$,*

$$\sum_{t=1}^{l} \lambda_{k-t+1}\left[\left(\sum_{j=1}^{m} A_j\right)/\alpha\right] \geq \sum_{t=1}^{l}\sum_{j=1}^{m} \lambda_{k-t+1}(A_j/\alpha)$$

and

$$\prod_{t=1}^{l} \lambda_{k-t+1}^{p/l}\left[\left(\sum_{j=1}^{m} A_j\right)/\alpha\right] \geq \sum_{j=1}^{m}\prod_{t=1}^{l} \lambda_{k-t+1}^{p/l}(A_j/\alpha).$$

The next theorem presents a sum-product to product-sum inequality on Schur complements. For this purpose, we recall the Hölder inequality [22]: Let $x_1, \ldots, x_n$ and $y_1, \ldots, y_n$ be nonnegative numbers, let p be a nonzero number, $p < 1$, and let $\frac{1}{p} + \frac{1}{p'} = 1$. Then, assuming x, $y > 0$ if $p < 0$,

$$\sum_{t=1}^{n} x_i y_i \geq \left(\sum_{t=1}^{n} x_i^p\right)^{\frac{1}{p}} \left(\sum_{t=1}^{n} y_i^{p'}\right)^{\frac{1}{p'}}.$$

We note here that if we take in the following theorem $\alpha = \{1\}$ or $\{1, \ldots, n\}$ and embed matrices A in

$$\tilde{A} = \begin{pmatrix} 1 & 0 \\ 0 & A \end{pmatrix} \quad \text{or} \quad \tilde{A} = \begin{pmatrix} A & 0 \\ 0 & \operatorname{tr} A \end{pmatrix}$$

respectively, we may arrive at many matrix (trace) and scalar inequalities.

Theorem 2.11 *Let A_{pq}, $p = 1, 2, \ldots, \mu$, $q = 1, 2, \ldots, \nu$, be $n \times n$ positive semidefinite matrices. Let $\alpha \subset \{1, 2, \ldots, n\}$, $k = n - |\alpha|$, and l be an integer, $1 \le l \le k$. Then for any nonzero real $r < 1$ and ω, $0 < \omega \le l$, conventionally assuming that all A_{pq} are positive definite if $r < 0$, we have*

$$\left(\sum_{p=1}^{\mu} \left\{ \prod_{t=1}^{l} \lambda_{k-t+1} \left[\left(\sum_{q=1}^{\nu} A_{pq} \right) / \alpha \right] \right\}^{r/\omega} \right)^{1/r}$$

$$\ge \sum_{q=1}^{\nu} \left\{ \sum_{p=1}^{\mu} \left[\prod_{t=1}^{l} \lambda_{k-t+1}(A_{pq}/\alpha) \right]^{r/\omega} \right\}^{1/r}.$$

Proof. Let s be the number so that $1/r + 1/s = 1$. Then $(r-1)s = r$. Set

$$C_{pq} \equiv \left[\prod_{t=1}^{l} \lambda_{k-t+1}(A_{pq}/\alpha) \right]^{1/\omega}$$

and

$$B_p \equiv \left\{ \prod_{t=1}^{l} \lambda_{k-t+1} \left[\left(\sum_{q=1}^{\nu} A_{pq} \right) / \alpha \right] \right\}^{(r-1)/\omega}.$$

Then we need to show

$$\left(\sum_{p=1}^{\mu} B_p^s \right)^{1/r} \ge \sum_{q=1}^{\nu} \left(\sum_{p=1}^{\mu} C_{pq}^r \right)^{1/r}.$$

Note that $1/\omega \ge 1/l$. By Corollary 2.8, we have

$$\left\{ \prod_{t=1}^{l} \lambda_{k-t+1} \left[\left(\sum_{q=1}^{\nu} A_{pq} \right) / \alpha \right] \right\}^{1/\omega} \ge \sum_{q=1}^{\nu} \left[\prod_{t=1}^{l} \lambda_{k-t+1}(A_{pq}/\alpha) \right]^{1/\omega} = \sum_{q=1}^{\nu} C_{pq},$$

from which, and by the Hölder inequality, we have

$$\begin{aligned}
\sum_{p=1}^{\mu} B_p^s &= \sum_{p=1}^{\mu} \left\{ \prod_{t=1}^{l} \lambda_{k-t+1} \left[\left(\sum_{q=1}^{\nu} A_{pq} \right) / \alpha \right] \right\}^{r/\omega} \\
&= \sum_{p=1}^{\mu} \left\{ \prod_{t=1}^{l} \lambda_{k-t+1} \left[\left(\sum_{q=1}^{\nu} A_{pq} \right) / \alpha \right] \right\}^{1/\omega} B_p \\
&\geq \sum_{p=1}^{\mu} \left(\sum_{q=1}^{\nu} C_{pq} \right) B_p \\
&= \sum_{q=1}^{\nu} \sum_{p=1}^{\mu} C_{pq} B_p \\
&\geq \sum_{q=1}^{\nu} \left[\left(\sum_{p=1}^{\mu} C_{pq}^r \right)^{1/r} \left(\sum_{p=1}^{\mu} B_p^s \right)^{1/s} \right] \\
&= \left[\sum_{q=1}^{\nu} \left(\sum_{p=1}^{\mu} C_{pq}^r \right)^{1/r} \right] \left(\sum_{p=1}^{\mu} B_p^s \right)^{1/s}.
\end{aligned}$$

Since $1 - \frac{1}{s} = \frac{1}{r}$, dividing by $\left(\sum_{p=1}^{\mu} B_p^s \right)^{1/s}$ yields the desired result. ∎

If we set $\omega = 1$ in the theorem, we then obtain

$$\left(\sum_{p=1}^{\mu} \left\{ \prod_{t=1}^{l} \lambda_{k-t+1} \left[\left(\sum_{q=1}^{\nu} A_{pq} \right) / \alpha \right] \right\}^r \right)^{1/r}$$

$$\geq \sum_{q=1}^{\nu} \left\{ \sum_{p=1}^{\mu} \left[\prod_{t=1}^{l} \lambda_{k-t+1}(A_{pq}/\alpha) \right]^r \right\}^{1/r}.$$

Theorem 2.12 *Let A_{pq}, $p = 1, 2, \ldots, \mu$, $q = 1, 2, \ldots, \nu$, be $n \times n$ positive semidefinite matrices. Let $\alpha \subset \{1, 2, \ldots, n\}$ and denote $k = n - |\alpha|$. Let l be an integer such that $1 \leq l \leq k$ and $c_1, c_2, \ldots, c_\mu$ be positive numbers such that $c_1 + c_2 + \cdots + c_\mu \geq 1/l$. Then*

$$\sum_{q=1}^{\nu} \sum_{p=1}^{\mu} \left[\prod_{t=1}^{l} \lambda_{k-t+1}(A_{pq}/\alpha) \right]^{c_p} \leq \prod_{p=1}^{\mu} \left\{ \prod_{t=1}^{l} \lambda_{k-t+1} \left[\left(\sum_{q=1}^{\nu} A_{pq} \right) / \alpha \right] \right\}^{c_p}.$$

Proof. All we need to show is that

$$L \equiv \frac{\sum_{q=1}^{\nu} \prod_{p=1}^{\mu} \left[\prod_{t=1}^{l} \lambda_{k-t+1}(A_{pq}/\alpha) \right]^{c_p}}{\prod_{p=1}^{\mu} \left\{ \prod_{t=1}^{l} \lambda_{k-t+1} \left[\left(\sum_{q=1}^{\nu} A_{pq} \right) / \alpha \right] \right\}^{c_p}} \leq 1.$$

Let $c = \sum_{p=1}^{\mu} c_p$, $c'_p = c_p/c$, $p = 1, 2, \ldots, \mu$. Then $c \geq 1/l$ and

$$\sum_{p=1}^{\mu} c'_p = \sum_{p=1}^{\mu} \frac{c_p}{c} = \frac{1}{c} \sum_{p=1}^{\mu} c_p = 1.$$

By the weighted arithmetic-geometric mean inequality and Corollary 2.8,

$$\begin{aligned}
L &= \sum_{q=1}^{\nu} \prod_{p=1}^{\mu} \frac{\left[\prod_{t=1}^{l} \lambda_{k-t+1}(A_{pq}/\alpha) \right]^{c_p}}{\left\{ \prod_{t=1}^{l} \lambda_{k-t+1} \left[\left(\sum_{q=1}^{\nu} A_{pq} \right) / \alpha \right] \right\}^{c_p}} \\
&= \sum_{q=1}^{\nu} \prod_{p=1}^{\mu} \left(\frac{\left[\prod_{t=1}^{l} \lambda_{k-t+1}(A_{pq}/\alpha) \right]^{c}}{\left\{ \prod_{t=1}^{l} \lambda_{k-t+1} \left[\left(\sum_{q=1}^{\nu} A_{pq} \right) / \alpha \right] \right\}^{c}} \right)^{c'_p} \\
&\leq \sum_{q=1}^{\nu} \sum_{p=1}^{\mu} c'_p \frac{\left[\prod_{t=1}^{l} \lambda_{k-t+1}(A_{pq}/\alpha) \right]^{c}}{\left\{ \prod_{t=1}^{l} \lambda_{k-t+1} \left[\left(\sum_{q=1}^{\nu} A_{pq} \right) / \alpha \right] \right\}^{c}} \\
&= \sum_{p=1}^{\mu} c'_p \frac{\sum_{q=1}^{\nu} \left[\prod_{t=1}^{l} \lambda_{k-t+1}(A_{pq}/\alpha) \right]^{c}}{\left\{ \prod_{t=1}^{l} \lambda_{k-t+1} \left[\left(\sum_{q=1}^{\nu} A_{pq} \right) / \alpha \right] \right\}^{c}} \\
&\leq \sum_{p=1}^{\mu} c'_p \quad \text{[by Corollary 2.8]} \\
&= 1. \blacksquare
\end{aligned}$$

2.5 The Hermitian case

In the previous sections, we presented some eigenvalue inequalities for the Schur complements of positive semidefinite matrices. In particular, we paid attention to the matrices in the form BAB^*, where A is positive semidefinite. We now study the inequalities for the Hermitian case of matrix A. Unless otherwise stated, we arrange the eigenvalues of $A \in \mathbb{H}_n$ in the order

$$\lambda_1(A) \geq \lambda_2(A) \geq \cdots \geq \lambda_n(A).$$

Theorem 2.13 *Let $A \in \mathbb{H}_n$, $B \in \mathbb{C}^{m\times n}$, and $\alpha \subset \{1, 2, \ldots, m\}$. Denote $k = m - |\alpha|$. Then for every $t = 1, 2, \ldots, k$,*

$$\lambda_t[(BAB^*)/\alpha] \geq \max_{t\leq r\leq k} \{\lambda_{n-r+t}(A)\lambda_r[(BB^*)/\alpha] : \lambda_{n-r+t}(A) \geq 0\}$$

and

$$\lambda_t[(BAB^*)/\alpha] \leq \min_{1\leq r\leq t} \{\lambda_r(A)\lambda_{k+r-t}[(BB^*)/\alpha] : \lambda_r(A) \leq 0\}.$$

Proof. Without loss of generality, we assume that $\alpha = \{k+1, \ldots, m\}$. Let

$$X = -[(BAB^*)[\alpha^c, \alpha]][(BAB^*)[\alpha]]^\dagger, \quad C = (I_k, X)B.$$

On one hand, for any integer r, $1 \leq r \leq k$, we have

$$CAC^* = C[A - \lambda_{n-r+t}(A)I_n]C^* + \lambda_{n-r+t}(A)CC^*,$$

where $A - \lambda_{n-r+t}(A)I_n$ is $n \times n$ Hermitian and $\lambda_{n-r+t}(A)CC^*$ is $k \times k$ Hermitian. Thus, there exists an $n \times n$ unitary matrix U such that

$$A - \lambda_{n-r+t}(A)I_n = U \operatorname{diag}(\lambda_1(A) - \lambda_{n-r+t}(A), \ldots, \lambda_n(A) - \lambda_{n-r+t}(A))U^*.$$

On the other hand, putting $P = CU$, we have

$$\begin{aligned} &C[A - \lambda_{n-r+t}(A)I_n]C^* \\ &\quad = P \operatorname{diag}(\lambda_1(A) - \lambda_{n-r+t}(A), \ldots, \lambda_n(A) - \lambda_{n-r+t}(A))P^* \\ &\quad \geq P \begin{pmatrix} 0 & 0 \\ 0 & [\lambda_n(A) - \lambda_{n-r+t}(A)]I_{r-t} \end{pmatrix} P^* \equiv D. \end{aligned}$$

Since $-D$ is $k \times k$ positive semidefinite and $\operatorname{rank}(-D) \leq r - t$, we see that

$$-\lambda_{k-r+t}(D) = \lambda_{r-t+1}(-D) = 0$$

and

$$\lambda_{k-r+t}[C(A - \lambda_{n-r+t}(A)I_n)C^*] \geq \lambda_{k-r+t}(D) = 0.$$

Thus

$$\begin{aligned}
\lambda_t[(BAB^*)/\alpha] &= \lambda_t[(I_k, X)BAB^*(I_k, X)^*] \\
&= \lambda_t(CAC^*) \\
&= \lambda_t[C(A - \lambda_{n-r+t}(A)I_n)C^* + \lambda_{n-r+t}(A)CC^*] \\
&\geq \max_{r+s=k+t} \{\lambda_s[C(A - \lambda_{n-r+t}(A)I_n)C^*] \\
&\quad +\lambda_r[\lambda_{n-r+t}(A)CC^*]\} \ \text{[by (2.0.9)]} \\
&= \max_{t\leq r\leq k} \{\lambda_{k-r+t}[C(A - \lambda_{n-r+t}(A)I_n)C^*] \\
&\quad +\lambda_r[\lambda_{n-r+t}(A)CC^*]\} \\
&\geq \max_{t\leq r\leq k} \{\lambda_{n-r+t}(A)\lambda_r(CC^*)\}.
\end{aligned}$$

It follows that, if $\lambda_{n-r+t}(A) \geq 0$, by (2.2.17), we have

$$\begin{aligned}
\lambda_t[(BAB^*)/\alpha] &\geq \max_{t\leq r\leq k} \{\lambda_{n-r+t}(A)\lambda_r(CC^*)\} \\
&= \max_{t\leq r\leq k} \{\lambda_{n-r+t}(A)\lambda_r[(I_k, X)BB^*(I_k, X)^*]\} \\
&\geq \max_{t\leq r\leq k} \{\lambda_{n-r+t}(A) \min_{Z\in\mathbb{C}^{k\times(n-k)}} \lambda_r[(I_k, Z)BB^*(I_k, Z)^*]\} \\
&= \max_{t\leq r\leq k} \{\lambda_{n-r+t}(A)\lambda_r[(BB^*)/\alpha]\}.
\end{aligned}$$

This completes the proof of the first inequality. The second inequality on the minimum can be similarly dealt with by substituting $-A$ for A. ∎

As an application of the theorem, setting $B = I_n, r = k$ and $B = I_n, r = t$, respectively, we see an interlacing-like result for the Hermitian case:

$$\lambda_t(A/\alpha) \geq \lambda_{n-k+t}(A), \quad \text{if } \lambda_{n-k+t}(A) \geq 0$$

and

$$\lambda_t(A/\alpha) \leq \lambda_t(A), \quad \text{if } \lambda_t(A) \leq 0.$$

In the following two theorems, Theorem 2.14 and Theorem 2.15, for a Hermitian $A \in \mathbb{H}_n$, we arrange and label the eigenvalues of A in the order so that $|\lambda_1(A)| \geq |\lambda_1(A)| \geq \cdots \geq |\lambda_n(A)|$. Our next theorem, like Theorem 2.8, gives lower bounds for the eigenvalues of the Schur complement of matrix product in terms of those of the individual matrices.

Theorem 2.14 *Let $A \in \mathbb{H}_n, B \in \mathbb{C}^{m\times n}$, and $\alpha \subset \{1, 2, \ldots, m\}$. Denote $k = m - |\alpha|$. Let the rank of A be r. Then for each $l = 1, 2, \ldots, k$,*

$$|\lambda_l[(BAB^*)/\alpha]| \geq \max_{\substack{t=1,\ldots,r-l+1\\ u=1,\ldots,t}} \begin{cases} \{\lambda_{l+t-1+n-r}[(BB^*)/\alpha]\lambda_{n-u+1}[(BB^*)/\alpha]\}^{\frac{1}{2}}|\lambda_{r+u-t}(A)| \\ \{\lambda_{l+t-1+n-r}[(BB^*)/\alpha]\lambda_{n-t+u}[(BB^*)/\alpha]\}^{\frac{1}{2}}|\lambda_{r-u+1}(A)| \\ \{\lambda_{n-t+1}[(BB^*)/\alpha]\lambda_{n-u+1}[(BB^*)/\alpha]\}^{\frac{1}{2}}|\lambda_{l+t+u-2}(A)| \\ \{\lambda_{n-t+1}[(BB^*)/\alpha]\lambda_{l+t+u-2+n-r}[(BB^*)/\alpha]\}^{\frac{1}{2}}|\lambda_{r-u+1}(A)|. \end{cases}$$

Proof. We may assume that $\alpha = \{k+1, \ldots, m\}$. Since $A \in \mathbb{H}_n$ and $\operatorname{rank}(A) = r$, there exists a unitary matrix $U \in \mathbb{C}^{n\times n}$ such that $UAU^* = \operatorname{diag}(D, 0)$, where $D = \operatorname{diag}(\lambda_1(A), \ldots, \lambda_r(A))$, and D is nonsingular. Let

$$X = -[(BAB^*)[\alpha^c, \alpha]][(BAB^*)[\alpha]]^\dagger.$$

Then

$$\begin{aligned} |\lambda_l[(BAB^*)/\alpha]| &= |\lambda_l[(I_k, X)BAB^*(I_k, X)^*]| \\ &= |\lambda_l[AB^*(I_k, X)^*(I_k, X)B]| \\ &= |\lambda_l(A\tilde{B})|, \end{aligned}$$

where $\tilde{B} = B^*(I_k, X)^*(I_k, X)B$. Partition

$$U\tilde{B}U^* = \begin{pmatrix} B_1 & B_2 \\ B_2^* & B_3 \end{pmatrix},$$

where B_1 is $r \times r$ positive semidefinite. Take

$$L = \begin{pmatrix} I_r & 0 \\ -B_2^* B_1^\dagger & I_{n-r} \end{pmatrix}.$$

Then

$$LU\tilde{B}U^*L^* = \operatorname{diag}(B_1, B_3 - B_2^* B_1^\dagger B_2)$$

and

$$L^{*-1}UAU^*L^{-1} = \operatorname{diag}(D, 0).$$

Thus

$$\begin{aligned} (L^{*-1}U)(A\tilde{B})(L^{*-1}U)^{-1} &= L^{*-1}UAU^*L^{-1}LU\tilde{B}U^*L^* \\ &= \operatorname{diag}(D, 0)\operatorname{diag}(B_1, B_3 - B_2^* B_1^\dagger B_2) \\ &= \operatorname{diag}(DB_1, 0). \end{aligned}$$

That is, $A\tilde{B}$ and $\operatorname{diag}(DB_1, 0)$ have the same set of eigenvalues. Thus

$$|\lambda_l[(BAB^*)/\alpha]| = |\lambda_l(A\tilde{B})| = |\lambda_l(\operatorname{diag}(DB_1, 0))|.$$

It follows that, for $l > r$,

$$|\lambda_l[(BAB^*)/\alpha]| = 0$$

and that, for $l = 1, 2, \ldots, r$,

$$|\lambda_l[(BAB^*)/\alpha]| = |\lambda_l(DB_1)|.$$

Notice that the eigenvalue interlacing theorem shows that, for $i = 1, 2, \ldots, r$,

$$\lambda_i(B_1) \geq \lambda_{i+n-r}(U\tilde{B}U^*) = \lambda_{i+n-r}(\tilde{B}).$$

We have, for $t = 1, \ldots, r-l+1$ and $u = 1, \ldots, t$,

$$\begin{aligned}
&|\lambda_l[(BAB^*)/\alpha]| \\
&= \lambda_l(B_1^{\frac{1}{2}} D B_1^{\frac{1}{2}}) \\
&\geq \lambda_{l+t-1}(B_1^{\frac{1}{2}})\lambda_{r-t+1}(DB_1^{\frac{1}{2}}) \quad \text{[by (2.0.3)]} \\
&\geq \lambda_{l+t-1}(B_1^{\frac{1}{2}})\lambda_{r-t+u}(D)\lambda_{r-u+1}(B_1^{\frac{1}{2}}) \quad \text{[by (2.0.3)]} \\
&\geq [\lambda_{l+t-1+n-r}(\tilde{B})\lambda_{n-u+1}(\tilde{B})]^{\frac{1}{2}}|\lambda_{r+u-t}(A)| \\
&= \Big\{\lambda_{l+t-1+n-r}[(I_k\ X)BB^*(I_k\ X)^*] \\
&\quad \cdot\lambda_{n-u+1}[(I_k\ X)BB^*(I_k\ X)^*]\Big\}^{\frac{1}{2}}|\lambda_{r+u-t}(A)| \\
&\geq \{\min_{Z\in\mathbb{C}^{k\times(m-k)}} \lambda_{l+t-1+n-r}[(I_k\ Z)BB^*(I_k\ Z)^*] \\
&\quad \cdot \min_{Z\in\mathbb{C}^{k\times(m-k)}} \lambda_{n-u+1}[(I_k\ Z)BB^*(I_k\ Z)^*]\}^{\frac{1}{2}}|\lambda_{r+u-t}(A)| \\
&= \{\lambda_{l+t-1+n-r}[(BB^*)/\alpha]\lambda_{n-u+1}[(BB^*)/\alpha]\}^{\frac{1}{2}} \\
&\quad \cdot|\lambda_{r+u-t}(A)| \quad \text{[by (2.2.17)]}.
\end{aligned}$$

The other remaining inequalities can be proved similarly. ∎

Setting $r = n$ in Theorem 2.14 yields the following

$$|\lambda_l[(BAB^*)/\alpha]| \geq$$

$$\max_{\substack{t=1,\ldots,n-l+1\\ u=1,\ldots,t}} \left\{ \begin{array}{l}
\{\lambda_{l+t-1}[(BB^*)/\alpha]\lambda_{n-u+1}[(BB^*)/\alpha]\}^{\frac{1}{2}}|\lambda_{n-t+u}(A)|, \\
\{\lambda_{l+t-1}[(BB^*)/\alpha]\lambda_{n-t+u}[(BB^*)/\alpha]\}^{\frac{1}{2}}|\lambda_{n-u+1}(A)|, \\
\{\lambda_{n-t+1}[(BB^*)/\alpha]\lambda_{n-u+1}[(BB^*)/\alpha]\}^{\frac{1}{2}}|\lambda_{l+t+u-2}(A)|, \\
\{\lambda_{n-t+1}[(BB^*)/\alpha]\lambda_{l+t+u-2}[(BB^*)/\alpha]\}^{\frac{1}{2}}|\lambda_{n-u+1}(A)|.
\end{array}\right.$$

Our next theorem is a version of Theorem 2.9 for Hermitian matrices.

Theorem 2.15 *Let $A \in \mathbb{H}_n$ with* $\operatorname{rank} A = r$ *and $B \in \mathbb{C}^{m\times n}$. Let $\alpha \subset \{1, 2, \ldots, m\}$ and denote $k = m - |\alpha|$. Then for any $1 \le i_1 < \cdots < i_u \le k$,*

$$\prod_{t=1}^{u} |\lambda_t[(BAB^*)/\alpha]| \ge \prod_{t=1}^{u} \{\lambda_{n-i_t+1}[(BB^*)/\alpha]\lambda_{n-r+i_t}[(BB^*)/\alpha]\}^{\frac{1}{2}} |\lambda_{r-t+1}(A)|.$$

Proof. Following the line of the proof of Theorem 2.14, we have

$$\begin{aligned}
&\prod_{t=1}^{u} |\lambda_t[(BAB^*)/\alpha]| \\
&= \prod_{t=1}^{u} \lambda_t(B_1^{\frac{1}{2}} D B_1^{\frac{1}{2}}) \\
&\ge \prod_{t=1}^{u} \lambda_{r-i_t+1}(B_1^{\frac{1}{2}})\sigma_{i_t}(DB_1^{\frac{1}{2}}) \quad \text{[by (2.0.1)]} \\
&\ge \prod_{t=1}^{u} \lambda_{r-i_t+1}(B_1^{\frac{1}{2}})\lambda_{i_t}(B_1^{\frac{1}{2}})\lambda_{r-t+1}(D) \quad \text{[by (2.0.2)]} \\
&\ge \prod_{t=1}^{u} \lambda_{r-i_t+1}(\tilde{B}^{\frac{1}{2}})\lambda_{n-r+i_t}(\tilde{B}^{\frac{1}{2}})\lambda_{r-t+1}(A) \\
&= \prod_{t=1}^{u} [\lambda_{n-i_t+1}(\tilde{B})\lambda_{n-r+i_t}(\tilde{B})]^{\frac{1}{2}} |\lambda_{r-t+1}(A)| \\
&= \prod_{t=1}^{u} \{\lambda_{n-i_t+1}[(I_k\ X)BB^*(I_k\ X)^*] \\
&\quad \cdot \lambda_{n-r+i_t}[(I_k\ X)BB^*(I_k\ X)^*]\}^{\frac{1}{2}} |\lambda_{r-t+1}(A)| \\
&\ge \prod_{t=1}^{u} \{\min_{Z\in\mathbb{C}^{k\times(m-k)}} \lambda_{n-i_t+1}[(I_k\ Z)BB^*(I_k\ Z)^*] \\
&\quad \cdot \min_{Z\in\mathbb{C}^{k\times(m-k)}} \lambda_{n-r+i_t}[(I_k\ Z)BB^*(I_k\ Z)^*]\}^{\frac{1}{2}} |\lambda_{r-t+1}(A)| \\
&= \prod_{t=1}^{u} \{\lambda_{n-i_t+1}[(BB^*)/\alpha]\lambda_{n-r+i_t}[(BB^*)/\alpha]\}^{\frac{1}{2}} \\
&\quad \cdot |\lambda_{r-t+1}(A)| \quad \text{[by (2.2.17)]}. \blacksquare
\end{aligned}$$

Setting $r = n$ in Theorem 2.15, we arrive at

$$\prod_{t=1}^{u} |\lambda_t[(BAB^*)/\alpha]| \ge \prod_{t=1}^{u} \{\lambda_{n-i_t+1}[(BB^*)/\alpha]\lambda_{i_t}[(BB^*)/\alpha]\}^{\frac{1}{2}} |\lambda_{n-t+1}(A)|.$$

Theorem 2.16 *Let $A \in \mathbb{H}_n$, $B \in \mathbb{C}^{n\times n}$, and $\alpha \subset \{1, 2, \ldots, n\}$. Denote $k = n - |\alpha|$. Then for any integer l with $1 < l < k$,*

$$\sum_{t=1}^{l} \lambda_{k-t+1}[(BAB^*)/\alpha] \le \sum_{t=1}^{l} \lambda_t(A)\lambda_{i_t}[(BB^*)/\alpha] \tag{2.5.28}$$

and

$$\sum_{t=1}^{l} \lambda_{i_t}[(BB^*)/\alpha]\lambda_{n-t+1}(A) \le \sum_{t=1}^{l} \lambda_t[(BAB^*)/\alpha]. \tag{2.5.29}$$

Proof. If $A \ge 0$, the inequalities follow immediately from Theorem 2.6. So we consider the case where A has negative eigenvalues. Let $\lambda_n(A) < 0$. Without loss of generality, we take $\alpha = \{k+1, \ldots, m\}$. Let

$$X = -[(BB^*)[\alpha^c, \alpha]][(BB^*)[\alpha]]^\dagger.$$

By (2.2.19), (2.0.6), we have

$$\begin{aligned}
&\sum_{t=1}^{l} \lambda_{i_t}\{[B(A - \lambda_n(A)I_n)B^*]/\alpha\} \\
= &\min_{Z\in\mathbb{C}^{k\times(n-k)}} \sum_{t=1}^{l} \lambda_{i_t}[(I_k, Z)B(A - \lambda_n(A)I_n)B^*(I_k, Z)^*] \\
= &\min_{Z\in\mathbb{C}^{k\times(n-k)}} \sum_{t=1}^{l} \lambda_{i_t}[(A - \lambda_n(A)I_n)B^*(I_k, Z)^*(I_k, Z)B] \\
\le &\min_{Z\in\mathbb{C}^{k\times(n-k)}} \sum_{t=1}^{l} \lambda_t[A - \lambda_n(A)]\lambda_{i_t}[B^*(I_k, Z)^*(I_k, Z)B] \\
= &\min_{Z\in\mathbb{C}^{k\times(n-k)}} \sum_{t=1}^{l} [\lambda_t(A) - \lambda_n(A)]\lambda_{i_t}[(I_k, Z)BB^*(I_k, Z)^*] \\
\le &\sum_{t=1}^{l} [\lambda_t(A) - \lambda_n(A)]\lambda_{i_t}[(I_k, X)BB^*(I_k, X)^*] \\
= &\sum_{t=1}^{l} [\lambda_t(A) - \lambda_n(A)]\lambda_{i_t}[(BB^*)/\alpha] \\
= &\sum_{t=1}^{l} \lambda_t(A)\lambda_{i_t}[(BB^*)/\alpha] - \lambda_n(A)\sum_{t=1}^{l} \lambda_{i_t}[(BB^*)/\alpha].
\end{aligned} \tag{2.5.30}$$

By (2.2.19) and (2.0.8), noticing that $-\lambda_n(A) \geq 0$, we have

$$\begin{aligned}
&\sum_{t=1}^{l} \lambda_{i_t}\{[B(A - \lambda_n(A)I_n)B^*]/\alpha\} \\
&= \min_{Z\in\mathbb{C}^{k\times(n-k)}} \sum_{t=1}^{l} \lambda_{i_t}[(I_k, Z)B(A - \lambda_n(A)I_n)B^*(I_k, Z)^*] \\
&= \min_{Z\in\mathbb{C}^{k\times(n-k)}} \sum_{t=1}^{l} \lambda_{i_t}[(I_k, Z)BAB^*(I_k, Z)^* - \lambda_n(A)(I_k, Z)BB^*(I_k, Z)^*] \\
&\geq \min_{Z\in\mathbb{C}^{k\times(n-k)}} \Big\{\sum_{t=1}^{l} \lambda_{k-t+1}[(I_k, Z)BAB^*(I_k, Z)^*] \\
&\quad + \sum_{t=1}^{l} \lambda_{i_t}[-\lambda_n(A)(I_k, Z)BB^*(I_k, Z)^*]\Big\} \text{ [by (2.0.8)]} \\
&= \min_{Z\in\mathbb{C}^{k\times(n-k)}} \Big\{\sum_{t=1}^{l} \lambda_{k-t+1}[(I_k, Z)BAB^*(I_k, Z)^*] \\
&\quad -\lambda_n(A)\sum_{t=1}^{l} \lambda_{i_t}[(I_k, Z)BB^*(I_k, Z)^*]\Big\} \\
&\geq \min_{Z\in\mathbb{C}^{k\times(n-k)}} \sum_{t=1}^{l} \lambda_{k-t+1}[(I_k, Z)BAB^*(I_k, Z)^*] \\
&\quad -\lambda_n(A) \min_{Z\in\mathbb{C}^{k\times(n-k)}} \sum_{t=1}^{l} \lambda_{i_t}[(I_k, Z)BB^*(I_k, Z)^*] \\
&= \sum_{t=1}^{l} \lambda_{k-t+1}[(BAB^*)/\alpha] - \lambda_n(A)\sum_{t=1}^{l} \lambda_{i_t}[(BB^*)/\alpha]. \qquad (2.5.31)
\end{aligned}$$

Combining (2.5.30) and (2.5.31) reveals (2.5.28). Likewise, by making use of (2.0.8) and (2.2.19) in the proof of (2.5.31), we have

$$\begin{aligned}
&\sum_{t=1}^{l} \lambda_{i_t}\{[B(A - \lambda_n(A)I_n)B^*]/\alpha\} \leq \\
&\qquad \sum_{t=1}^{l} \lambda_t[(BAB^*)/\alpha] - \lambda_n(A)\sum_{t=1}^{l} \lambda_{i_t}[(BB^*)/\alpha]. \qquad (2.5.32)
\end{aligned}$$

Using (2.2.19), (2.0.7) and as in the proof of (2.5.30), we have

$$\sum_{t=1}^{l} \lambda_{i_t}\{[B(A-\lambda_n(A)I_n)B^*]/\alpha\} \geq$$
$$\sum_{t=1}^{l} \lambda_{n-t+1}(A)\lambda_{i_t}[(BB^*)/\alpha] - \lambda_n(A)\sum_{t=1}^{l} \lambda_{i_t}[(BB^*)/\alpha]. \quad (2.5.33)$$

Combining (2.5.32) and (2.5.33), we obtain the inequality (2.5.29). ∎

2.6 Singular values of the Schur complement of product

Singular values are, in many aspects, as important as the eigenvalues for matrices. This section, based upon [285], is devoted to the inequalities on singular values of the Schur complements of products of general matrices.

Theorem 2.17 *Let $A \in \mathbb{C}^{m\times n}$ and $B \in \mathbb{C}^{n\times p}$. Let $\alpha \subset \{1,2,\ldots,l\}$, where $l = \min\{m,n,p\}$. If B^*B is nonsingular, then for $s = 1,2,\ldots,l-|\alpha|$,*

$$\sigma_s^2[(AB)/\alpha] \geq \max_{1\leq i\leq m-|\alpha|+p-n} \lambda_{p-|\alpha|-i+s}\left[(B^*B)/\alpha\right]\lambda_{n-p+i}\left[(AA^*)/\alpha\right].$$

Proof. We first claim that we may take $\alpha = \{1,2,\ldots,|\alpha|\}$. To see this, let $\alpha^c = \{1,\ldots,m\} - \alpha$, $\beta^c = \{1,\ldots,n\} - \alpha$, and $\gamma^c = \{1,\ldots,p\} - \alpha$. There exist permutation matrices $U \in \mathbb{C}^{m\times m}$, $V \in \mathbb{C}^{n\times n}$, $W \in \mathbb{C}^{p\times p}$ such that

$$UAV = \begin{pmatrix} A[\alpha] & A[\alpha,\beta^c] \\ A[\alpha^c,\alpha] & A[\alpha^c,\beta^c] \end{pmatrix},$$

$$V^TBW = \begin{pmatrix} B[\alpha] & B[\alpha,\gamma^c] \\ B[\beta^c,\alpha] & B[\beta^c,\gamma^c] \end{pmatrix},$$

and

$$(UAV)(V^TBW) = UABW = \begin{pmatrix} (AB)[\alpha] & (AB)[\alpha,\gamma^c] \\ (AB)[\alpha^c,\alpha] & (AB)[\alpha^c,\gamma^c] \end{pmatrix}.$$

Let $\bar{\alpha} = \{1,2,\ldots,|\alpha|\}$. Then

$$(AB)/\alpha = (UABW)/\bar{\alpha},$$
$$(B^*B)/\alpha = (W^*B^*BW)/\bar{\alpha},$$
$$(AA^*)/\alpha = (UAA^*U^*)/\bar{\alpha}.$$

So we may replace A with UAV and B with V^TBW in the theorem so that the submatrices indexed by α are now located in the upper left corners. Thus, without loss of generality, we may assume that $\alpha = \{1,2,\ldots,|\alpha|\}$.

The idea of the proof of the inequality is to obtain two quantities, one of which bounds $\lambda_i\{[AB(B^*B)^{-1}B^*A^*]/\alpha\}$ from above, and the other from below; combining the two inequalities will yield the desired inequality.

We shall make heavy use of Corollary 2.6. Let

$$C = AB(B^*B)^{-1}B^*A^*,$$

$$X = -C[\alpha^c,\ \alpha][C[\alpha]]^\dagger,$$

$$Y = -[(AB)[\alpha^c,\alpha]][(AB)[\alpha]]^\dagger,$$

and

$$Z = -[(AA^*)[\alpha^c,\alpha]][(AA^*)[\alpha]]^\dagger.$$

Using (2.0.3) and upon computation, we have

$$\begin{aligned}
&\lambda_i\{[AB(B^*B)^{-1}B^*A^*]/\alpha\}\\
&= \lambda_i[C[\alpha^c] + XC[\alpha,\ \alpha^c]]\\
&= \lambda_i\{(Y,\ I_{m-|\alpha|})C(Y,\ I_{m-|\alpha|})^* - Y(C[\alpha,\ \alpha^c])\\
&\quad -(C[\alpha^c,\ \alpha])Y^* - Y(C[\alpha])Y^* + XC[\alpha,\ \alpha^c]\}\\
&= \lambda_i\{(Y,\ I_{m-|\alpha|})C(Y,\ I_{m-|\alpha|})^*\\
&\quad +(X-Y)(C[\alpha])(C[\alpha])^\dagger(C[\alpha,\ \alpha^c])\\
&\quad -(C[\alpha^c,\ \alpha])(C[\alpha])^\dagger(C[\alpha])Y^* - Y(C[\alpha])Y^*\}\\
&= \lambda_i\{(Y,\ I_{m-|\alpha|})C(Y,\ I_{m-|\alpha|})^* - (X-Y)(C[\alpha])X^*\\
&\quad +X(C[\alpha])Y^* - Y(C[\alpha])Y^*\}\\
&= \lambda_i\{(Y,\ I_{m-|\alpha|})C(Y,\ I_{m-|\alpha|})^*\\
&\quad -(X-Y)(C[\alpha])(X-Y)^*\} \qquad (2.6.34)\\
&\le \lambda_i\left[(Y,\ I_{m-|\alpha|})C(Y,\ I_{m-|\alpha|})^*\right]\\
&= \lambda_i\{[(Y,\ I_{m-|\alpha|})AB](B^*B)^{-1}[(Y,\ I_{m-|\alpha|})AB]^*\}\\
&= \lambda_i\left[(0,\ (AB)/\alpha)(B^*B)^{-1}(0,\ (AB)/\alpha)^*\right]\\
&= \lambda_i\{[(AB)/\alpha][(B^*B)^{-1}[\gamma^c]][(AB)^*/\alpha]\}\\
&= \lambda_i\{[(B^*B)^{-1}[\gamma^c]][(AB)/\alpha]^*[(AB)/\alpha]\} \qquad (2.6.35)\\
&\le \min_{\substack{t+s=i+1\\ t=1,\ldots,p-|\alpha|\\ s=1,\ldots,p-|\alpha|}} \lambda_t[(B^*B)^{-1}[\gamma^c]]\lambda_s\{[(AB)/\alpha]^*[(AB)/\alpha]\}\\
&\le \min_{\substack{t+s=i+1\\ t=1,\ldots,l-|\alpha|\\ s=1,\ldots,l-|\alpha|}} \lambda_t[(B^*B)^{-1}[\gamma^c]]\sigma_s^2[(AB)/\alpha]. \qquad (2.6.36)
\end{aligned}$$

On the other hand, by (2.0.6), for every $i = 1, 2, \ldots, m - |\alpha| + p - n$,

$$\begin{aligned}
&\lambda_i\{[AB(B^*B)^{-1}B^*A^*]/\alpha\} \\
&= \lambda_i\left[(X,\ I_{m-|\alpha|})AB(B^*B)^{-1}B^*A^*(X,\ I_{m-|\alpha|})^*\right] \\
&= \lambda_i\{[B(B^*B)^{-1}B^*][A^*(X,\ I_{m-|\alpha|})^*(X,\ I_{m-|\alpha|})A]\} \\
&\geq \max_{t+s=n+i} \lambda_t[B(B^*B)^{-1}B^*]\lambda_s[A^*(X,\ I_{m-|\alpha|})^*(X,\ I_{m-|\alpha|})A] \\
&\geq \lambda_p[B(B^*B)^{-1}B^*]\lambda_{n-p+i}[(X,\ I_{m-|\alpha|})AA^*(X,\ I_{m-|\alpha|})^*] \\
&= \lambda_p[(B^*B)^{-1}B^*B]\lambda_{n-p+i}[(X,\ I_{m-|\alpha|})AA^*(X,\ I_{m-|\alpha|})^*] \\
&= \lambda_{n-p+i}[(X,\ I_{m-|\alpha|})AA^*(X,\ I_{m-|\alpha|})^*] \\
&= \lambda_{n-p+i}\{(Z,\ I_{m-|\alpha|})AA^*(Z,\ I_{m-|\alpha|})^* \\
&\quad +(X-Z)[(AA^*)[\alpha]](X-Z)^*\} \ \text{[by (2.6.35)]} \qquad (2.6.37) \\
&\geq \lambda_{n-p+i}\left[(Z,\ I_{m-|\alpha|})AA^*(Z,\ I_{m-|\alpha|})^*\right] \\
&= \lambda_{n-p+i}[(AA^*)/\alpha]. \qquad (2.6.38)
\end{aligned}$$

By Theorem 1.2, $(B^*B)^{-1}[\gamma^c] = [(B^*B)/\alpha]^{-1}$, so for $t = 1, 2, \ldots, p - |\alpha|$,

$$\lambda_t^{-1}\{[(B^*B)/\alpha]^{-1}\} = \lambda_{p-|\alpha|-t+1}[(B^*B)/\alpha],$$

it follows that, by using (2.6.36) and (2.6.38), for $s = 1, 2, \ldots, l - |\alpha|$,

$$\begin{aligned}
\sigma_s^2[(AB)/\alpha] &\geq \max_{\substack{1\leq i\leq m-|\alpha|+p-n \\ t=i-s+1}} \lambda_{p-|\alpha|-t+1}[(B^*B)/\alpha]\lambda_{n-p+i}[(AA^*)/\alpha] \\
&= \max_{1\leq i\leq m-|\alpha|+p-n} \lambda_{p-|\alpha|-i+s}[(B^*B)/\alpha]\lambda_{n-p+i}[(AA^*)/\alpha]. \ \blacksquare
\end{aligned}$$

Corollary 2.9 *Let $A \in \mathbb{C}^{m\times n}$, $l = \min\{m,\ n\}$, $\alpha \subset \{1, 2, \ldots, l\}$. Then*

$$\sigma_i(A/\alpha) \geq \lambda_i^{\frac{1}{2}}[(AA^*)/\alpha] \geq \sigma_{i+|\alpha|}(A), \quad i = 1, 2, \ldots, l - |\alpha|.$$

Proof. Set $B = I$, $i = s$ in Theorem 2.17 and use Corollary 2.4. ∎

Corollary 2.10 *If $A \in \mathbb{C}^{m\times n}$, $B \in \mathbb{C}^{n\times p}$, $l = \min\{m, n, p\}$, and let $\alpha \subset \{1, 2, \ldots, l\}$. If B^*B is nonsingular, then for $s = 1, 2, \ldots, l - |\alpha|$,*

$$\sigma_s^2[(AB)/\alpha] \geq \max_{i=1,2,\ldots,m-|\alpha|+p-n} \lambda_{n-p+i}[(AA^*)/\alpha]\sigma_{p-i+s}^2(B).$$

Proof. This follows from (2.1.12) and Theorem 2.17. ∎

Setting $m = n = p$ in Corollary 2.10 shows, for each $s = 1, 2, \ldots, n - |\alpha|$,

$$\sigma_s[(AB)/\alpha] \geq \max_{i+j=n+s} \lambda_i^{\frac{1}{2}}[(AA^*)/\alpha]\sigma_j(B).$$

In a similar manner, we obtain lower bounds for products of singular values of Schur complements of matrix products.

Theorem 2.18 *Let the assumptions of Theorem 2.17 be satisfied. Let u be an integer, $1 \le u \le l - |\alpha|$, and $n - p + 1 \le i_1 < \cdots < i_u \le l - |\alpha|$. Then*

$$\prod_{t=1}^{u} \sigma_t[(AB)/\alpha] \ge \prod_{t=1}^{u} \lambda_{i_t}^{\frac{1}{2}}[(AA^*)/\alpha]\lambda_{p-|\alpha|-t+1}^{\frac{1}{2}}[(B^*B)/\alpha].$$

Proof. Following the proof of Theorem 2.17, by (2.6.35) and using (2.0.1), we have, for every $n \ge i_t \ge n - p + 1$, $t = 1, 2, \ldots, u$,

$$\begin{aligned}
&\prod_{t=1}^{u} \lambda_t\{[AB(B^*B)^{-1}B^*A^*]/\alpha\} \\
&\quad = \prod_{t=1}^{u} \lambda_t \left[(X,\ I_{m-|\alpha|})AB(B^*B)^{-1}B^*A^*(X,\ I_{m-|\alpha|})^*\right] \\
&\quad \ge \prod_{t=1}^{u} \lambda_{i_t}[A^*(X,\ I_{m-|\alpha|})^*(X,\ I_{m-|\alpha|})A] \\
&\quad \cdot \lambda_{n-i_t+1}[B(B^*B)^{-1}B^*] \ \text{[by 2.0.1)]} \\
&\quad = \prod_{t=1}^{u} \lambda_{i_t}\{(Z,\ I_{m-|\alpha|})AA^*(Z,\ I_{m-|\alpha|})^* \\
&\quad +(X - Z)[(AA^*)[\alpha]](X - Z)^*\} \ \text{[see (2.6.35)]} \\
&\quad \ge \prod_{t=1}^{u} \lambda_{i_t} \left[(Z,\ I_{m-|\alpha|})AA^*(Z,\ I_{m-|\alpha|})^*\right] \\
&\quad = \prod_{t=1}^{u} \lambda_{i_t}[(AA^*)/\alpha].
\end{aligned} \tag{2.6.39}$$

On the other hand, by (2.6.35) and (2.0.2), we have

$$\begin{aligned}
&\prod_{t=1}^{u} \lambda_t \left\{[AB(B^*B)^{-1}B^*A^*]/\alpha\right\} \\
&\quad = \prod_{t=1}^{u} \lambda_t\{(Y,\ I_{m-|\alpha|})C(Y,\ I_{m-|\alpha|})^* \\
&\quad -(X - Y)[C[\alpha]](X - Y)^*\} \ \text{[by (2.6.35)]} \\
&\quad \le \prod_{t=1}^{u} \lambda_t \left[(Y,\ I_{m-|\alpha|})C(Y,\ I_{m-|\alpha|})^*\right] \\
&\quad \le \prod_{t=1}^{u} \lambda_t[(B^*B)^{-1}(\gamma^c)]\sigma_t^2[(AB)/\alpha] \ \text{[by (2.6.35)]}.
\end{aligned} \tag{2.6.40}$$

Combining (2.6.39) and (2.6.40) we obtain the desired inequality. ∎

The proof of the next theorem is similar to the above, thus omitted.

Theorem 2.19 *Let $A \in \mathbb{C}^{m\times n}$ and B be $n \times n$ nonsingular. Let $l = \min\{m, n\}$, $\alpha \subset \{1, 2, \ldots, l\}$, and u be an integer with $1 \le u \le l - |\alpha|$. Then for $1 \le i_1 < \cdots < i_u \le l - |\alpha|$,*

$$\prod_{t=1}^{u} \sigma_{i_t}[(AB)/\alpha] \ge \prod_{t=1}^{u} \lambda_{i_t}^{\frac{1}{2}}[(AA^*)/\alpha]\lambda_{n-|\alpha|-t+1}^{\frac{1}{2}}[(B^*B)/\alpha]$$

and

$$\prod_{t=1}^{u} \sigma_t[(AB)/\alpha] \ge \prod_{t=1}^{u} \lambda_{i_t}^{\frac{1}{2}}[(AA^*)/\alpha]\lambda_{n-|\alpha|-i_t+1}^{\frac{1}{2}}[(B^*B)/\alpha].$$

Corollary 2.11 *Let $A \in \mathbb{C}^{m\times n}$ and $B \in \mathbb{C}^{n\times p}$. Let $l = \min\{m, n, p\}$, $\alpha \subset \{1, 2, \ldots, l\}$, and u be an integer such that $1 \le u \le l - |\alpha|$. Then for $n - p + 1 \le i_1 < \cdots < i_u \le l - |\alpha|$,*

$$\prod_{t=1}^{u} \sigma_t[(AB)/\alpha] \ge \prod_{t=1}^{u} \lambda_{i_t}^{\frac{1}{2}}[(AA^*)/\alpha]\sigma_{p-t+1}(B) \tag{2.6.41}$$

and if $n = p$, then

$$\prod_{t=1}^{u} \sigma_{i_t}[(AB)/\alpha] \ge \prod_{t=1}^{u} \lambda_{i_t}^{\frac{1}{2}}[(AA^*)/\alpha]\sigma_{n-t+1}(B). \tag{2.6.42}$$

Proof. If B^*B is singular, for $t = 1$ we have $\sigma_{p-t+1}(B) = \sigma_p(B) = 0$. The first inequality holds. If B^*B is nonsingular, then Theorem 2.18, together with Theorem 2.2, yields the first inequality again. The second inequality can be obtained in a similar manner. ∎

Corollary 2.12 *Let all the assumptions of Theorem 2.19 be satisfied. Then*

$$\prod_{t=1}^{u} \sigma_t[(AB)/\alpha] \ge \prod_{t=1}^{u} \lambda_{i_t}^{\frac{1}{2}}[(AA^*)/\alpha]\sigma_{n-i_t+1}(B)$$

and

$$\prod_{t=1}^{u} \sigma_t[(AB)/\alpha] \ge \prod_{t=1}^{\mu} \sigma_{i_t+|\alpha|}(A)\lambda_{n-|\alpha|-i_t+1}^{\frac{1}{2}}[(B^*B)/\alpha].$$

All the inequalities we obtained so far in this section present lower bounds for the singular values. It is tempting to obtain analogous results on upper bounds. For instance, we may ask if an analog of (2.0.2)

$$\prod_{t=1}^{l} \sigma_{i_t}[(AB)/\alpha] \le \min\left\{\prod_{t=1}^{l} \sigma_{i_t}(A/\alpha)\sigma_t(B),\ \prod_{t=1}^{l} \sigma_t(A/\alpha)\sigma_{i_t}(B)\right\}$$

or an analog of (2.0.3)

$$\sigma_t[(AB)/\alpha] \le \min_{i+j=t+1} \{\sigma_i(A/\alpha)\sigma_j(B),\ \sigma_i(B/\alpha)\sigma_j(A)\}$$

holds. The answer is negative as the following example shows: Take

$$A = I_2, \quad B = \begin{pmatrix} 1 & 2 \\ -2 & 1 \end{pmatrix}, \quad \alpha = \{2\}.$$

Then $\sigma_1[(AB)/\alpha] = \sigma_1(5) = 5$, $\sigma_1(A/\alpha)\sigma_1(B) = \sigma_1(B) = \sqrt{5}$. Thus

$$\sigma_1[(AB)/\alpha] \ge \sigma_1(A/\alpha)\sigma_1(B).$$

This says neither of the above two inequalities holds. This comes as no surprise if one reinspects the signs of the second summands in (2.6.35) and (2.6.38). Furthermore, invalidity remains true even if one replaces the pair $\sigma(A/\alpha)$ and $\sigma(B/\alpha)$ or $\sigma(A)$ and $\sigma(B)$ by the pair $\lambda^{\frac{1}{2}}[(AA^*)/\alpha]$ and $\lambda^{\frac{1}{2}}[(BB^*)/\alpha]$.

Finally, we apply the theorems of this section to obtain some new upper bounds for eigenvalues of the Schur complements of BAB^*, where A is $n \times n$ positive semidefinite and B is any $m \times n$ matrix.

Theorem 2.20 *Let A be $n \times n$ positive semidefinite and B be $m \times n$. Let $l = \min\{m, n\}$, $\alpha \subset \{1, 2, \ldots, l\}$, and $\alpha^c = \{1, 2, \ldots, n\} - \alpha$. Then for every $i = 1, 2, \ldots, m - |\alpha|$ and every $t = 1, 2, \ldots, l - |\alpha|$,*

$$\lambda_i[(BAB^*)/\alpha] \le \min_{s+t=i+1} \lambda_s(A[\alpha^c])\sigma_t{}^2(B/\alpha).$$

Proof. Without loss of generality, assume $\alpha = \{1, 2, \ldots, |\alpha|\}$. By Theorem 2.17, since $(A^{\frac{1}{2}})^* = A^{\frac{1}{2}}$, $(A^{-\frac{1}{2}})^* = A^{-\frac{1}{2}}$, for $t = 1, 2, \ldots, l - |\alpha|$, we have

$$\begin{aligned} \sigma_t{}^2(B/\alpha) &= \sigma_t{}^2\left[(BA^{\frac{1}{2}}A^{-\frac{1}{2}})/\alpha\right] \\ &\ge \max_{i=1,2,\cdots,m-|\alpha|} \lambda_{n-|\alpha|-i+t}(A^{-1}/\alpha)\lambda_i[(BAB^*)/\alpha]. \end{aligned}$$

By Theorem 1.2, we have

$$A^{-1}/\alpha = [(A^{-1}/\alpha)^{-1}]^{-1} = (A[\alpha^c])^{-1}.$$

So, for any $1 \le j \le n$, we have

$$\lambda_j^{-1}(A^{-1}/\alpha) = \lambda_{n-|\alpha|-j+1}(A[\alpha^c]).$$

It follows that, for every $i = 1, 2, \ldots, m - |\alpha|$ and $t = 1, 2, \ldots, l - |\alpha|$,

$$\begin{aligned} \lambda_i[(BAB^*)/\alpha] &\le \min_{t=1,2,\ldots,l-|\alpha|} \sigma_t^2(B/\alpha)\lambda_{i-t+1}(A[\alpha^c]) \\ &= \min_{t+s=i+1} \lambda_s(A[\alpha^c])\sigma_t^2(B/\alpha). \blacksquare \end{aligned}$$

Setting $B = I$ in Theorem 2.20 results in eigenvalue inequalities that may be compared with the ones in Section 2.1: For $i = 1, 2, \ldots, n - |\alpha|$,

$$\lambda_i(A/\alpha) \le \lambda_i(A[\alpha^c]).$$

By Theorems 2.18 and 2.19, one gets the following result which is proven in a manner similar to that of Theorem 2.20.

Theorem 2.21 *Let all the assumptions of Theorem 2.20 be satisfied. Let u be an integer such that $1 \le u \le l - |\alpha|$. Then for $1 \le i_1 < \cdots < i_t \le l - |\alpha|$,*

$$\prod_{t=1}^{u} \lambda_{i_t}[(BAB^*)/\alpha] \le \min\left\{ \prod_{t=1}^{u} \lambda_{i_t}(A[\alpha^c])\sigma_t{}^2(B/\alpha), \prod_{t=1}^{u} \lambda_t(A[\alpha^c])\sigma_{i_t}{}^2(B/\alpha) \right\}.$$

Chapter 3

Block Matrix Techniques

3.0 Introduction

This chapter is an expository study of matrix inequalities by means of the techniques on block matrices; usually they are 2×2 in most applications. The 2×2, ordinary or partitioned, matrices play an important role in various matrix problems, particularly in deriving matrix inequalities. We begin by showing a few examples that often appear in the literature, in which the block matrix techniques are used to obtain the desired results.

Example 1 on the eigenvalues of AB and BA. For $m \times n$ matrix A and $n \times m$ matrix B, AB and BA have the same set of nonzero eigenvalues, including multiplicity. Here is a neat and simple proof by similarity:

$$\begin{pmatrix} I & A \\ 0 & I \end{pmatrix}^{-1} \begin{pmatrix} AB & 0 \\ B & 0 \end{pmatrix} \begin{pmatrix} I & A \\ 0 & I \end{pmatrix} = \begin{pmatrix} 0 & 0 \\ B & BA \end{pmatrix}.$$

Example 2 on the singular values of matrices. To extend a result on Hermitian matrices to a general matrix A, say, an eigenvalue or singular value inequality, one sometimes needs to form the block Hermitian matrix

$$H = \begin{pmatrix} 0 & A \\ A^* & 0 \end{pmatrix}.$$

If $\sigma_1(A), \ldots, \sigma_r(A)$ are the nonzero singular values of the (square or rectangular) matrix A, then $\sigma_1(A), \ldots, \sigma_r(A), 0, \cdots, 0, -\sigma_r(A), \ldots, -\sigma_1(A)$ are the eigenvalues of the Hermitian matrix H. Thus the properties of the Hermitian H may pass to A, which was embedded in H as a submatrix.

For instance, given that $\lambda_{\max}(M+N) \leq \lambda_{\max}(M) + \lambda_{\max}(N)$ for Hermitian matrices M and N, where $\lambda_{\max}(X)$ denotes the largest eigenvalue of the Hermitian matrix X, setting $K = \begin{pmatrix} 0 & B \\ B^* & 0 \end{pmatrix}$, we obtain $\sigma_{\max}(A+B) \leq \sigma_{\max}(A) + \sigma_{\max}(B)$, where $\sigma_{\max}(X)$ is the largest singular value of X.

Example 3 on the convexity of the numerical range of a complex matrix. The well-known Toeplitz–Hausdorff Theorem asserts that the numerical range of a complex matrix is convex in $\mathbb{C}$. Its proof is by a reduction through unitary similarity to the 2×2 case (see, e.g., [230, p. 18]).

Example 4 on the solution to the matrix equation $AX - XB = C$. A classical result in solving matrix equations states that the matrix equation $AX - XB = C$ has a solution X if and only if the block matrices

$$\begin{pmatrix} A & C \\ 0 & B \end{pmatrix} \quad \text{and} \quad \begin{pmatrix} A & 0 \\ 0 & B \end{pmatrix}$$

are similar. As a consequence, if $s(A) \cap s(B) = \emptyset$; that is, A and B have no eigenvalues in common, then the matrix equation has a solution X for every matrix C of appropriate size (see, e.g., [55]).

Example 5 on the numerical radius of a square matrix. For an $n \times n$ matrix A, the *numerical radius* of A is defined and denoted by $w(A) := \max\{x^*Ax : \|x\| = 1\}$. That $w(A) \leq 1$ can be characterized by the positive semidefiniteness of certain block matrices [19]. To be precise, a matrix A has $w(A) \leq 1$ if and only if there exists a Hermitian matrix Z so that

$$\begin{pmatrix} I + Z & A \\ A^* & I - Z \end{pmatrix} \geq 0.$$

The purpose of this chapter is to present, mainly through demonstrating examples, a variety of matrix inequalities by the techniques on 2×2 matrices. Much of the material is in the spirit of the Schur complement. We have made no attempt to cover all the methods on the 2×2 block matrices nor to include all the results on the matrices; either is an impossible task.

In addition to the notations $A \geq 0$ for the positive semidefiniteness of the matrix A and $|A|$ for the matrix absolute value of A, we denote by $\langle u, v \rangle$ the inner product of vectors u and v in a vector space. In particular, for x, $y \in \mathbb{C}^n$, $\langle x, y \rangle = y^*x$, and for A, B in the unitary space $\mathbb{C}^{m \times n}$ of all $m \times n$ complex matrices, $\langle A, B \rangle = \operatorname{tr}(B^*A)$. A norm $\|\cdot\|$ on the matrix space $\mathbb{C}^{m \times n}$ is said to be *unitarily invariant* if $\|UAV\| = \|A\|$ for all $A \in \mathbb{C}^{m \times n}$ and for all unitary matrices $U \in \mathbb{C}^{m \times m}$ and $V \in \mathbb{C}^{n \times n}$. Throughout the chapter all norms are assumed to be unitarily invariant.

Unitarily invariant norms of matrices can be characterized by singular values and symmetric gauge functions. A celebrated theorem due to von Neumann and Fan [301, p. 264] reveals an important connection between the inequalities on singular values and unitarily invariant norms.

For vectors $x = (x_1, x_2, \ldots, x_n)$ and $y = (y_1, y_2, \ldots, y_n)$ with nonnegative entries in decreasing order, we write $x \prec_w y$ to mean

$$\sum_{i=1}^{k} x_i \le \sum_{i=1}^{k} y_i, \quad k = 1, 2, \ldots, n,$$

and we say that y *weakly majorizes* x. If, in addition, the above inequality becomes an equality for $k = n$, we write $x \prec y$ and say that y *majorizes* x.

Let A be an $n \times n$ matrix. As usual, we designate the vectors of the main diagonal entries, eigenvalues, and singular values of A respectively by

$$d(A) = (a_{11}, \ldots, a_{nn}),$$

$$\lambda(A) = (\lambda_1(A), \ldots, \lambda_n(A)),$$

$$\sigma(A) = (\sigma_1(A), \ldots, \sigma_n(A)).$$

The von Neumann-Fan theorem states that if $A,\ B \in \mathbb{C}^{m\times n}$ then

$$\sigma(A) \prec_w \sigma(B) \quad \Leftrightarrow \quad \|A\| \le \|B\| \text{ for all } \|\cdot\| \text{ on } \mathbb{C}^{m\times n}. \tag{3.0.1}$$

A theorem of I. Schur asserts that if H is a Hermitian matrix then

$$d(H) \prec \lambda(H). \tag{3.0.2}$$

It follows that for every Hermitian matrix H and unitarily invariant norm

$$\|\operatorname{diag}(H)\| = \|H \circ I\| \le \|H\|. \tag{3.0.3}$$

For more results on majorization inequalities, we refer the reader to [301].

3.1 Embedding approach

While the sets of values x^*Ax and y^*Ax with some constraints on vectors x and y have been extensively studied as numerical ranges or fields of values [230, Chapter 1], we shall inspect a number of matrix inequalities that involve the quadratic terms x^*Ax and x^*Ay through the standpoint of *embedding*. Namely, we will embed x^*Ax and x^*Ay in 2×2 matrices of the forms $\begin{pmatrix} x^*Ax & \star \\ \star & \star \end{pmatrix}$ and $\begin{pmatrix} \star & x^*Ay \\ \star & \star \end{pmatrix}$, respectively, where $\star$ stands for some entries irrelevant to our discussions, so that the results on 2×2 matrices can be utilized to derive equalities or inequalities of x^*Ax and x^*Ay. This idea is further used to "couple" matrices A and X in the form $\begin{pmatrix} \star & \langle A,X\rangle \\ \star & \star \end{pmatrix}$ when a trace inequality involving $\operatorname{tr}(AX^*) = \langle A, X\rangle$ is to be studied.

To begin with, consider a 2×2 Hermitian matrix $A = \begin{pmatrix} a & b \\ \bar{b} & c \end{pmatrix}$. It is obvious that $A \ge 0$ if and only if $a,\ c \ge 0$, and $|b|^2 \le ac$. If we denote the eigenvalues of A by α and β, then $\alpha+\beta = a+c$ and $\alpha\beta = ac - |b|^2$. Furthermore, we have the following elementary results as a lemma in which lower bounds of certain expressions involving the eigenvalues of A are given in terms of those of the entries of A.

Lemma 3.4 *Let* $A = \begin{pmatrix} a & b \\ \bar{b} & c \end{pmatrix}$ *be a* 2×2 *Hermitian matrix and let* α *and* β *be the (necessarily real) eigenvalues of* A *with* $\alpha \geq \beta$. *Then*

$$2|b| \leq \alpha - \beta. \tag{3.1.4}$$

If A *is further positive definite; that is,* $\alpha \geq \beta > 0$, *then*

$$\frac{|b|}{\sqrt{ac}} \leq \frac{\alpha-\beta}{\alpha+\beta}, \tag{3.1.5}$$

$$\frac{|b|}{a} \leq \frac{\alpha-\beta}{2\sqrt{\alpha\beta}}, \tag{3.1.6}$$

and

$$\frac{|b|}{\sqrt{c}} \leq \sqrt{\alpha} - \sqrt{\beta}. \tag{3.1.7}$$

Proof. The inequalities (3.1.4) and (3.1.5) follow from the observation that

$$\alpha,\ \beta = \frac{(a+c) \pm \sqrt{(a-c)^2 + 4|b|^2}}{2}.$$

To show (3.1.6), notice that for any real parameter t,

$$(\alpha+\beta)t - \alpha\beta \leq \frac{(\alpha+\beta)^2}{4\alpha\beta}t^2.$$

Put $t = a$. The replacements of $\alpha+\beta$ with $a+c$ and $\alpha\beta$ with $ac - |b|^2$ on the left hand side lead to (3.1.6). To prove (3.1.7), use, for all $t > 0$,

$$(\alpha+\beta) - \alpha\beta t \leq t^{-1} + (\sqrt{\alpha} - \sqrt{\beta})^2,$$

then set $t = c^{-1}$ and $\alpha+\beta = a+c$, $\alpha\beta = ac - |b|^2$ on the left hand side. ∎

We shall frequently use an equivalent form of (3.1.5):

$$|b|^2 \leq \left(\frac{\alpha-\beta}{\alpha+\beta}\right)^2 ac. \tag{3.1.8}$$

In addition, (3.1.6) holds for c in place of a. This reveals the inequality

$$4|b| \leq \frac{\alpha^2-\beta^2}{\sqrt{\alpha\beta}}.$$

In a similar way, from (3.1.7), one obtains

$$\sqrt{2}|b| \leq (\sqrt{\alpha} - \sqrt{\beta})\sqrt{\alpha + \beta}.$$

Using embedding techniques, we proceed to inspect some matrix equalities and inequalities that have often made their appearance; that is, we formulate a matrix inequality in terms of a sesquilinear form involving $\langle Ax, x\rangle$ or $\langle Ax, y\rangle$ as an inequality involving the entries of a matrix or a submatrix of the original matrix. We will then extend our studies to the matrix absolute values and Ky Fan singular value majorization theorem [282].

The Cauchy-Schwarz inequality. The classic Cauchy-Schwarz inequality (see, e.g., [228, p. 261]) states that for any vectors $x,\ y \in \mathbb{C}^n$,

$$|\langle x, y\rangle|^2 \leq \langle x, x\rangle\langle y, y\rangle.$$

Proof. This inequality is traditionally proved by examining the discriminant of the quadratic function $\langle x + ty, x + ty\rangle$ in t. We now notice that

$$(x, y)^*(x, y) = \begin{pmatrix} x^* \\ y^* \end{pmatrix} (x, y) = \begin{pmatrix} x^*x & x^*y \\ y^*x & y^*y \end{pmatrix} \geq 0.$$

The inequality follows at once by taking the determinant for the 2×2 matrix. Equality occurs if and only if the $n \times 2$ matrix (x, y) has rank 1; that is, x and y are linearly dependent. ∎

The Wielandt inequality. Let A be a nonzero $n \times n$ positive semidefinite matrix having eigenvalues $\lambda_1 \geq \lambda_2 \geq \cdots \geq \lambda_n$. The Wielandt inequality (see, e.g., [228, p. 443]) asserts that for all orthogonal $x,\ y \in \mathbb{C}^n$,

$$|x^*Ay|^2 \leq \left(\frac{\lambda_1 - \lambda_n}{\lambda_1 + \lambda_n}\right)^2 (x^*Ax)(y^*Ay). \tag{3.1.9}$$

Proof. (3.1.9) involves the quadratic terms x^*Ax, x^*Ay, and y^*Ay. It is natural for us to think of the 2×2 matrix

$$M = \begin{pmatrix} x^*Ax & x^*Ay \\ y^*Ax & y^*Ay \end{pmatrix}.$$

If $\lambda_n = 0$, (3.1.9) follows immediately by taking the determinant of M. Let $\lambda_n > 0$. Then $M = (x, y)^*A(x, y)$ is bounded from below by $\lambda_n(x, y)^*(x, y)$ and from above by $\lambda_1(x, y)^*(x, y)$. We may assume that x and y are orthonormal by scaling both sides of (3.1.9). Then $\lambda_n I_2 \leq M \leq \lambda_1 I_2$ and thus the eigenvalues λ and μ of M with $\lambda \geq \mu$ are contained in $[\lambda_n, \lambda_1]$. Therefore $\frac{\lambda-\mu}{\lambda+\mu} \leq \frac{\lambda_1-\lambda_n}{\lambda_1+\lambda_n}$ since $\frac{t-1}{t+1}$ is monotone in t.

An application of (3.1.8) to M results in

$$|x^*Ay|^2 \leq \left(\frac{\lambda-\mu}{\lambda+\mu}\right)^2 (x^*Ax)(y^*Ay) \leq \left(\frac{\lambda_1-\lambda_n}{\lambda_1+\lambda_n}\right)^2 (x^*Ax)(y^*Ay). \blacksquare$$

In a similar manner, since $t-\frac{1}{t}$ is an increasing function in t, by (3.1.6),

$$|x^*Ay| \leq \frac{\lambda_1-\lambda_n}{2\sqrt{\lambda_1\lambda_n}} \min\{x^*Ax,\ y^*Ay\}$$

and, by (3.1.7),

$$|x^*Ay|^2 \leq (\sqrt{\lambda_1}-\sqrt{\lambda_n})^2 \min\{x^*Ax,\ y^*Ay\}.$$

A theorem of Mirsky. The spreads of Hermitian matrices have been studied by many authors, especially by R. C. Thompson (see, e.g., [439]). Recall that the spread of a Hermitian matrix A is defined to be $\text{Spread}(A) = \lambda_{\max} - \lambda_{\min}$, where $\lambda_{\max}$ and $\lambda_{\min}$ are the largest and smallest eigenvalues of A, respectively. It is shown in [315] and [316] that

$$\text{Spread}(A) = 2\sup_{u,v} |u^*Av|, \tag{3.1.10}$$

where the "sup" is taken with respect to all pairs of orthonormal u, v.

Proof. To show (3.1.10), place u^*Av in a 2×2 matrix as follows: Let

$$M = (u,v)^*A(u,v) = \begin{pmatrix} u^*Au & u^*Av \\ v^*Au & v^*Av \end{pmatrix}.$$

Let U be a unitary matrix with u, v as the 1st and 2nd columns. Then

$$U^*AU = \begin{pmatrix} M & \star \\ \star & \star \end{pmatrix}.$$

By the interlacing eigenvalue theorem, $|\lambda_{\max} - \lambda_{\min}| \geq |\lambda-\mu|$, where λ and μ are the eigenvalues of M. On the other hand $|\lambda-\mu| \geq 2|u^*Av|$ by (3.1.4). It follows that $\text{Spread}(A) \geq 2|u^*Av|$. For the other direction of the inequality, that $\text{Spread}(A) \leq 2\sup_{u,v}|u^*Av|$ is seen, as in [316], by taking $u = \frac{1}{\sqrt{2}}(x+iy)$ and $v = \frac{1}{\sqrt{2}}(x-iy)$, where x and y are the orthonormal eigenvectors belonging to the eigenvalues $\lambda_{\max}$ and $\lambda_{\min}$, respectively. $\blacksquare$

We remark that (3.1.10) is proved in two separate papers. The inequality "$\geq$" follows from a discussion on the results of normal matrices in [315, Equation (6)], while "$\leq$" is shown in [316].

Positivity and inner product. Let $X \in \mathbb{C}^{n\times n}$. It is well-known that $X \geq 0 \Leftrightarrow \langle X, Y\rangle \geq 0$ for every $n \times n$ positive semidefinite matrix Y. It would be tempting to generalize the statement for partitioned (block) matrices X, namely when X is given in a partitioned form. An existing result (see [228, p. 473]) has shed light on this: With $A > 0$, $C > 0$,

$$\begin{pmatrix} A & B \\ B^* & C \end{pmatrix} \geq 0 \quad \Leftrightarrow \quad |x^*By|^2 \leq (x^*Ax)(y^*Cy) \text{ for all } x,\ y.$$

Equivalently, by writing $x^*By = \operatorname{tr}(yx^*B) = \langle B, xy^*\rangle$, we get

$$\begin{pmatrix} A & B \\ B^* & C \end{pmatrix} \geq 0 \quad \Leftrightarrow \quad \begin{pmatrix} \langle A, xx^*\rangle & \langle B, xy^*\rangle \\ \langle B^*, yx^*\rangle & \langle C, yy^*\rangle \end{pmatrix} \geq 0. \tag{3.1.11}$$

This generalizes to the following:

$$\begin{pmatrix} A & B \\ B^* & C \end{pmatrix} \geq 0 \quad \Leftrightarrow \quad \begin{pmatrix} \langle A, P\rangle & \langle B, Q\rangle \\ \langle B^*, Q^*\rangle & \langle C, R\rangle \end{pmatrix} \geq 0, \tag{3.1.12}$$

whenever the conformally partitioned matrix

$$\begin{pmatrix} P & Q \\ Q^* & R \end{pmatrix} \geq 0.$$

Proof. Let $A \in \mathbb{C}^{m\times m}$, $B \in \mathbb{C}^{m\times n}$, and $C \in \mathbb{C}^{n\times n}$. Then (3.1.11) ensures "$\Leftarrow$" in (3.1.12) by taking $P = xx^*$, $Q = xy^*$, and $R = yy^*$. For the other direction, let W be an $(m+n) \times (m+n)$ matrix such that

$$\begin{pmatrix} P & Q \\ Q^* & R \end{pmatrix} = WW^*.$$

Let U and V be the matrices consisting of the first m rows of W and the remaining rows of W, respectively, and denote the ith column of U by u_i and the ith column of V by v_i for each i. Then

$$P = \sum_{i=1}^{m+n} u_i u_i^*, \quad Q = \sum_{i=1}^{m+n} u_i v_i^*, \quad R = \sum_{i=1}^{m+n} v_i v_i^*.$$

By using (3.1.11) again, the block matrix with the inner products in (3.1.12), when written as a sum of $(m+n)$ positive semidefinite 2×2 matrices, is positive semidefinite. ∎

Note that the positive semidefiniteness of the block matrix with inner products as entries in (3.1.12) implies the trace inequality

$$|\operatorname{tr}(BQ^*)|^2 \leq \operatorname{tr}(AP)\operatorname{tr}(CR). \tag{3.1.13}$$

This trace inequality will be extensively used later. As an application, for any positive definite $X \in \mathbb{C}^{m\times m}$ and $A, B \in \mathbb{C}^{m\times n}$, since block matrices

$$\begin{pmatrix} X & I \\ I & X^{-1} \end{pmatrix} \quad \text{and} \quad \begin{pmatrix} AA^* & AB^* \\ BA^* & BB^* \end{pmatrix}$$

are both positive semidefinite, we obtain the known trace inequality

$$|\operatorname{tr}(A^*B)|^2 \le \operatorname{tr}(A^*XA)\operatorname{tr}(B^*X^{-1}B).$$

In what follows we make use of (3.1.12) to the block matrix

$$\begin{pmatrix} |A^*| & A \\ A^* & |A| \end{pmatrix} \ge 0. \tag{3.1.14}$$

Notice that (3.1.13), applied to (3.1.14), shows the trace inequality

$$|\operatorname{tr}(AQ^*)|^2 \le \operatorname{tr}(|A^*|P)\ \operatorname{tr}(|A|R) \tag{3.1.15}$$

whenever

$$\begin{pmatrix} P & Q \\ Q^* & R \end{pmatrix} \ge 0.$$

Following are immediate consequences of (3.1.15) for a square A.

Case 1). Setting $P = Q = R = I$ yields $|\operatorname{tr} A| \le \operatorname{tr}|A|$ ([468, p. 260]).

Case 2). Putting $P = Q = R = J$, the matrix all whose entries are 1,

$$|\Sigma(A)|^2 \le \Sigma(|A^*|)\ \Sigma(|A|), \tag{3.1.16}$$

where $\Sigma(X) = \sum_{ij} x_{ij}$ is the sum of all entries of matrix $X = (x_{ij})$. Note that (3.1.16) implies $|\Sigma(A)| \le \Sigma(|A|)$ if A is normal (not conversely).

Case 3). Letting $P = |A|$, $Q = A^*$, and $R = |A^*|$, we obtain

$$|\operatorname{tr} A^2| \le \operatorname{tr}(|A|\ |A^*|).$$

Note that $\operatorname{tr}(AA^*) \le \operatorname{tr}(|A|\ |A^*|)$ is not generally true while it is valid that

$$|\operatorname{tr} A^2| \le \operatorname{tr}(AA^*).$$

Case 4). Replacing P, Q, and R with yy^*, yx^*, and xx^*, respectively,

$$|\langle Ax, y\rangle|^2 \le \langle |A|x, x\rangle \langle |A^*|y, y\rangle. \tag{3.1.17}$$

Setting $x = y$ in (3.1.17) leads to the statement on matrix normality:

$$|\langle Ax, x\rangle| \le \langle |A|x, x\rangle \ \text{ for all } x \in \mathbb{C}^n \quad \Leftrightarrow \quad A \text{ is normal} \tag{3.1.18}$$

or

$$|\langle Ax, x\rangle| \le \langle |A^*|x, x\rangle \text{ for all } x \in \mathbb{C}^n \quad \Leftrightarrow \quad A \text{ is normal} \tag{3.1.19}$$

which have appeared in [247] and [157].

The sufficiency of (3.1.18) or (3.1.19) is immediate from (3.1.17), while the necessity is done in the same way as in [247] or [157] by an induction on n by assuming A to be upper triangular and by taking $x = (1, 0, \ldots, 0)^T$.

If we use (3.1.13) and take the positive semidefinite matrices

$$\begin{pmatrix} \sigma_{\max}(A)I & A \\ A^* & \sigma_{\max}(A)I \end{pmatrix} \quad \text{and} \quad \begin{pmatrix} yy^* & yx^* \\ xy^* & xx^* \end{pmatrix},$$

we may obtain a representation of the spectral norm (see, e.g, [468, p. 91])

$$\sigma_{\max}(A) = \sup_{\|x\|=\|y\|=1} |\langle Ax, y\rangle|,$$

where the "sup" is attained by taking x and y to be the first columns of the unitary matrices V and U, here $A = U\operatorname{diag}(\sigma_{\max}(A), \ldots, \sigma_n(A))V^*$.

More generally if we take, in (3.1.15), the positive semidefinite matrix

$$\begin{pmatrix} P & Q \\ Q^* & R \end{pmatrix} = \begin{pmatrix} XX^* & XY^* \\ YX^* & YY^* \end{pmatrix},$$

we get, for $A \in \mathbb{C}^{m\times n}$, $X \in \mathbb{C}^{m\times k}$, $Y \in \mathbb{C}^{n\times k}$,

$$|\operatorname{tr}(X^*AY)|^2 \le \operatorname{tr}(X^*|A^*|X)\operatorname{tr}(Y^*|A|Y),$$

with which we can derive the representation (see, e.g., [230, p. 195])

$$\sum_{i=1}^{k} \sigma_i(A) = \max_{X\in\mathbb{C}^{m\times k}, Y\in\mathbb{C}^{n\times k}} \{|\operatorname{tr}(X^*AY)| : X^*X = I_k = Y^*Y\}. \tag{3.1.20}$$

To show (3.1.20), note that if P is $n \times n$ positive semidefinite having eigenvalues $\lambda_1 \ge \cdots \ge \lambda_n$, and $U \in \mathbb{C}^{n\times k}$ is such that $U^*U = I_k$, then $\operatorname{tr}(U^*PU) \le \lambda_1 + \cdots + \lambda_k$. Noticing that $|A^*|$ and $|A|$ have the same eigenvalues, we have for $X \in \mathbb{C}^{m\times k}$ and $Y \in \mathbb{C}^{n\times k}$ with $X^*X = I_k = Y^*Y$,

$$\operatorname{tr}(X^*|A^*|X) \text{ and } \operatorname{tr}(Y^*|A|Y) \le \sigma_1(A) + \cdots + \sigma_k(A).$$

Thus

$$|\operatorname{tr}(X^*AY)| \le \sigma_1(A) + \cdots + \sigma_k(A).$$

For the other direction, let $A = VDW$ be a singular value decomposition of A with the ith largest singular value of A in the (i, i)-position of D for each i, where V and W are $m \times m$ and $n \times n$ unitary matrices, respectively. Then the extremal value is attained by taking X and Y to be the first k columns of V and W^*, respectively.

The representation (3.1.20) is traditionally and commonly proved through stochastic matrix theory [301, p. 511] or by eigenvalue and singular value inequalities for matrix product [230, p. 196]. The case $k = 1$ was discussed previously. For the case $k = m = n$ (see, e.g., [228, p. 430]), we have

$$\sigma_1(A) + \sigma_2(A) + \cdots + \sigma_n(A) = \max_{\text{unitary } U \in \mathbb{M}_n} |\operatorname{tr}(UA)|$$

which is also proved by taking the positive semidefinite matrix in (3.1.15)

$$\begin{pmatrix} P & Q \\ Q^* & R \end{pmatrix} = \begin{pmatrix} I & U^* \\ U & I \end{pmatrix}.$$

The representation (3.1.20) implies at once (see, e.g., [301, p. 243])

Theorem 3.1 (Ky Fan) *Let $A, B \in \mathbb{C}^{m \times n}$ and $1 \leq k \leq \min\{m, n\}$. Then*

$$\sum_{i=1}^{k} \sigma_i(A + B) \leq \sum_{i=1}^{k} \sigma_i(A) + \sum_{i=1}^{k} \sigma_i(B),$$

that is, in the notation of majorization,

$$\sigma(A + B) \prec_w \sigma(A) + \sigma(B).$$

This theorem reveals that $N_k(A) := \sum_{i=1}^{k} \sigma_i(A)$ is a unitarily invariant norm for any $k \leq \min\{m, n\}$; it is usually refered to as *Ky Fan k-norm.*

3.2 A matrix inequality and its consequences

In this section we demonstrate how to use the Schur complement to obtain matrix inequalities. Much of the material of this section is taken from [469].

Let $A > 0$. It is easy to see that for any matrix X of appropriate size,

$$\begin{pmatrix} A & X \\ X^* & X^*A^{-1}X \end{pmatrix} \geq 0 \tag{3.2.21}$$

and that $X^*A^{-1}X$ is the smallest matrix such that (3.2.21) holds; namely,

$$\begin{pmatrix} A & X \\ X^* & Z \end{pmatrix} \geq 0 \quad \Rightarrow \quad Z \geq X^*A^{-1}X.$$

In case where $X^*A^{-1}X$ is nonsingular, taking the Schur complement of the (2,2)-block in the block matrix in (3.2.21) shows, with A^{-1} replaced by A,

$$X(X^*AX)^{-1}X^* \le A^{-1}.$$

In particular, if $X^*X = I$, then, by pre- and post-multiplying X^* and X,

$$(X^*AX)^{-1} \le X^*A^{-1}X.$$

Many matrix inequalities may be obtained similarly. For instance, since

$$\begin{pmatrix} I+A^*A & A^*+B^* \\ A+B & I+BB^* \end{pmatrix} = \begin{pmatrix} I & A^* \\ B & I \end{pmatrix} \begin{pmatrix} I & B^* \\ A & I \end{pmatrix} \ge 0$$

for any matrices A and B of the same size, and since $I + A^*A$ is always positive definite, by taking the Schur complement of $I + A^*A$, we arrive at

$$I + BB^* \ge (A+B)(I+A^*A)^{-1}(A+B)^*,$$

which yields, when A and B are square,

$$\det(I + A^*A)\det(I + BB^*) \ge |\det(A+B)|^2.$$

Recall Theorem 1.20 of Chapter 1 that if A is an $n \times n$ positive semidefinite matrix and X is an $n \times m$ matrix such that the column space of X is contained in that of A, with $A^\dagger$ for the Moore-Penrose inverse of A, then

$$\begin{pmatrix} A & X \\ X^* & X^*A^\dagger X \end{pmatrix} \ge 0.$$

Theorem 3.2 *Let A_i be $n \times n$ positive semidefinite matrices and X_i be $n \times m$ matrices such that the column space of X_i is contained in that of A_i, for $i = 1, 2, \dots, k$. Then*

$$\left(\sum_{i=1}^k \alpha_i X_i\right)^* \left(\sum_{i=1}^k \alpha_i A_i\right)^\dagger \left(\sum_{i=1}^k \alpha_i X_i\right) \le \sum_{i=1}^k \alpha_i X_i^* A_i^\dagger X_i. \qquad (3.2.22)$$

Proof. Put

$$M_i = \begin{pmatrix} A_i & X_i \\ X_i^* & X_i^*A_i^\dagger X_i \end{pmatrix} \ge 0$$

and let $\alpha_1 + \alpha_2 + \cdots + \alpha_k = 1$, where all $\alpha_i \ge 0$. By taking the sum, we have

$$\sum_{i=1}^k \alpha_i M_i = \begin{pmatrix} \sum \alpha_i A_i & \sum \alpha_i X_i \\ \sum \alpha_i X_i^* & \sum \alpha_i X_i^* A_i^\dagger X_i \end{pmatrix} \ge 0.$$

Taking the Schur complement, we obtain the desired inequality. ∎

Theorem 3.2 is discussed in [259] for the nonsingular case of k matrices, in [213] for the nonsingular case of two matrices, in [284] in the Hilbert space setting, in [179] for the singular case in terms of generalized inverses, and in [106] for the generalized inverse case of two matrices. We notice from the proof that the theorem obviously holds for the Hadamard product.

As consequences of (3.2.22), we have the following results that have appeared in a fragmentary literature: For any $n \times n$ positive semidefinite matrices A and B (nonsingular if inverses are present), $n \times m$ matrices X and Y, and scalars α and β such that $\alpha + \beta = 1$, $\alpha, \beta \geq 0$,

$$(X+Y)^*(A+B)^{-1}(X+Y) \leq X^*A^{-1}X + Y^*B^{-1}Y;$$

$$(\alpha X + \beta Y)^*(\alpha X + \beta Y) \leq \alpha X^*X + \beta Y^*Y;$$

$$(\alpha A + \beta B)^{-1} \leq \alpha A^{-1} + \beta B^{-1};$$

$$(\alpha A + \beta B)^2 \leq \alpha A^2 + \beta B^2;$$

$$(\alpha A + \beta B)^{1/2} \geq \alpha A^{1/2} + \beta B^{1/2}.$$

We now present an explicit matrix inequality and discuss the related ones. Even though the inequality is obtainable from a later theorem, it is in a simple and elegant form and has many interesting applications. For this reason, we prefer to give a direct and elementary proof of it. Notice that if $M \geq 0$ and α is an index set, then the principal submatrix $M[\alpha] \geq 0$.

Theorem 3.3 *Let $M \geq 0$ and let $M[\alpha]$ be nonsingular. If $\beta \subseteq \alpha^c$, then*

$$M[\beta] \geq M[\beta,\alpha]M[\alpha]^{-1}M[\alpha,\beta]. \tag{3.2.23}$$

Proof. This is because the block matrix, a principal submatrix of M,

$$\begin{pmatrix} M[\alpha] & M[\alpha,\beta] \\ M[\beta,\alpha] & M[\beta] \end{pmatrix}$$

is positive semidefinite. Taking the Schur complement reveals (3.2.23). ∎

The following are immediate consequences of the above theorem.

Corollary 3.13 *Let $\begin{pmatrix} A & B \\ B^* & C \end{pmatrix} \geq 0$, in which $A \in \mathbb{C}^{n \times n}$ and $C \in \mathbb{C}^{m \times m}$. Let $\alpha \subseteq \{1, 2, \ldots, n\}$, $\beta \subseteq \{1, 2, \ldots, m\}$, and $A[\alpha]$ be nonsingular. Then*

$$C[\beta] \geq B^*[\beta,\alpha]A[\alpha]^{-1}B[\alpha,\beta].$$

Corollary 3.14 *Let A be an $n \times n$ positive definite matrix. Then for any $m \times n$ matrix X and any index sets $\alpha \subseteq \{1, 2, \ldots, n\}$ and $\beta \subseteq \{1, 2, \ldots, m\}$,*

$$X^*[\beta, \alpha]A[\alpha]^{-1}X[\alpha, \beta] \le (X^*A^{-1}X)[\beta]. \tag{3.2.24}$$

In particular, for any $A > 0$ and any matrix X, both $n \times n$,

$$X^*[\alpha]A[\alpha]^{-1}X[\alpha] \le (X^*A^{-1}X)[\alpha]. \tag{3.2.25}$$

Proof. By (3.2.21) and the above argument, it is sufficient to note that

$$\begin{pmatrix} A[\alpha] & X[\alpha, \beta] \\ X^*[\beta, \alpha] & (X^*A^{-1}X)[\beta] \end{pmatrix} \ge 0. \blacksquare$$

We now show some applications of the theorem.

Application 1. Let A be $n \times n$ positive definite and $\alpha = \{1, 2, \ldots, k\}$, where $1 \le k \le n$. With $\tilde{A} = (1) \oplus A$, $\tilde{X} = I_{n+1}$, $\tilde{\alpha} = \{1, 2, \ldots, k+1\}$, and $\tilde{\beta} = \{2, 3, \ldots, n+1\}$ in (3.2.24), by computation, one gets

$$\begin{pmatrix} A[\alpha]^{-1} & 0 \\ 0 & 0 \end{pmatrix} \le A^{-1}. \tag{3.2.26}$$

Inequality (3.2.26) may generalize for any α by permutation similarity. As a result of (3.2.26) (or from (3.2.25) with $X = I$), one gets

$$A[\alpha]^{-1} \le A^{-1}[\alpha]. \tag{3.2.27}$$

Application 2. For $A, B \in \mathbb{C}^{n\times n}$, since the Hadamard product $A \circ B$ is a principal submatrix of the Kronecker product $A \otimes B$, we may write

$$A \circ B = (A \otimes B)[\alpha]. \tag{3.2.28}$$

It is immediate from (3.2.27) that for any $n \times n$ positive definite A, B,

$$\begin{aligned} A^{-1} \circ B^{-1} &= (A^{-1} \otimes B^{-1})[\alpha] \\ &= (A \otimes B)^{-1}[\alpha] \\ &\ge ((A \otimes B)[\alpha])^{-1} \\ &= (A \circ B)^{-1}. \end{aligned}$$

In a similar manner, noticing that for any X, $Y \in \mathbb{C}^{n\times n}$,

$$(X^*A^{-1}X) \otimes (Y^*B^{-1}Y) = (X^* \otimes Y^*)(A \otimes B)^{-1}(X \otimes Y).$$

By (3.2.25), we have (see, e.g., [468, p. 198])

$$(X^*A^{-1}X) \circ (Y^*B^{-1}Y) \geq (X \circ Y)^*(A \circ B)^{-1}(X \circ Y).$$

Application 3. Taking $A = I$ in (3.2.25) yields for any matrix X,

$$X^*[\alpha]X[\alpha] \leq (X^*X)[\alpha] \tag{3.2.29}$$

In particular, if $X^* = X$, then

$$X[\alpha]^2 \leq X^2[\alpha].$$

As an application of (3.2.29), if H is a positive definite matrix, we write $H = CC^*$, a *Cholesky factorization* of H, where C is lower triangular with positive diagonal entries. Since $C \circ C^{-T} = I$, where $C^{-T} = (C^{-1})^T$, and

$$H \otimes H^{-T} = (C \otimes C^{-T})(C^* \otimes \bar{C}^{-1}),$$

we have

$$H \circ H^{-T} \geq (C \circ C^{-T})(C^* \circ \bar{C}^{-1}) = I.$$

Inequality (3.2.29) yields at once: For any $X \in \mathbb{C}^{m\times n}$ and $Y \in \mathbb{C}^{n\times m}$

$$(Y^*X^*)[\alpha](XY)[\alpha] \leq (Y^*X^*XY)[\alpha].$$

We next present a matrix inequality that generates a large family of inequalities, old and new, including (3.2.23) and many more on the Hadamard product. Such a technique by embedding the Hadamard product into a Kronecker product as a principal submatrix is often used for deriving matrix inequalities on the Hadamard product.

Let $A = (A_{ij})_{i,j=1}^2$ and $B = (B_{ij})_{i,j=1}^2$ be 2×2 block matrices. We write the 2×2 block matrix with the Kronecker products of the corresponding blocks of A, B as

$$A \boxtimes B = (A_{ij} \otimes B_{ij}) = \begin{pmatrix} A_{11} \otimes B_{11} & A_{12} \otimes B_{12} \\ A_{21} \otimes B_{21} & A_{22} \otimes B_{22} \end{pmatrix}.$$

It is easy to see that $A \boxtimes B$ is a submatrix of $A \otimes B$. Moreover $A \boxtimes B \geq 0$ if $A, B \geq 0$ and each diagonal block of A and B is square [233].

Theorem 3.4 *Let H, $R \in \mathbb{C}^{m\times m}$ and K, $S \in \mathbb{C}^{n\times n}$ be positive definite matrices. Then for any A, $C \in \mathbb{C}^{p\times m}$, B, $D \in \mathbb{C}^{q\times n}$, U, $V \in \mathbb{C}^{r\times m}$ with* $\operatorname{rank}(U) = r$ *or* $\operatorname{rank}(V) = r$, *and for any real numbers a and b*

$$a^2(AH^{-1}A^*) \otimes (BK^{-1}B^*) + b^2(CR^{-1}C^*) \otimes (DS^{-1}D^*) \geq$$

$$(aAU^* \otimes B + bCV^* \otimes D)(UHU^* \otimes K + VRV^* \otimes S)^{-1}(aUA^* \otimes B^* + bVC^* \otimes D^*).$$

Proof. Note that if $T > 0$, then for any matrix X of appropriate size,

$$\begin{pmatrix} T & X^* \\ X & XT^{-1}X^* \end{pmatrix} \geq 0.$$

Thus, by pre- and post-multiplying the first row and the first column of the 2×2 block matrix by Y and Y^*, respectively, we have

$$\begin{pmatrix} YTY^* & YX^* \\ XY^* & XT^{-1}X^* \end{pmatrix} \geq 0.$$

Therefore the following block matrix M, the sum of two block Kronecker products, is positive semidefinite:

$$\begin{aligned} M &= \begin{pmatrix} UHU^* & aUA^* \\ aAU^* & a^2AH^{-1}A^* \end{pmatrix} \boxtimes \begin{pmatrix} K & B^* \\ B & BK^{-1}B^* \end{pmatrix} \\ &+ \begin{pmatrix} VRV^* & bVC^* \\ bCV^* & b^2CR^{-1}C^* \end{pmatrix} \boxtimes \begin{pmatrix} S & D^* \\ D & DS^{-1}D^* \end{pmatrix}. \end{aligned}$$

The (1, 1)-block of M is $UHU^* \otimes K + VRV^* \otimes S$, which is nonsingular since $\operatorname{rank}(U) = r$ or $\operatorname{rank}(V) = r$ implies the invertibility of UHU^* or VRV^*. Taking the Schur complement in M gives the desired inequality. ∎

In what follows we show that many existing inequalities (mainly on the Hadamard product) are in fact consequences of Theorem 3.4 by making special choices of the following numbers and matrices:

$$m,\ n,\ a,\ b,\ H,\ R,\ K,\ S,\ A,\ B,\ C,\ D,\ U,\ V.$$

Case 1. Take $n = 1$, $K = S = B = D = (1)$, $U = V = I_m$. Then

$$a^2AH^{-1}A^* + b^2CR^{-1}C^* \geq (aA + bC)(H + R)^{-1}(aA^* + bC^*). \quad (3.2.30)$$

Setting $a = b = 1$ in (7.6.9) reveals the Haynsworth's inequality [213]

$$AH^{-1}A^* + CR^{-1}C^* \geq (A + C)(H + R)^{-1}(A + C)^*.$$

The case where A and C are vectors were discussed in [294] and [47]. Setting $A = C = I$ in (7.6.9) results in

$$a^2H^{-1} + b^2R^{-1} \geq (a + b)^2(H + R)^{-1},$$

which is equivalent to the matrix inverse convexity inequality: For $t \in [0, 1]$,

$$tH^{-1} + (1 - t)R^{-1} \geq (tH + (1 - t)R)^{-1}.$$

Case 2. Set $a = 1$, $b = 0$, $U = I$, $V = 0$, and use (3.2.25) to get

$$AH^{-1}A^* \circ BK^{-1}B^* \geq (A \circ B)(H \circ K)^{-1}(A \circ B)^*.$$

It is immediate that, by taking $H = K = I$,

$$AA^* \circ BB^* \geq (A \circ B)(A \circ B)^* \tag{3.2.31}$$

and that, by setting $A = B = I$,

$$H^{-1} \circ K^{-1} \geq (H \circ K)^{-1}.$$

Replacing K with H^{-1} yields

$$H \circ H^{-1} \geq I.$$

It is immediate from (3.2.31) that for any $n \times n$ matrices $A, B \geq 0$,

$$A^2 \circ B^2 \geq (A \circ B)^2.$$

Thus for $A, B \geq 0$,

$$(A \circ B)^{\frac{1}{2}} \geq A^{\frac{1}{2}} \circ B^{\frac{1}{2}}.$$

Case 3. Let $H = K = R = S = U = V = I$. Then

$$\begin{aligned} a^2 AA^* &\otimes BB^* + b^2 CC^* \otimes DD^* \\ &\geq \frac{1}{2}(aA \otimes B + bC \otimes D)(aA^* \otimes B^* + bC^* \otimes D^*) \\ &= \frac{1}{2}(a^2 AA^* \otimes BB^* + abAC^* \otimes BD^* \\ &\quad + abCA^* \otimes DB^* + b^2 CC^* \otimes DD^*). \end{aligned}$$

Using (3.2.28) and (3.2.29), we have

$$\begin{aligned} a^2 AA^* &\circ BB^* + b^2 CC^* \circ DD^* \\ &\geq \frac{1}{2}(a^2 AA^* \circ BB^* + abAC^* \circ BD^* \\ &\quad + abCA^* \circ DB^* + b^2 CC^* \circ DD^*) \\ &\geq \frac{1}{2}(aA \circ B + bC \circ D)(aA^* \circ B^* + bC^* \circ D^*). \end{aligned}$$

Take $C = B$, $D = A$, and multiply through by 2. Then

$$\begin{aligned} 2(a^2 &+ b^2)(AA^* \circ BB^*) \\ &\geq (a^2 + b^2)(AA^* \circ BB^*) + 2abAB^* \circ BA^* \\ &\geq (a + b)^2 (A \circ B)(A^* \circ B^*). \end{aligned}$$

Setting $k = 2ab/(a^2 + b^2)$, then $k \in [-1, 1]$. The first inequality yields

$$AA^* \circ BB^* \geq kAB^* \circ BA^*$$

and the second one implies [448]

$$AA^* \circ BB^* + kAB^* \circ BA^* \geq (1 + k)(A \circ B)(A \circ B)^*.$$

In particular,

$$AA^* \circ BB^* \geq \pm AB^* \circ BA^*$$

and

$$AA^* \circ BB^* \geq \frac{1}{2}(AA^* \circ BB^* + AB^* \circ BA^*) \geq (A \circ B)(A \circ B)^*.$$

Case 4. To show that Theorem 3.4 implies Theorem 3.3, we notice that for any $m \times n$ matrix M, $\alpha \subseteq \{1, 2, \ldots, m\}$, and $\beta \subseteq \{1, 2, \ldots, n\}$,

$$M[\alpha, \beta] = S_\alpha M S_\beta^T,$$

where S_α is the $|\alpha| \times n$ matrix whose jth row is the n-vector with 1 in the α_jth position and 0 elsewhere. It follows that for any A, $B \in \mathbb{C}^{m \times n}$,

$$A \circ B = (A \otimes B)[\alpha, \beta] = S_\alpha (A \otimes B) S_\beta^T, \tag{3.2.32}$$

where $\alpha = \{1, m+2, 2m+3, \ldots, m^2\}$ and $\beta = \{1, n+2, 2n+3, \ldots, n^2\}$.

Now we take $a = 1$, $b = 0$, $B = K = (1)$, $V = 0$, and $U = S_\alpha$ to get

$$AH^{-1}A^* \geq (AS_\alpha^T)(S_\alpha H S_\alpha^T)^{-1}(S_\alpha A^*).$$

Pre- and post-multiplying respectively by S_β and S_β^T yields inequality (3.2.24).

Note that S_α and S_β are the only matrices of zeros and ones that satisfy (3.2.32) for all $m \times n$ matrices A and B. It can be shown that $S_\alpha S_\alpha^T = I$ and $0 \leq S_\alpha^T S_\alpha \leq I$ and that there exists a matrix Q_α of zeros and ones such that the augmented matrix (S_α, Q_α) is a permutation matrix [448].

3.3 A technique by means of 2×2 block matrices

In this section we continue to present matrix inequalities involving the sum and the Hadamard product of positive semidefinite matrices through 2×2 block matrices. Two facts which we shall use repeatedly are A, $B \geq 0 \Rightarrow A + B$, $A \circ B \geq 0$. In addition, for $x = (x_1, \ldots, x_n)$, we denote $|x| = (|x_1|, \ldots, |x_n|)$.

Theorem 3.5 *Let H and K be $n \times n$ Hermitian matrices. Then*

$$\begin{pmatrix} H & K \\ K & H \end{pmatrix} \geq 0 \Leftrightarrow \pm K \leq H \Rightarrow |\lambda(K)| \prec_w \lambda(H) \Rightarrow \|K\| \leq \|H\|. \quad (3.3.33)$$

Proof. The implications "$\Leftrightarrow$" follow from the identity via *-congruence:

$$\left[\frac{1}{\sqrt{2}}\begin{pmatrix} I & -I \\ I & I \end{pmatrix}\right]\begin{pmatrix} H & K \\ K & H \end{pmatrix}\left[\frac{1}{\sqrt{2}}\begin{pmatrix} I & I \\ -I & I \end{pmatrix}\right] = \begin{pmatrix} H-K & 0 \\ 0 & H+K \end{pmatrix}.$$

The second "$\Rightarrow$" is immediate from (3.0.1). (If $H \geq 0$, "$\Leftarrow$" also holds.) For the first "$\Rightarrow$", note that the singular values of K are the absolute values of the eigenvalues of K. Since $\pm K \leq H$ yields $\pm U^*KU \leq U^*HU$ for every $n \times n$ unitary matrix U, without loss of generality, we may assume that $K = \operatorname{diag}(k_1, \ldots, k_n)$ is a diagonal matrix with $k_1 \geq \cdots \geq k_n$. Thus $\pm K \leq H$ implies that $\pm k_i \leq h_{ii}$, i.e., $|k_i| \leq h_{ii}$ for $i = 1, 2, \ldots, n$, where h_{ii} are corresponding diagonal entries of H.

On the other hand, by the eigenvalue interlacing theorem,

$$\sum_{i=1}^{m} \lambda_i(H_m) \leq \sum_{i=1}^{m} \lambda_i(A),$$

where H_m denotes the $m \times m$ leading principal submatrix of H, $m \leq n$. So

$$\sum_{i=1}^{m} |k_i| \leq \sum_{i=1}^{m} h_{ii} = \sum_{i=1}^{m} \lambda_i(H_m) \leq \sum_{i=1}^{m} \lambda_i(H).$$

It follows that

$$|\lambda(K)| \prec_w \lambda(H). \blacksquare$$

The eigenvalues of the block matrix $\left(\begin{smallmatrix} H & K \\ K & H \end{smallmatrix}\right)$ are those of $H \pm K$. A proof of "$\Leftrightarrow$" in (3.3.33) for the real case of H and K is given in [168] via quadratic forms, and a characterization of the matrices K, which comprise a convex set for the given H by trace inequalities, is presented in [42]. Some similar or stronger results are seen in [53, 231, 467, 470]. Moreover, we note that $H \geq \pm K$ and $H \geq |K|$ are not equivalent. Take, for example,

$$H = \begin{pmatrix} 4 & 2 \\ 2 & 4 \end{pmatrix} \quad \text{and} \quad K = \begin{pmatrix} 3 & 0 \\ 0 & -3 \end{pmatrix}.$$

Then

$$\pm K \leq H \quad \text{but} \quad |K| \not\leq H.$$

Notice that $H \geq |K| \Rightarrow H \geq \pm K$. It is also worth noting that

$$\begin{pmatrix} H & K \\ K & H \end{pmatrix} = \begin{pmatrix} H-K & 0 \\ 0 & H-K \end{pmatrix} + \begin{pmatrix} K & K \\ K & K \end{pmatrix}.$$

We now show our basic inequalities that easily follow from (3.3.33).

Theorem 3.6 *Let A, B, and C be $n \times n$ matrices such that*

$$\begin{pmatrix} A & B \\ B^* & C \end{pmatrix} \geq 0.$$

Then, with $\star$ for sum $+$ or the Hadamard product $\circ$,

$$\pm(B^* \star B) \leq A \star C \tag{3.3.34}$$

and, for any unitarily invariant norm $\|\cdot\|$ on $\mathbb{C}^{n\times n}$,

$$\|B^* \star B\| \leq \|A \star C\|. \tag{3.3.35}$$

Proof. Since the partitioned positive semidefinite matrix through a permutation congruence transformation is also positive semidefinite, we have

$$\begin{pmatrix} A & B \\ B^* & C \end{pmatrix} \star \begin{pmatrix} C & B^* \\ B & A \end{pmatrix} = \begin{pmatrix} A\star C & B^*\star B \\ B^*\star B & A\star C \end{pmatrix} \geq 0.$$

Thus both (3.3.34) and (3.3.35) follow from Theorem 3.5. ∎

Note that $B^* \star B \leq A \star C$ alone does not imply $\|B^* \star B\| \leq \|A \star C\|$.

Applications of Theorem 3.6 to some 2×2 block positive semidefinite matrices that we frequently encounter result in certain interesting inequalities. We present some as examples. Assume in the following that matrices A, B, and C are all $n \times n$. (Some results also hold for the rectangular case.)

Application 1. Since $\begin{pmatrix} A & A^{\frac{1}{2}} \\ A^{\frac{1}{2}} & I \end{pmatrix}$ for any $A \geq 0$, we have

$$2A^{\frac{1}{2}} \leq A + I \quad \text{and} \quad A^{\frac{1}{2}} \circ A^{\frac{1}{2}} \leq A \circ I.$$

In particular, if A is a correlation matrix (all $a_{ii} = 1$), then $A^{\frac{1}{2}} \circ A^{\frac{1}{2}} \leq I$.

For a pair of positive semidefinite matrices A, B of the same size,

$$\begin{pmatrix} A & A^{\frac{1}{2}}B^{\frac{1}{2}} \\ B^{\frac{1}{2}}A^{\frac{1}{2}} & B \end{pmatrix} \geq 0 \;\Rightarrow\; \pm(A^{\frac{1}{2}}B^{\frac{1}{2}} \star B^{\frac{1}{2}}A^{\frac{1}{2}}) \leq A \star B.$$

Application 2. Note that $\begin{pmatrix} A^*A & A^* \\ A & I \end{pmatrix} \geq 0$ for all square matrices A. So

$$A + A^* \leq I + A^*A, \quad A \circ A^* \leq I \circ A^*A.$$

More generally, since $\begin{pmatrix} |A|^{2\alpha} & A^* \\ A & |A^*|^{2(1-\alpha)} \end{pmatrix} \geq 0$ for any $\alpha \in [0,1]$, we obtain

$$\pm(A \star A^*) \leq |A|^{2\alpha} \star |A^*|^{2(1-\alpha)}.$$

In particular, taking $\alpha = \frac{1}{2}$, we have

$$\pm(A + A^*) \leq |A| + |A^*|, \quad \pm(A \circ A^*) \leq |A| \circ |A^*|$$

and for any unitarily invariant norm

$$\|A + A^*\| \leq \| \, |A| + |A^*| \, \|, \quad \|A \circ A^*\| \leq \| \, |A| \circ |A^*| \, \|.$$

We point out that $|A + A^*| \leq |A| + |A^*|$ does not hold in general as the following example, due to G. Visick, shows (there is an error in 3i) in [470])

$$A = \begin{pmatrix} 1 & 0 & 1 \\ 1 & 0 & -1 \\ 0 & 1 & 0 \end{pmatrix}.$$

Application 3. Let A, B be $n \times n$ and both positive definite. Then

$$\begin{pmatrix} A & I \\ I & A^{-1} \end{pmatrix} \circ \begin{pmatrix} B^{-1} & I \\ I & B \end{pmatrix} = \begin{pmatrix} A \circ B^{-1} & I \\ I & A^{-1} \circ B \end{pmatrix} \geq 0.$$

It follows that

$$A \circ B^{-1} + B \circ A^{-1} \geq 2I.$$

Taking $B = A$ shows

$$A \circ A^{-1} \geq I.$$

Application 4. Let A, B be $m \times n$ matrices. Then $\begin{pmatrix} A^*A & A^*B \\ B^*A & B^*B \end{pmatrix} \geq 0$. So

$$\pm(A^*B \star B^*A) \leq A^*A \star B^*B.$$

By (3.3.35), we have [53]

$$\|A^*B + B^*A\| \leq \|A^*A + B^*B\|$$

and

$$\|A^*B \circ B^*A\| \leq \|A^*A \circ B^*B\|.$$

Application 5. Let A, B be $m \times n$ matrices. Then

$$\begin{pmatrix} A^*A & A^* \\ A & I \end{pmatrix} + \begin{pmatrix} I & B^* \\ B & BB^* \end{pmatrix} = \begin{pmatrix} I + A^*A & A^* + B^* \\ A + B & I + BB^* \end{pmatrix} \geq 0.$$

It follows that

$$(A+B) \circ (A+B)^* \leq (I + A^*A) \circ (I + BB^*).$$

In a similar fashion

$$\begin{pmatrix} A^*A & A^* \\ A & I \end{pmatrix} + \begin{pmatrix} B^*B & B^* \\ B & I \end{pmatrix} = \begin{pmatrix} A^*A \circ B^*B & A^* \circ B^* \\ A \circ B & I \end{pmatrix} \geq 0.$$

This reveals

$$A \circ B + A^* \circ B^* \leq I + A^*A \circ B^*B$$

and

$$A^* \circ A \circ B^* \circ B \leq \operatorname{diag}(A^*A \circ B^*B).$$

We end this section by presenting a set of singular value inequalities concerning matrices A^*B, AA^*, and BB^* [52, 53].

Recall that for any $n \times n$ Hermitian matrices H and K (see (2.0.9)),

$$\lambda_t(H+K) \leq \lambda_i(H) + \lambda_j(K), \quad \text{if } i+j = t+1$$

and

$$\lambda_t(H+K) \geq \lambda_i(H) + \lambda_j(K), \quad \text{if } i+j = t+n,$$

here the eigenvalues λs are enumerated in decreasing order. Thus, we have

$$\lambda_t(H+K) \leq \lambda_t(H) + \lambda_1(K)$$

and

$$\lambda_t(H+K) \geq \lambda_{2t-1}(H) + \lambda_{n-t+1}(K).$$

Lemma 3.5 *Let H and K be $n \times n$ positive semidefinite. Then for $t \leq n$,*

$$-\lambda_t(K) \leq \lambda_t(H-K) \leq \lambda_t(H).$$

Proof. The second inequality follows by writing $H - K = H + (-K)$:

$$\lambda_t(H-K) \leq \lambda_t(H) + \lambda_1(-K) = \lambda_t(H) - \lambda_n(K) \leq \lambda_t(H).$$

Likewise,

$$\lambda_t(H-K) \geq \lambda_{2t-1}(H) + \lambda_{n-t+1}(-K) \geq \lambda_{2t-1}(H) - \lambda_t(K) \geq -\lambda_t(K). \blacksquare$$

Theorem 3.7 *Let A, B be $m \times n$ complex matrices and denote by $\sigma_i(Z)$ the ith largest singular value of matrix Z. Then for $i = 1, 2, \ldots, \min\{m, n\}$,*

$$2\sigma_i(A^*B) \leq \sigma_i(AA^* + BB^*).$$

Proof. Let $X = \begin{pmatrix} A & B \\ 0 & 0 \end{pmatrix}$ and $U = \begin{pmatrix} I & 0 \\ 0 & -I \end{pmatrix}$. Upon computation, we have

$$XX^* = \begin{pmatrix} AA^* + BB^* & 0 \\ 0 & 0 \end{pmatrix}, \quad X^*X = \begin{pmatrix} A^*A & A^*B \\ B^*A & B^*B \end{pmatrix}$$

and

$$X^*X - U(X^*X)U^* = \begin{pmatrix} 0 & 2A^*B \\ 2B^*A & 0 \end{pmatrix} := Y.$$

Notice that the first n eigenvalues of Y are the singular values of $2A^*B$. On the other hand, by Lemma 3.5, we see for each $i = 1, 2, \ldots, \min\{m, n\}$,

$$-\lambda_i(U(X^*X)U^*) \le \lambda_i(Y) \le \lambda_i(X^*X).$$

It follows that

$$-\lambda_i(X^*X) \le \lambda_i(Y) \le \lambda_i(X^*X)$$

or

$$\sigma_i(Y) \le \lambda_i(X^*X) = \lambda_i(XX^*) = \sigma_i(AA^* + BB^*). \blacksquare$$

The following result on unitarily invariant norms is immediate [53].

Corollary 3.15 *For any unitarily invariant norm on* $\mathbb{C}^{n\times n}$, A, $B \in \mathbb{C}^{n\times n}$

$$\|A^*B + B^*A\| \le 2\|A^*B\| \le \|AA^* + BB^*\|.$$

More generally [52], for arbitrary $n \times n$ matrices A, B, X,

$$2\|A^*XB\| \le \|AA^*X + XBB^*\|.$$

3.4 Liebian functions

We have seen many inequalities on trace, eigenvalues, and unitarily invariant norms in the previous sections. We now study a larger class of matrix functions and continue to derive inequalities through block matrices.

If A, B, and C are $n \times n$ complex matrices such that

$$\begin{pmatrix} A & B \\ B^* & C \end{pmatrix} \ge 0,$$

then, by taking the Schur complement, we have $C \ge B^*A^\dagger B$, and thus

$$|\det B|^2 \le \det A \det C. \tag{3.4.36}$$

There are many other inequalities resembling (3.4.36). For example, (3.4.36) holds for trace, the spectral norm, as well as unitarily invariant norms; all of these are members of a larger class of matrix functions.

A continuous complex-valued function $\mathcal{L}$ on a space of matrices is said to be a *Liebian function* if it is positive, i.e., $A \geq 0 \Rightarrow \mathcal{L}(A) \geq 0$, and if

$$\begin{pmatrix} A & B \\ B^* & C \end{pmatrix} \geq 0 \quad \Rightarrow \quad |\mathcal{L}(B)|^2 \leq \mathcal{L}(A)\mathcal{L}(C).$$

Each of the following is a Liebian function [51, p. 269], [420, p. 99]:

- trace $\operatorname{tr}(A)$;
- determinant $\det(A)$;
- spectral radius $\rho(A) = \max_i\{|\lambda_i(A)|\}$;
- spectral norm $\sigma_{\max}(A)$;
- unitarily invariant norm $\|A\|$;
- unitarily invariant norm $\|A\|^p$ for ;
- product of first k largest eigenvalues in absolute values $\prod_{i=1}^k |\lambda_i(A)|$;
- product of first k largest singular values $\prod_{i=1}^k \sigma_i(A)$.

As is well known, $M = \left(\begin{smallmatrix} A & B \\ B^* & C \end{smallmatrix}\right) \geq 0 \Rightarrow M/A \geq 0$ and the converse is true if $M = M^*$, $A \geq 0$, and if the space spanned by the columns of B is contained in that of A (Theorem 1.19). Thus one may verify that for any $A \geq 0$ and $z \in \mathbb{C}$, the following 2×2 block matrices are positive semidefinite:

$$\begin{pmatrix} A & A \\ A & A \end{pmatrix}, \quad \begin{pmatrix} |z|A & zA \\ \bar{z}A & |z|A \end{pmatrix}, \quad \begin{pmatrix} A & I \\ I & A^{-1} \end{pmatrix} \text{ if } A > 0. \tag{3.4.37}$$

In what follows we exhibit some examples of matrix inequalities on sum, Hadamard product, and on determinant, unitarily invariant norm; some are Cauchy-Schwarz type. These examples or results have appeared in the literature and have been employed to deduce many other advanced and sophisticated inequalities by a number of authors (see, e.g., [51, 230]).

Example 1. Since

$$\begin{pmatrix} A^* \\ B^* \end{pmatrix} (A, B) = \begin{pmatrix} A^*A & A^*B \\ B^*A & B^*B \end{pmatrix} \geq 0,$$

we have

$$|\mathcal{L}(A^*B)|^2 \leq \mathcal{L}(A^*A)\mathcal{L}(B^*B).$$

In particular, for any unitarily invariant norm $\|\cdot\|$ (see [232] or [230, p. 212])

$$\|A^*B\|^2 \le \|A^*A\|\ \|B^*B\|.$$

More generally, let A_i and B_i be $m_i \times n$ matrices, $i = 1, 2, \ldots, k$. Then

$$\begin{pmatrix} A_i^* \\ B_i^* \end{pmatrix} (A_i, B_i) = \begin{pmatrix} A_i^*A_i & A_i^*B_i \\ B_i^*A_i & B_i^*B_i \end{pmatrix} \ge 0.$$

It follows that

$$\begin{pmatrix} \sum A_i^*A_i & \sum A_i^*B_i \\ \sum B_i^*A_i & \sum B_i^*B_i \end{pmatrix} \ge 0.$$

Thus [51, p. 270]

$$\left\|\sum_{i=1}^k A_i^*B_i\right\|^2 \le \left\|\sum_{i=1}^k A_i^*A_i\right\| \left\|\sum_{i=1}^k B_i^*B_i\right\|.$$

Example 2. Let $A,\ B \ge 0$. Then for any $z \in \mathbb{C}$, by (3.4.37),

$$\begin{pmatrix} A & A \\ A & A \end{pmatrix} + \begin{pmatrix} |z|B & zB \\ \bar{z}B & |z|B \end{pmatrix} = \begin{pmatrix} A+|z|B & A+zB \\ A+\bar{z}B & A+|z|B \end{pmatrix} \ge 0.$$

For any unitarily invariant norm, we have (see [54, 470])

$$\|A + zB\| \le \|A + |z|B\|.$$

More generally, let $\lambda_i \in \mathbb{C}$ and $A_i \ge 0$, $i = 1, 2, \ldots, k$. Then

$$\sum \begin{pmatrix} |\lambda_i|A_i & \lambda_i A_i \\ \bar{\lambda}_i A_i & |\lambda_i|A_i \end{pmatrix} = \begin{pmatrix} \sum |\lambda_i|A_i & \sum \lambda_i A_i \\ \sum \bar{\lambda}_i A_i & \sum |\lambda_i|A_i \end{pmatrix} \ge 0.$$

For determinant, we have [354, p. 144]

$$\left|\det\left(\sum_{i=1}^k \lambda_i A_i\right)\right| \le \det\left(\sum_{i=1}^k |\lambda_i| A_i\right).$$

Similarly, one can obtain the singular value majorization inequality:

$$\sigma\left(\sum_{i=1}^k \lambda_i A_i\right) \prec_w \sum_{i=1}^k |\lambda_i(A)|\sigma(A_i).$$

Example 3. Notice that for any matrix A,

$$\begin{pmatrix} A \\ I \end{pmatrix} \begin{pmatrix} A \\ I \end{pmatrix}^* = \begin{pmatrix} A \\ I \end{pmatrix} (A^*, I) = \begin{pmatrix} AA^* & A \\ A^* & I \end{pmatrix} \ge 0.$$

We obtain that

$$\prod_{i=1}^{k} |\lambda_i(A)|^2 \le \prod_{i=1}^{k} \lambda_i(AA^*)$$

for each k, that is,

$$\prod_{i=1}^{k} |\lambda_i(A)| \le \prod_{i=1}^{k} \sigma_i(A),$$

from which we may derive [230, p. 171]

$$|\lambda(A)| \prec_w \sigma(A).$$

Example 4. Since for any matrices A and B of the same size,

$$\begin{pmatrix} AA^* & A \\ A^* & I \end{pmatrix} \circ \begin{pmatrix} I & B \\ B^* & B^*B \end{pmatrix} = \begin{pmatrix} (AA^*)\circ I & A\circ B \\ (A\circ B)^* & I\circ(B^*B) \end{pmatrix} \ge 0,$$

we have, by (3.0.3), for every unitarily invariant norm $\|\cdot\|$ [230, p. 212],

$$\|A\circ B\|^2 \le \|(AA^*)\circ I\| \, \|(B^*B)\circ I\| \le \|AA^*\| \, \|B^*B\|,$$

which is a Cauchy-Schwarz inequality for the Hadamard product, and

$$\prod_{i=1}^{k} \sigma_i^2(A\circ B) \le \prod_{i=1}^{k} \sigma_i(AA^*\circ I) \prod_{i=1}^{k} \sigma_i(I\circ BB^*).$$

This implies

$$\sigma^2(A\circ B) \prec_w \lambda(AA^*\circ I)\circ \lambda(I\circ B^*B)$$

which yields, by (3.0.2),

$$\sigma(A\circ B) \prec_w \sigma(A)\circ\sigma(B).$$

Example 5. For any matrices A and B of the same size,

$$\begin{pmatrix} |A| & A^* \\ A & |A^*| \end{pmatrix} + \begin{pmatrix} |B| & B^* \\ B & |B^*| \end{pmatrix} = \begin{pmatrix} |A|+|B| & (A+B)^* \\ A+B & |A^*|+|B^*| \end{pmatrix} \ge 0.$$

It follows that for any unitarily invariant norm [227]

$$\begin{aligned} \|A+B\| &\le \| \, |A|+|B| \, \|^{1/2} \, \| \, |A^*|+|B^*| \, \|^{1/2} \\ &\le \frac{1}{2}\left(\| \, |A|+|B| \, \| + \| \, |A^*|+|B^*| \, \|\right), \end{aligned}$$

which is the refined norm triangle inequality for normal matrices A and B:

$$\|A+B\| \le \| \, |A|+|B| \, \| \le \|A\| + \|B\|. \tag{3.4.38}$$

The first inequality in (3.4.38), though it is very tempting, does not hold in general for nonnormal matrices A and B by taking the spectral norm, $A = \left(\begin{smallmatrix} 1 & 1 \\ 0 & 0 \end{smallmatrix}\right)$ and $B = \left(\begin{smallmatrix} 1 & -1 \\ 0 & 0 \end{smallmatrix}\right)$, while the second inequality holds for all A and B (the triangular inequality). Note that $\| \, |X| \, \| = \|X\|$. The analog of (3.4.38) for the Hadamard product of normal matrices should read

$$\|A \circ B\| \leq \| \, |A| \circ |B| \, \| \leq \|A\| \, \|B\|. \qquad (3.4.39)$$

Its proof is parallel to that of (3.4.38), and likewise the first inequality in (3.4.39) is false for nonnormal A and B in general [231], whereas the second inequality, which is essentially the same as $\|A \circ B\| \leq \|A\| \, \|B\|$, $A, B \geq 0$, has been known to hold for all positive semidefinite A and B of the same size. We note here that the Hadamard submultiplicativity of a unitarily invariant norm is equivalent to the conventional submultiplicativity of the norm; that is, $\|X \circ Y\| \leq \|X\| \, \|Y\|$ for all X and Y if and only if $\|XY\| \leq \|X\| \, \|Y\|$ for all X and Y (see, e.g., [229] or [230, p. 335]).

3.5 Positive linear maps

Let M be an $n \times n$ matrix and let $\alpha \subseteq \{1, 2, \ldots, n\}$ be a nonempty index subset. Two elementary but interesting known results on Hermitian matrices (see, e.g., [468, p. 177]) are

$$A^2[\alpha] \geq A[\alpha]^2, \quad \text{if } A \text{ is Hermitian}$$

and

$$A^{-1}[\alpha] \geq A[\alpha]^{-1}, \quad \text{if } A \text{ is positive definite.}$$

Generalizations of these beautiful inequalities to primary matrix functions involve matrix monotonicity and convexity and have attracted a lot of attentions [121]. Notice that the map $A \mapsto A[\alpha]$ is linear and positive; namely, $A[\alpha] \geq 0$ if $A \geq 0$. This idea of "extracting a principal submatrix" works well for more general positive linear maps; a map Φ from the $n \times n$ complex matrix space to a matrix space is *positive linear* if Φ is linear and preserves positivity, that is, $X \geq 0 \Rightarrow \Phi(X) \geq 0$.

Let A be a normal matrix and let $A = \sum_{i=1}^n \lambda_i u_i u_i^*$ be the spectral decomposition of A, where λ_i are the eigenvalues of A and u_i are the orthonormal eigenvectors belonging to λ_i [468, p. 241]. Then $\sum_{i=1}^n u_i u_i^* = I$ and $|A|^\alpha = \sum_{i=1}^n |\lambda_i|^\alpha u_i u_i^*$ for any real α. (Conventionally, we assume $0^0 = 0$.) Let Φ be a positive linear map. If A is Hermitian, then so is $\Phi(A)$.

Notice that for any $\lambda \in \mathbb{C}$, $\alpha \in [0, 1]$, and $B \geq 0$, the 2×2 block matrix

$$\begin{pmatrix} |\lambda|^{2\alpha} B & \lambda B \\ \bar{\lambda} B & |\lambda|^{2(1-\alpha)} B \end{pmatrix} \geq 0. \qquad (3.5.40)$$

Since Φ is positive linear, each $\Phi(u_i u_i^*) \geq 0$. By (3.5.40), for $i = 1, 2, \ldots, n$,

$$\begin{pmatrix} |\lambda_i|^{2\alpha}\Phi(u_i u_i^*) & \lambda_i \Phi(u_i u_i^*) \\ \bar{\lambda}_i \Phi(u_i u_i^*) & |\lambda_i|^{2(1-\alpha)}\Phi(u_i u_i^*) \end{pmatrix} \geq 0.$$

Thus

$$\sum_{i=1}^{n} \begin{pmatrix} |\lambda_i|^{2\alpha}\Phi(u_i u_i^*) & \lambda_i \Phi(u_i u_i^*) \\ \bar{\lambda}_i \Phi(u_i u_i^*) & |\lambda_i|^{2(1-\alpha)}\Phi(u_i u_i^*) \end{pmatrix} \geq 0,$$

thus,

$$\begin{pmatrix} \Phi\left(|A|^{2\alpha}\right) & \Phi(A) \\ \Phi(A)^* & \Phi\left(|A|^{2(1-\alpha)}\right) \end{pmatrix} \geq 0. \tag{3.5.41}$$

By taking the Schur complement,

$$\Phi\left(|A|^{2(1-\alpha)}\right) \geq \Phi(A)^* \Phi\left(|A|^{2\alpha}\right)^{\dagger} \Phi(A).$$

In particular,

$$\Phi\left(|A|\right) \geq \Phi(A)^* \Phi\left(|A|\right)^{\dagger} \Phi(A).$$

And also, for unitarily invariant norms,

$$\|\Phi(A)\|^2 \leq \left\|\Phi\left(|A|^{2\alpha}\right)\right\| \left\|\Phi\left(|A|^{2(1-\alpha)}\right)\right\|.$$

Putting $\alpha = \frac{1}{2}$,

$$\|\Phi(A)\| \leq \|\Phi(|A|)\|.$$

If V is unitary, then

$$\|\Phi(V)\| \leq \|\Phi(I)\|.$$

In addition, by (3.3.34), (3.5.41) shows

$$\Phi(A)^* \circ \Phi(A) \leq \Phi\left(|A|^{2\alpha}\right) \circ \Phi\left(|A|^{2(1-\alpha)}\right).$$

Similarly, noticing that $A^*A = \sum_{i=1}^n |\lambda_i|^2 u_i u_i^*$ and $A^{-1} = \sum_{i=1}^n \lambda_i^{-1} u_i u_i^*$ when A is nonsingular, one arrives at

$$\begin{pmatrix} \Phi(I) & \Phi(A) \\ \Phi(A)^* & \Phi(A^*A) \end{pmatrix} \geq 0 \quad \text{and} \quad \begin{pmatrix} \Phi(A) & \Phi(I) \\ \Phi(I) & \Phi(A^{-1}) \end{pmatrix} \geq 0.$$

Thus if Φ is further *normalized*; that is, it maps identity to identity, then

$$\Phi(A^*A) \geq \Phi(A)^*\Phi(A).$$

If A is Hermitian, in particular,

$$\Phi(A^2) \geq \Phi(A)^2, \tag{3.5.42}$$

and, if A is positive definite,

$$\Phi(A^{-1}) \geq \Phi(A)^{-1}. \tag{3.5.43}$$

(3.5.42) belongs to Kadison [251] in the setting of C^*-algebras. Both (3.5.42) and (3.5.43) are consequences of a result of Choi [120] which asserts that $\Phi(f(A)) \geq f(\Phi(A))$ for all Hermitian matrices A, operator convex functions f, and normalized positive linear maps Φ.

Chapter 4

Closure Properties

4.0 Introduction

Our purpose here is to understand matrix classes that enjoy closure properties under Schur complementation. In the process, certain other inheritance properties within classes are also encountered.

In this and the next two sections, we assume that all matrices are square, have entries from a designated field $\mathbb{F}$ and are *principally nonsingular* (PN), that is, all principal minors are nonzero. If A is $n \times n$ and $\alpha \subseteq N \equiv \{1, 2, \ldots, n\}$ is an index set, the Schur complement in A of the principal submatrix $A[\alpha]$ is, from (1.7) when $\beta = \alpha$,

$$A/A[\alpha] \equiv A[\alpha^c] - A[\alpha^c, \alpha]A[\alpha]^{-1}A[\alpha, \alpha^c].$$

Throughout this chapter, we assume that α is a proper subset of N.

In order to understand closure, we need to be able to refer generally to square matrix classes defined in all dimensions. To this end, let $\mathcal{C}_n$ be a set of $n \times n$ PN–matrices and let $\mathcal{C} = \bigcup_{i=1}^{n} \mathcal{C}_n$. By $\mathcal{C}^{-1}$ we mean $\{ A^{-1} : A \in \mathcal{C} \}$. We say that $\mathcal{C}$ is α *SC–closed* if for any $A \in \mathcal{C}$, $A/A[\alpha] \in \mathcal{C}$, when it is defined (i.e., A is $n \times n$ over $\mathbb{F}$, $\alpha \subseteq N$). Class $\mathcal{C}$ is α *hereditary* if $A \in \mathcal{C}$ implies $A[\alpha] \in \mathcal{C}$, when it is defined. If $\mathcal{C}$ is α SC–closed (α hereditary) for all α, we say that $\mathcal{C}$ is SC–closed (hereditary). As we shall see, there are interesting classes that are, for example, α SC–closed for some α and not others.

4.1 Basic theory

We recall certain background that will be needed.

Theorem 4.8 (Jacobi's identity [228, p. 21]) *Let A be an $n\times n$ matrix over $\mathbb{F}$, and α, $\beta \subseteq N$. Then,*

$$\det A^{-1}[\alpha^c, \beta^c] = \left[(-1)^{(\sum_{i\in\alpha} i+\sum_{j\in\beta} j)}\right] \frac{\det A[\beta,\alpha]}{\det A}.$$

Typically, we apply only the case $\alpha = \beta$, which becomes

$$\det A^{-1}[\alpha^c] = \frac{\det A[\alpha]}{\det A}. \tag{4.1.1}$$

The special case of Theorem 1.2 in which the block with respect to which the Schur complement is taken is principal will be of considerable use.

Theorem 4.9 (The Schur complement and blocks of the inverse) *If A is a nonsingular $n \times n$ matrix over $\mathbb{F}$, $\alpha \subseteq N$ and $A[\alpha]$ is nonsingular, then*

$$A^{-1}[\alpha^c] = (A/A[\alpha])^{-1}.$$

A key property of the Schur complement is its nested sequential nature that comes from Gaussian elimination. The following is equivalent to the quotient formula, Theorem 1.4.

Theorem 4.10 (Sequential property of Schur complements) *Let $A \in M_n(F)$ be PN and suppose that α is a proper subset of N and that β is a proper subset of α^c. Then,*

$$(A/A[\alpha])/(A/A[\alpha])[\beta] = A/A[\alpha \cup \beta].$$

Proof. Identify the β of Theorem 1.4 with $\alpha \cup \beta$ above and note that $(A/A[\alpha])[\beta] = A[\alpha \cup \beta]/A[\alpha]$. ∎

Theorem 4.3 has a natural connection with SC–closure. It shows, for example, that SC–closure follows from α SC–closure for arbitrary index sets α of cardinality 1.

We may now make several simple, but far-reaching, observations that substantially clarify questions of SC–closure. The first is the key.

Theorem 4.11 *The class $\mathcal{C}$ is α SC–closed if and only if the class $\mathcal{C}^{-1}$ is α^c hereditary.*

Proof. Use the identity of Theorem 4.2. First, suppose that $\mathcal{C}$ is α SC–closed and let $A \in \mathcal{C}$. Then $A/A[\alpha] \in \mathcal{C}$, so that $(A/A[\alpha])^{-1} \in \mathcal{C}^{-1}$. Thus, $A^{-1}[\alpha^c] \in \mathcal{C}^{-1}$ and $\mathcal{C}^{-1}$ is α^c hereditary.

Conversely, if $\mathcal{C}^{-1}$ is α^c hereditary and $A \in \mathcal{C}$, then $A^{-1}[\alpha^c] \in \mathcal{C}^{-1}$. Then, $(A/A[\alpha])^{-1} \in \mathcal{C}^{-1}$, which means that $(A/A[\alpha]) \in \mathcal{C}$ and $\mathcal{C}$ is α SC–closed, completing the proof. ∎

It follows readily that

Corollary 4.16 *The class $\mathcal{C}$ is SC–closed if and only if the class $\mathcal{C}^{-1}$ is hereditary.*

It is not immediately clear whether Theorem 4.4 and Corollary 4.1 is more useful to exhibit SC–closure of one class or heredity in another class. In fact, depending upon the class, they could be useful in either way. However, two further observations are worth making.

Corollary 4.17 *If a class $\mathcal{C}$ is (α) hereditary, then its inverse class $\mathcal{C}^{-1}$ is (α) SC–closed.*

As, here, we are primarily interested in SC–closure and many classes are obviously hereditary, many examples of SC–closed classes may be readily verified.

In case a class is inverse–closed ($\mathcal{C}^{-1} = \mathcal{C}$), the two qualities coincide.

Corollary 4.18 *If a class $\mathcal{C}$ is inverse–closed, then $\mathcal{C}$ is SC–closed if and only if $\mathcal{C}$ is hereditary.*

It can happen that an inverse–closed class is neither (see the discussion of circulants to follow), but often, as we shall see, both do occur. As one may be easier to verify than the other, the above observation is useful.

As some important classes involve diagonal products with matrices of another class, it is useful that diagonal multiplication commutes, in a sense, with the Schur complement. An easy calculation verifies that

Lemma 4.6 *Suppose that D and E are invertible diagonal matrices and A is PN. Then,*

$$DA/(DA)[\alpha] = D[\alpha^c](A/A[\alpha]),$$

$$AE/(AE)[\alpha] = (A/A[\alpha])E[\alpha^c],$$

and

$$DAE/(DAE)[\alpha] = D[\alpha^c](A/A[\alpha])E[\alpha^c].$$

Proof. It suffices to verify the final statement and, for this, it suffices to consider the case in which α consists of the first k indices. Let $D = \begin{pmatrix} D_1 & 0 \\ 0 & D_2 \end{pmatrix}$, $E = \begin{pmatrix} E_1 & 0 \\ 0 & E_2 \end{pmatrix}$, and $A = \begin{pmatrix} A_{11} & A_{12} \\ A_{21} & A_{22} \end{pmatrix}$, in which the upper left block is $k \times k$ in each case. Then,

$$\begin{aligned} DAE/D_1A_{11}E_1 &= D_2A_{22}E_2 - D_2A_{21}E_1(D_1A_{11}E_1)^{-1}D_1A_{12}E_2 \\ &= D_2(A_{22} - A_{21}A_{11}^{-1}A_{12})E_2 \\ &= D_2(A/A_{11})E_2, \end{aligned}$$

as required. ∎

Now, let $\mathcal{D}$ be the set of nonsingular diagonal matrices and $\mathcal{D}_+$ be the set of positive (real) diagonal matrices. For a given class $\mathcal{C}$, we may form such classes $\mathcal{C}'$ as $\mathcal{DC}$, $\mathcal{D}_+\mathcal{C}$, $\mathcal{CD}$, $\mathcal{CD}_+$, $\mathcal{DCD}$, and $\mathcal{D}_+\mathcal{CD}_+$. Call such classes *diagonally derived* from $\mathcal{C}$.

Corollary 4.19 *Let $\mathcal{C}'$ be a class of matrices diagonally derived from the class $\mathcal{C}$. Then, $\mathcal{C}'$ is SC–closed (α SC–closed) if and only if $\mathcal{C}$ is SC–closed (α SC–closed).*

If $\mathcal{C}$ is a class of PN matrices, then so is $-\mathcal{C} = \{-\mathcal{A} : \mathcal{A} \in \mathcal{C}\}$. It is clear that $(-\mathcal{C})^{-1} = -\mathcal{C}^{-1}$ and by direct appeal to the formula for Schur complement we have

Lemma 4.7 *The class $-\mathcal{C}$ is (α) SC–closed ((α) hereditary, inverse–closed) if and only if $\mathcal{C}$ is (α) SC–closed ((α) hereditary, inverse–closed).*

It follows that whatever closure properties hold in such classes as the positive definite or stable matrices also hold for the negatives of these classes (the negative definite or positive stable matrices).

We close by noting that for a class $\mathcal{C}$ that is permutationally similarity invariant ($P^T\mathcal{C}P = \mathcal{C}$ whenever P is a permutation matrix), the inverse class is as well, and the class is $\{i\}$ SC–closed if and only if it is $\{1\}$ SC–closed (α SC–closed if and only if β SC–closed, whenever $|\alpha| = |\beta|$). Because of the sequential property of Schur complements (Theorem 4.3), it then follows that SC–closure is equivalent to $\{1\}$ SC–closure. Of course, many familiar classes are permutationally similarity invariant, but some, such as the totally positive matrices, are not.

Theorem 4.12 *Suppose that $\mathcal{C}$ is a permutationally similarity invariant class. Then, $\mathcal{C}$ is SC–closed if and only if $\mathcal{C}$ is $\{1\}$ SC–closed.*

More generally, if $\mathcal{C}' = \bigcup_P P^T\mathcal{C}P$ in which the union is over all permutation matrices P, the set of all permutation similarities of matrices in $\mathcal{C}$, then $\mathcal{C}'$ is SC–closed (hereditary, inverse–closed) if $\mathcal{C}$ is SC–closed (hereditary, inverse–closed). Of course, $\mathcal{C}' = \mathcal{C}$ when $\mathcal{C}$ is permutationally similarity invariant, and, in general, $\mathcal{C}'$ itself is permutationally similarity invariant, so that Theorem 4.5 applies to it.

4.2 Particular classes

Of course, matrix classes of interest are not usually viewed as the union of sets from each dimension, but have a natural definition that transcends dimension. Using the tools derived in the previous section, we consider here a wide variety of familiar classes. We continue to consider only subclasses of the PN matrices, and for convenience, we group the classes as follows: (a) "structured" matrices; (b) inverse–closed classes; (c) classes based upon dominance; (d) further positivity classes; and (e) other classes.

a) Structured Matrices

Matrices with special "structure" arise in many parts of mathematics and its applications. A natural question is which particular structures are preserved by Schur complementation. In this subsection, we answer this and other questions for a variety of familiar classes defined by some form of entry-defined structure. As in subsequent subsections, we give, at the end, a table summarizing our conclusions for each type of matrix. In most cases, the structure does not guarantee principal nonsingularity (PN), but this is assumed, in keeping with earlier comments. In each case, the observations of the first section will be used to give efficient discussions and to exploit or extend the observations to the inverse class. In particular, we note that, by Corollary 4.1, heredity and SC–closure for the inverse class are entirely determined by these properties for the class itself and that, by Corollary 4.3, in the event that the class is inverse–closed, it suffices to check either heredity or SC–closure (and not both).

• **Diagonal matrices**

A matrix is *diagonal* if its off-diagonal entries vanish. It is obvious that the diagonal matrices $\mathcal{D}$ are hereditary and inverse–closed. Thus, it follows from Corollary 4.3 that $\mathcal{D}$ is also SC–closed.

• **Triangular matrices**

A matrix is *lower (upper) triangular* if its entries above (below) the main diagonal vanish. In either case, the matrix is said to be *triangular*. It is immediate that lower (upper) triangular matrices are hereditary and inverse–closed. Therefore, applying Corollary 4.3, we see that triangular matrices are also SC–closed.

We note also, by permutation similarity (see comment at end of Section 4.1), that the essentially triangular matrices (those permutationally similar to some triangular matrix) are SC–closed and hereditary, as well as inverse–closed.

• **Reducible matrices**

A matrix A is *reducible* if there is a permutation matrix P such that

$$PAP^T = \begin{pmatrix} B & C \\ 0 & D \end{pmatrix}.$$

It is apparent that a matrix is reducible if and only if its inverse is. Thus, the reducible matrices are inverse–closed.

Consider the reducible matrix

$$A = \begin{pmatrix} 1 & 0 & 0 \\ 1 & 1 & 2 \\ 1 & 1 & 1 \end{pmatrix}.$$

Since $A[\{2,3\}] = \left(\begin{smallmatrix} 1 & 2 \\ 1 & 1 \end{smallmatrix}\right)$ is not reducible, it follows from Corollary 4.3 that the class of reducible matrices are neither hereditary nor SC–closed. Reducibility will, however, occur for certain α specific to the particular A.

- **Irreducible matrices**

A matrix is *irreducible* if it is not reducible. Irreducible matrices are inverse–closed since reducible matrices are.

Consider the irreducible matrix

$$A = \begin{pmatrix} 1 & 2 & 2 \\ 1 & 1 & 2 \\ 1 & 0 & 1 \end{pmatrix}.$$

Since $A[\{2,3\}] = \left(\begin{smallmatrix} 1 & 2 \\ 0 & 1 \end{smallmatrix}\right)$ is reducible, it follows from Corollary 4.3 that the class of irreducible matrices is neither hereditary nor SC–closed.

- **Symmetric matrices; Hankel matrices**

An $n \times n$ matrix $A = (a_{ij})$ is *symmetric* if $a_{ij} = a_{ji}$ for all $i, j \in N$ or, equivalently, if $A = A^T$. Since $A^{-1} = (A^T)^{-1} = (A^{-1})^T$ for a symmetric matrix A, symmetric matrices are inverse–closed. Obviously, symmetric matrices are hereditary and so, by Corollary 4.3, they are SC–closed.

We say that a matrix $A = (a_{ij})$ is *Hankel* if, for some sequence a_2, $a_3, \ldots, a_{2n}$, $a_{ij} = a_{i+j}$ for all $i, j \in N$. That is, the entries of A are constant down the "backward" (upper right to lower left) diagonal and also down the diagonals parallel to the backward diagonal. Hankel matrices are particular symmetric matrices.

Consider the Hankel matrix

$$A = \begin{pmatrix} 1 & 2 & 0 & 1 \\ 2 & 0 & 1 & 3 \\ 0 & 1 & 3 & -1 \\ 1 & 3 & -1 & 1 \end{pmatrix}.$$

Since $A(\{2\}) = \begin{pmatrix} 1 & 0 & 1 \\ 0 & 3 & -1 \\ 1 & -1 & 1 \end{pmatrix}$ does not have constant backward diagonal, we see that Hankel matrices are not hereditary, and, since

$$A/a_{11} = \begin{pmatrix} -4 & 1 & 1 \\ 1 & 3 & -1 \\ 1 & -1 & 0 \end{pmatrix},$$

they are not SC–closed as well. Lastly,

$$A^{-1} = \begin{pmatrix} 34 & -6 & -5 & -21 \\ -6 & 1 & 1 & 4 \\ -5 & 1 & 1 & 3 \\ -21 & 4 & 3 & 13 \end{pmatrix}$$

is not constant down the backward diagonal, illustrating the fact that Hankel matrices are not inverse–closed.

• Persymmetric matrices; Toeplitz matrices

A matrix $A = (a_{ij})$ is called *persymmetric* if $a_{ij} = a_{n+1-j,n+1-i}$ for all $i, j \in N$ or, equivalently, if $A = KA^TK$ in which K is the *backward identity* (ones on the backward diagonal, zeros elsewhere). That is, A is persymmetric if it is symmetric with respect to the backward diagonal. Since $A^{-1} = K(A^T)^{-1}K = K(A^{-1})^TK$ for each persymmetric matrix A, persymmetric matrices are inverse–closed.

We say a matrix $A = (a_{ij})$ is *Toeplitz* if, for some sequence $a_{-(n-1)}, \ldots,$ $a_{-1}, a_0, a_1, \ldots, a_{n-1}$, $a_{ij} = a_{j-i}$ for all $i, j \in N$. Equivalently, a Toeplitz matrix is one that is constant down the main diagonal and down the diagonals parallel to the main diagonal and thus is persymmetric. Since B is Toeplitz if and only if $B = AK$ in which A is Hankel and K is the backward identity, one can use the Hankel matrix above to illustrate that Toeplitz (persymmetric) matrices are neither hereditary nor SC–closed and use its inverse to show that Toeplitz matrices are not inverse–closed.

• Bisymmetric matrices

We say that a matrix A is *bisymmetric* if it is both symmetric and persymmetric. Since both symmetric and persymmetric matrices are inverse–closed, so are bisymmetric matrices. Consider the bisymmetric matrix

$$A = \begin{pmatrix} 2 & 2 & 3 \\ 2 & 1 & 2 \\ 3 & 2 & 2 \end{pmatrix}.$$

Since $A(\{1\}) = \left(\begin{smallmatrix} 1 & 2 \\ 2 & 2 \end{smallmatrix}\right)$ is not persymmetric, we see that bisymmetric matrices are not hereditary, and, hence, not SC–closed as well.

• Hermitian matrices; Skew-Hermitian matrices

A matrix $A = (a_{ij})$ is *Hermitian* if $a_{ij} = \overline{a_{ji}}$ for all $i, j \in N$ or, equivalently, if $A = A^*$ in which A^* denotes the conjugate transpose of A. Since $A^{-1} = (A^*)^{-1} = (A^{-1})^*$ for a Hermitian matrix A, Hermitian matrices are inverse–closed. Obviously, Hermitian matrices are hereditary and so, by Corollary 4.3, they are SC–closed as well.

A matrix $A = (a_{ij})$ is *skew-Hermitian* if $a_{ij} = -\overline{a_{ji}}$ for all $i, j \in N$ or, equivalently, if $A = -A^*$. By a similar line of reasoning, it follows that skew-Hermitian matrices are inverse–closed, hereditary, and SC–closed.

• **Centrosymmetric matrices**

Call a matrix $A = (a_{ij})$ *centrosymmetric* if $a_{ij} = a_{n+1-i,n+1-j}$ for all $i, j \in N$ or, equivalently, if $A = KAK$ in which K is the backward identity. (Essentially, such matrices are symmetric about their geometric center.) Since $A^{-1} = KA^{-1}K$ for each centrosymmetric matrix A, centrosymmetric matrices are inverse–closed.

Consider the centrosymmetric matrix

$$A = \begin{pmatrix} 1 & 0 & 2 & 1 & 1 \\ 0 & 1 & 3 & 2 & 0 \\ 1 & 1 & 1 & 1 & 1 \\ 0 & 2 & 3 & 1 & 0 \\ 1 & 1 & 2 & 0 & 1 \end{pmatrix}.$$

Since $A(\{1,2\}) = \begin{pmatrix} 1 & 1 & 1 \\ 3 & 1 & 0 \\ 2 & 0 & 1 \end{pmatrix}$ is not centrosymmetric, we see that centrosymmetric matrices are not hereditary and, hence, not SC–closed as well, by Corollary 4.3.

• **Circulant matrices**

We say that a matrix $A = (a_{ij})$ is a *circulant* matrix if for some sequence $a_0, a_1, \ldots, a_{n-1}$, $a_{ij} = a_{j-i}$ for all $i, j \in N$. Here, the subscripts are taken modulo n. It is clear that each circulant is a polynomial in the basic circulant P whose first row is $[0\,1\,0 \cdots 0]$ and thus circulants are inverse–closed.

Consider the circulant matrix

$$A = \begin{pmatrix} 1 & 2 & 3 \\ 3 & 1 & 2 \\ 2 & 3 & 1 \end{pmatrix}.$$

Since $A(\{1\}) = \left(\begin{smallmatrix} 1 & 2 \\ 3 & 1 \end{smallmatrix}\right)$, one sees that circulant matrices are not hereditary and, thus, by Corollary 4.3, they are not SC–closed.

• **Normal matrices; Unitary matrices; Householder matrices**

A matrix A is *normal* (*unitary*) if $AA^* = A^*A$ $(A^{-1} = A^*)$. It is clear that normal (unitary) matrices are inverse–closed and that normal (unitary) matrices are not hereditary. Thus, by Corollary 4.3, they are not SC–closed as well. A matrix A is said to be a *Householder matrix* if $A = I - \frac{2}{x^T x} x x^T$ for some $0 \neq x \in \mathbb{R}^n$. Thus, Householder matrices are symmetric, orthogonal, and (hence) inverse–closed. Since this class is obviously not hereditary, it is not SC–closed as well.

• Hessenberg matrices; Tridiagonal matrices

We say a matrix $A = a_{ij}$ is *lower* (*upper*) *Hessenberg* if $a_{ij} = 0$ for $j - i > 1$ $(i - j > 1)$ or, equivalently, if the only nonzero entries lie in the lower triangular part or on the first super-diagonal (the upper triangular part or on the first sub-diagonal). In either case we say the matrix is *Hessenberg*. A matrix is *tridiagonal* if it is both lower and upper Hessenberg. It is known that a matrix is lower (upper) Hessenberg if and only if the 2×2 minors above (below) or touching the main diagonal in its inverse vanish. It then follows that Hessenberg and tridiagonal matrices are SC–closed (because the inverse classes are hereditary), and it is clear that they are hereditary. Lastly, since the inverse of a tridiagonal matrix is full when its super- and sub-diagonals are completely nonzero, neither Hessenberg nor tridiagonal matrices are inverse–closed.

Table for Subsection (a)

Class	SC Closure	Heredity	Inverse Closure	Inverse Class SC Closure	Inverse Class Heredity
Diagonal	Y	Y	Y	Y	Y
Triangular	Y	Y	Y	Y	Y
Reducible	N	N	Y	N	N
Irreducible	N	N	Y	N	N
Symmetric	Y	Y	Y	Y	Y
Hankel	N	N	N	N	N
Persymmetric	N	N	Y	N	N
Toeplitz	N	N	N	N	N
Bisymmetric	N	N	Y	N	N
Hermitian	Y	Y	Y	Y	Y
Skew-Hermitian	Y	Y	Y	Y	Y
Centrosymmetric	N	N	Y	N	N
Circulant	N	N	Y	N	N
Normal	N	N	Y	N	N
Unitary	N	N	Y	N	N
Householder	N	N	Y	N	N
Hessenberg	Y	Y	N	Y	Y
Tridiagonal	Y	Y	N	Y	Y

For completeness, consider closure properties of k, l–banded matrices. Say that an $n \times n$ matrix A over field $\mathbb{F}$ is *k,l–banded* if the nonzero entries are confined to the k bands below the main diagonal, the main diagonal and the l bands above the main diagonal. For example, tridiagonal is $1, 1$–banded, while Hessenberg is $1, n-1$ or $n-1, 1$–banded. Triangular is $0, n-1$ or $n-1, 0$–banded. It is straightforward that the k, l–banded matrices are hereditary, but, while the tridiagonal, triangular, and Hessenberg matrices are SC–closed, other k, l–banded classes are α SC–closed only for certain α. To understand this, we discuss the case $\alpha = \{i\}$, from which more general situations may be deduced. We determine k', l' for which $A/A[i]$ is k', l'–banded.

Now,

$$A/A[i] = A(i) - \frac{1}{a_{ii}} A(i, i]A[i, i).$$

Since $A(i)$ is k, l–banded and $a_{ii} \neq 0$, it suffices to determine the bandedness indices for the $(n-1) \times (n-1)$ matrix $A(i, i]A[i, i)$. The vector $A(i, i]$ has possibly nonzero entries stretching consecutively from position $\min\{i-l, 1\}$ to position $\max\{i+k-1, n-1\}$, while in $A[i, i)$ they stretch from $\min\{i-k, 1\}$ to $\max\{i+l-1, n-1\}$. The outer product is then a block whose nonzero entries are confined to a rectangular submatrix whose northeast corner lies at

$$\min\{i-l, 1\}, \max\{i+l-1, n-1\}$$

and southwest corner at

$$\max\{i+k-1, n-1\}, \min\{i-k, 1\}.$$

Since the bandedness indices for this block are the coordinate differences, we have that $A/A[i]$ is k', l'–banded for $k' = \max\{k, k''\}, l' = \max\{l, l''\}$, in which

$$k'' = \min\{i+k-1, n-1\} - \max\{i-k, 1\}$$

and

$$l'' = \min\{i+l-1, n-1\} - \max\{i-l, 1\}.$$

Note that the lower (upper) band index is determined by i and the lower (upper) band index for A; no interaction occurs. It follows that for $i = 1, 2, n-1, n$ (and, thus, for α any consecutive sequence containing one of these) we have α SC–closure, but there is not generally SC–closure. For example, when $n = 5$, $k = l = 2$, $i = 3$, there is not. There is SC–closure when $k, l \in \{0, 1, n-2, n-1\}$ and this is all, but for low-order cases. The case $k = l = 1$ is the tridiagonal matrices, already addressed

b) Inverse–Closed Classes.

In this subsection, prominent classes that happen to be inverse–closed and have not yet been discussed are addressed. By Corollary 4.3, such classes are either simultaneously hereditary and SC–closed, or neither. This abbreviates the necessary discussion. As we saw in the prior section, both possibilities can occur; the symmetric matrices are both hereditary and SC–closed while the circulant matrices are neither. As usual, we conclude this subsection with a table summarizing the results of the discussion of each case.

- **PN–matrices**

The PN (all principal submatrices nonsingular) –matrices themselves are inverse–closed by Jacobi's identity, Theorem 4.1. Since it is immediate that they are hereditary, it follows that they are SC–closed as well.

- **P–matrices**

A real square matrix is called a *P–matrix* if all its principal minors are positive. Again by Jacobi's identity, this property conveys to the inverse, so that the P–matrices are inverse–closed. As heredity is also immediate, P–matrices are also both SC–closed and hereditary.

- **Positive definite matrices**

A Hermitian matrix is *positive definite* (PD) if it is a P–matrix. (In the presence of Hermicity, this is equivalent to the positivity of the eigenvalues or the positivity of the quadratic form x^*Ax on nonzero vectors x.) Since the Hermitian and P–matrices are inverse–closed, the PD matrices are as well. Since the Hermitian and P–matrices are hereditary, the PD matrices are and thus are SC–closed as well.

- **Elliptic matrices**

An $n \times n$ Hermitian matrix is called *elliptic* (E) if it has exactly 1 positive eigenvalue. Here, for consistency, we also require that a matrix be PN to be elliptic. Thus, there are $n - 1$ negative eigenvalues (and no principal submatrix has a 0 eigenvalue). That the elliptic matrices are inverse–closed follows from the spectral theorem; the inverse is Hermitian and the eigenvalues of the inverse are the inverses of the eigenvalues. That the elliptic matrices are *not* hereditary follows from the simple example

$$A = \begin{pmatrix} 1 & 0 \\ 0 & -1 \end{pmatrix},$$

since $A[\{2\}]$ has no positive eigenvalues. By permutation similarity, there is α heredity for no α. Thus, the elliptic matrices are neither (α) SC–closed nor α hereditary.

- **Symmetric Sector Field Containment**

The *field of values* of an $n \times n$ complex matrix A is defined by

$$F(A) = \{\, x^*Ax : x \in \mathbb{C}^n, x^*x = 1 \,\}.$$

It is known that $F(A)$ is a compact convex set in the complex plane, that $F(A^*) = \overline{F(A)}$; and that, for a congruence C^*AC, $C \in \mathbb{C}^{n \times n}$ and nonsingular, $F(C^*AC)$ lies in the same angular sector anchored at the origin as $F(A)$ [230]. Now let S be an angular sector, anchored at the origin, that is symmetric with respect to the real axis.

Figure 4.1:

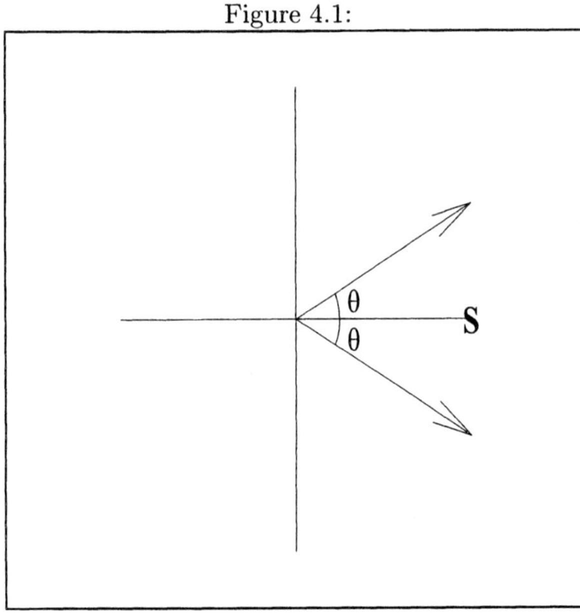

This sector is allowed to be either the positive or negative real axis. Since $F(A[\alpha]) \subseteq F(A)$, for any $\alpha \subseteq N$ [230], we see that the property that $F(A) \subseteq S$ is hereditary. But, this property (call it *symmetric sector field containment*, or $SSFC$ is inverse–closed, as $F(A^*A^{-1}A) = F(A^*) = \overline{F(A)}$ implies that $F(A^{-1})$ lies in the same sector S. Thus, an $SSFC$ class is both SC–closed and hereditary.

- **Positive definite Hermitian part**

An important special case of the property of symmetric sector field containment is that in which S is the open right half-plane. This is equivalent [230] to the requirement that the *Hermitian part* $H(A) = \frac{1}{2}(A + A^*)$ be positive definite. Thus, the matrices with positive definite Hermitian part ($PDHP$) are inverse–closed, SC–closed and hereditary.

• **Regional Field Containment**

We conclude this section with a class that does not exactly fit, as it is not inverse–closed; but it is defined via the field of values. Let R be any region of the complex plane, and define $\mathcal{R}$ to be the class of matrices A for which $F(A) \subseteq R$. We say such a class is a *regional field containment* (RFC) class. By the principal submatrix containment property mentioned above, $\mathcal{R}$ is hereditary. But, as $\mathcal{R}$ is not generally inverse–closed, $\mathcal{R}$ is not SC–closed.

Table for Subsection (b)

Class	SC Closure	Heredity	Inverse Closure	Inverse Class SC Closure	Inverse Class Heredity
PN	Y	Y	Y	Y	Y
P	Y	Y	Y	Y	Y
PD	Y	Y	Y	Y	Y
E	N	N	Y	N	N
$SSFC$	Y	Y	Y	Y	Y
$PDHP$	Y	Y	Y	Y	Y
RFC	N	Y	N	Y	N

c) Dominance Based Classes.

In this subsection we consider classes based upon some form of diagonal dominance. In each case the table profile will be the same. The SC–closure is not so apparent (either directly or indirectly), but is based upon one calculation that we make. The SC–closure then follows from Corollary 4.5, as each class is permutationally similarity invariant. Several of these classes arise frequently, are very important in computation and the SC–closure is key to their analysis.

• **Diagonal Dominance Classes**

An $n \times n$ complex matrix $A = (a_{ij})$ is called (strictly) *row diagonally dominant* (RDD) if, for each i, $1 \leq i \leq n$,

$$|a_{ii}| > \sum_{j \neq i} |a_{ij}| = R_i'(A)$$

It is straightforward [228] to show that an RDD matrix is nonsingular, and, as a moment's reflection reveals, that the property RDD is hereditary, and that RDD matrices are PN. We may analogously define CDD to be the (strictly) column diagonally dominant matrices (i.e., A is CDD if and only if A^T is RDD), and, if both requirements hold for the same matrix, we use $RCDD$ (row and column diagonally dominant). For both CDD and $RCDD$, heredity (and nonsingularity) are equally clear. What about SC–closure, however? As heredity in the inverse classes is not immediate, we use the observation that these classes are permutationally similarity invariant, so that SC–closure may be checked by checking $\{1\}$ SC–closure. We make this calculation only once, as it is similar for each class. Suppose that $A = (a_{ij})$ is RDD, and we calculate $A/A[\{1\}]$. It suffices to check that the first row of the result satisfies the row dominance criterion (though other rows would be similar). That first row is

$$a_{22} - a_{21}(\frac{a_{12}}{a_{11}}),\ a_{23} - a_{21}(\frac{a_{13}}{a_{11}}),\ \ldots,\ a_{2n} - a_{21}(\frac{a_{1n}}{a_{11}}).$$

Then, we need to verify that the absolute first entry beats the sum of the absolute values of the remaining entries, or

$$\begin{aligned}\triangle &= |a_{22} - a_{21}(\frac{a_{12}}{a_{11}})| - |a_{23} - a_{21}(\frac{a_{13}}{a_{11}})| - \cdots - |a_{2n} - a_{21}(\frac{a_{1n}}{a_{11}})| \\ &> 0.\end{aligned}$$

But,

$$\begin{aligned}\triangle &\geq |a_{22}| - |a_{21}||\frac{a_{12}}{a_{11}}| - |a_{23}| - |a_{21}||\frac{a_{13}}{a_{11}}| - \cdots - |a_{2n}| - |a_{21}||\frac{a_{1n}}{a_{11}}| \\ &= |a_{22}| - |a_{21}|\frac{\sum_{j\neq 1}|a_{1j}|}{|a_{11}|} - \sum_{j\neq 1,2}|a_{2j}| \\ &\geq |a_{22}| - R_2'(A) \\ &> 0,\end{aligned}$$

as was to be shown. The last weak inequality is because $\frac{R_1'(A)}{a_{11}} < 1$ by the row dominance in row 1 of A and the strict inequality follows from the row dominance in row 2 of A. We now have that each of the classes RDD, CDD and $RCDD$ is SC–closed and hereditary, so that the inverse classes are as well. For $n \geq 3$, it is easily seen by example that none of these dominance classes is inverse–closed.

- **H–matrices**

A square matrix A is called an *H–matrix* if there is a (nonsingular) diagonal matrix D such that DA is CDD. (It is equivalent to say that AD be RDD or DAE, E diagonal, be $RCDD$.) Variants upon this definition occur in the literature, and it is the same to say that $A \in \mathcal{D}(CDD)$ or

$\mathcal{D}_+(CDD)$, etc. By Corollary 4.4, we conclude from the RDD, CDD and $RCDD$ cases that the H–matrices are SC–closed and hereditary. Of course, they, again, need not be inverse–closed.

• Z–matrices

A *Z–matrix* is a matrix with nonpositive off-diagonal entries. The Z–matrices are obviously hereditary, but are neither inverse–closed nor SC–closed.

• M–matrices; $\mathbf{L}_k$–matrices

A very important special case of H–matrices is the *M–matrices*: the real H–matrices that are Z–matrices with positive diagonal. Many other descriptions are available (see, e.g., [230] or [49]). In this case, the inverse class, the *inverse M–matrices* (nonnegative matrices whose inverses are M–matrices) are of independent interest. It follows that if A is a Z–matrix, then $A/A[\alpha]$ is also *provided* $A[\alpha]$ is an M–matrix. That M–matrices are SC–closed follows since H–matrices are SC–closed. Further, since H–matrices and Z–matrices are hereditary, it follows that M–matrices are as well. (The M–matrices are permutationally similarity invariant.) Thus, the inverse M–matrices are also hereditary and SC–closed, a fact not so easily seen directly.

We mention two more classes without full discussion. The Z–matrices may be partitioned into a collection of classes L_k, $k = 0, \ldots, n$. In general, L_k consists of those Z–matrices in which every $t \times t$ principal minor, $t \leq k$, is nonnegative (so that the corresponding submatrices are in the closure of the M–matrices) and some $(k+1)\times(k+1)$ principal minor is negative. Thus, L_n is the closure of the M–matrices. Let $L'_k = L_k \cap PN$, $k = 0, 1, \ldots, n$. We have just seen that L'_n (the class of M–matrices) is hereditary and SC–closed, i.e., if $\alpha \subseteq N$ and $A[\alpha]$ is a principal submatrix of $A \in L'_n$, then $A[\alpha] \in L'_{|\alpha|}$ and $A/A[\alpha] \in L'_{n-|\alpha|}$. In general, L'_k, $k = 0, 1, \ldots, n-1$, is neither inverse–closed nor hereditary nor SC–closed. But, as we shall see, certain subclasses have α SC–closure. To see this, let $k \in \{1, \ldots, n-1\}$, $A \in L'_{n-k}$, and $\alpha \subseteq N$ in which $|\alpha| \leq n-k$ (so that $A[\alpha]$ is an M–matrix). Then it follows from Schur's identity (Theorem 1.1) that $A/A[\alpha] \in L'_{(n-|\alpha|)-k}$ if *every* $(n-k+1) \times (n-k+1)$ principal minor of A is negative (see [166]). So we have α SC–closure under these restrictions. In particular, we note that L'_{n-1} is always closed under Schur complementation since the only principal submatrix of order $n-1+1 = n$ is the matrix itself (which is not an M–matrix).

Let $L''_k = \bigcup_{i=k}^{n} L'_i$, $k = 1, \ldots, n$, i.e., the class of Z–matrices in which every $t \times t$ principal submatrix, $t \leq k$, is an M–matrix. In order to analyze α SC–closure in L''_k, assume that $A \in L''_k$, say $A \in L'_s$ for some $s \geq k$. If

$\alpha \subseteq N$ with $|\alpha| \leq k$, then $A/A[\alpha] \in L'_j$ for some $j \geq k - |\alpha|$ [166]. That is, $A/A[\alpha] \in L''_{k-|\alpha|} = \bigcup_{i=k-|\alpha|}^{n-|\alpha|} L'_i$. Thus, L''_k is α SC–closed (and its inverse class α^c hereditary) provided $|\alpha| \leq k$.

In order to analyze heredity in L''_k, again assume that $A \in L''_k$, say $A \in L'_s$ for some $s \geq k$, but, this time, assume that $\alpha \subseteq N$ with $|\alpha| \geq k$. Now, if $s \leq |\alpha|$ (so that $k \leq s \leq |\alpha|$), $A[\alpha] \in L'_j$ for some j, $s \leq j \leq |\alpha|$, and, hence, $A[\alpha] \in \bigcup_{i=k}^{|\alpha|} L'_i = L''_k$. On the other hand, if $s > |\alpha|$, then $A[\alpha]$ is an M–matrix, i.e., $A[\alpha] \in L'_{|\alpha|} = L''_{|\alpha|}$. So L''_k is α hereditary provided $|\alpha| \geq k$. Since the case $|\alpha| < k$ is satisfied vacuously, L''_k is hereditary (and its inverse class is SC–closed).

- **Cassini matrices**

If A is not necessarily RDD but, for each pair i, j,

$$|a_{ii}||a_{jj}| > R_i(A)R_j(A),$$

it is still the case that A is nonsingular. Because of the ovals of Cassini, as used by Brauer [228], we call the class defined by the above inequalities the *Cassini* class, denoted C. It is clearly hereditary, and it is also SC–closed. Like the class RDD, the Cassini class is not inverse-closed.

Table for Subsection (c)

Class	SC Closure	Heredity	Inverse Closure	Inverse Class SC Closure	Inverse Class Heredity
RDD	Y	Y	N	Y	Y
CDD	Y	Y	N	Y	Y
$RCDD$	Y	Y	N	Y	Y
H	Y	Y	N	Y	Y
Z	N	Y	N	Y	N
M	Y	Y	N	Y	Y
L_k	N	N	N	N	N
C	Y	Y	N	Y	Y

d) Further Positivity Classes.

We consider here several important classes involving positivity that have not been discussed in prior subsections. The most important of these is the very interesting case of totally positive (TP) matrices. In several other cases there is no α SC–closure, but clear heredity implies surprising SC–closure results for the inverse classes.

• Totally positive matrices

Though it is the most complicated case discussed, we begin with the case of *totally positive matrices.* A real matrix is TP if *all* its minors are positive. It is important to note that this class is not permutationally similarity invariant; in fact, besides the identity, the only permutation similarity that does not change the signs of some non-principal minors is the one by the backward identity K. The TP class is clearly hereditary, is not generally SC–closed but is α SC–closed for many α (and, in fact, for "other" α, $A/A[\alpha]$ *cannot* remain TP). To determine when it is α SC–closed, we observe that, by Jacobi's identity, the class TP, while not truly inverse–closed is almost so; an n–by–n matrix B is inverse TP if and only if it is of the form SCS, in which S is the alternating sign signature matrix, i.e.,

$$S = \begin{pmatrix} 1 & 0 & 0 & \cdots & \cdots & 0 \\ 0 & -1 & 0 & & & \cdots \\ 0 & 0 & 1 & \cdots & & \cdots \\ \cdots & \cdots & \cdots & \cdots & \cdots & \cdots \\ \cdots & \cdots & & \cdots & \cdots & 0 \\ 0 & 0 & \cdots & \cdots & 0 & (-1)^{n+1} \end{pmatrix}$$

and C is TP.

Thus, the entries of such a B have a "checkerboard" sign pattern. In fact, again by Jacobi, a submatrix of B is inverse TP if and only if it is checkerboard. Thus, to determine the α^c for which $(TP)^{-1}$ is α^c hereditary (and thus the for α for which TP is α SC–closed by Theorem 4.4) it suffices to determine the α^c for which $B[\alpha^c]$ remains checkerboard. Though straightforward, this is somewhat tedious to describe. It hinges upon the maximal consecutive subsets of α^c. Among these consecutive subsets, we distinguish the extremal ones (those containing 1 or n) and the "interior" ones. The extremal ones may be of any cardinality, but the interior ones must be of even cardinality to maintain the checkerboard pattern. We call an index set meeting these requirements *proper*. Now, it is important to reference index sets α relative to TP_n, rather than the class TP of matrices of unboundedly many rows and columns (in the latter case, no extremal consecutive subset from the "end" makes any sense). We then have that TP_n is α SC–closed ($(TP)^{-1}$ is α^c hereditary) if and only if α^c is proper. In fact, if α^c is not proper, $A/A[\alpha]$ is TP for *no* $A \in TP_n$.

• Nonnegative, doubly nonnegative, completely positive, and copositive matrices

We next consider four interesting positivity classes simultaneously, as the results are identical and the analysis is similar because of the role of nonnegative entries.

A matrix is (entry-wise) *nonnegative* (*positive*) if all its entries are nonnegative (positive). We denote the nonnegative matrices by NN. A matrix that is both positive semi-definite (positive definite for us) and nonnegative is called *doubly nonnegative* (DN), and if $A = BB^T$, with B nonnegative, A is called *completely positive* (CP) (we add the requirement that B have full row rank). Finally, a real symmetric A is called (strictly) *copositive* ($CoPos$) if $x \geq 0$, $0 \neq x \in R^n$ imply $x^T Ax > 0$. (We add the requirement here that A be PN.) Clearly, a positive symmetric matrix is $CoPos$, as is a PD matrix or the sum of a nonnegative and a PD matrix. Because the inverse of a nonnegative matrix is nonnegative if and only if it is monomial (diagonal times permutation), none of the first three classes (nonnegative, DN, CP) is inverse–closed. The class $CoPos$ is also not, as shown by such simple examples as

$$A = \begin{pmatrix} 1 & 2 \\ 2 & 1 \end{pmatrix}.$$

Each of these four classes is hereditary, as is known and may easily be verified. None, however, is SC–closed and, thus, not α SC–closed for any α (as all are permutationally similarity invariant). The example above illustrates this fact for copositive matrices since $A/a_{11} = -3$. For the other three classes, consider the completely positive matrix

$$A = \begin{pmatrix} 9 & 7 & 7 \\ 7 & 6 & 5 \\ 7 & 5 & 6 \end{pmatrix}.$$

$$\left(A = BB^T \quad \text{in which} \quad B = \begin{pmatrix} 2 & 1 & 2 \\ 2 & 1 & 1 \\ 1 & 1 & 2 \end{pmatrix}\right).$$

Since

$$A^{-1} = \begin{pmatrix} 11 & -7 & -7 \\ -7 & 5 & 4 \\ -7 & 4 & 5 \end{pmatrix},$$

it follows from Theorem 1.2 that

$$A/a_{11} = \begin{pmatrix} 5 & -4 \\ -4 & 5 \end{pmatrix}.$$

We conclude that the class of completely positive (nonnegative, doubly nonnegative) matrices is neither inverse–closed nor SC–closed. Thus, none of the inverse classes is hereditary, but, curiously, each inverse class is SC–closed.

• Stable and positive stable matrices

A square complex matrix is *positive stable* (stable) if all of its eigenvalues lie in the open right (left) complex half-plane, denoted PST (ST). Each class, stable and positive stable, is inverse–closed but neither is hereditary nor SC–closed, as seen by simple examples, such as

$$A = \begin{pmatrix} -1 & 4 \\ -1 & 3 \end{pmatrix}.$$

• Fischer matrices; Koteljanski matrices

A P–matrix $A = (a_{ij})$ is said to be a *Fischer matrix* if it satisfies Fischer's inequality, i.e.,

$$\det A[\alpha \cup \beta] \leq \det A[\alpha] \det A[\beta]$$

for all $\alpha, \beta \subseteq N$ in which $\alpha \cap \beta$ is empty and to be a *Koteljanski matrix* if it satisfies Koteljanski's inequality, i.e.,

$$\det A[\alpha \cup \beta] \det A[\alpha \cap \beta] \leq \det A[\alpha] \det A[\beta]$$

for all $\alpha, \beta \subseteq N$. Here, $\det A[\phi]$ is taken to be 1. It follows from their definitions that each of the classes, Fischer and Koteljanski, are hereditary. Also, upon applying Jacobi's identity to the defining inequality for Koteljanski matrices, one finds that Koteljanski matrices are inverse–closed and, hence, SC–closed. It is easily checked that

$$A = \begin{pmatrix} 4 & 1 & 2 \\ 2 & 1 & 1 \\ 5 & 3 & 4 \end{pmatrix}$$

is a Fischer matrix. By inspection of

$$A^{-1} = \frac{1}{3} \begin{pmatrix} 1 & 2 & -1 \\ -3 & 6 & 0 \\ 1 & -7 & 2 \end{pmatrix},$$

one sees that the determinant of $A^{-1}[\{1, 2\}]$ is greater than the product of its diagonal entries. So Fischer matrices are not inverse–closed. Moreover, since

$$A/a_{22} = \begin{pmatrix} 2 & 1 \\ -1 & 1 \end{pmatrix}$$

is not Fischer, Fischer matrices are not SC-closed either.

• The s, t section of positive definite matrices

We also mention the s,t section of the PD-matrices. Let $0 < s < t$ and define $PD_{[s,t]} = \{ A^* = A : sI \leq A \leq tI \}$. We mean to include the possibility of $t = \infty$, making $[s,t] = [s,\infty]$ semi-infinite. The following are simple exercises (using Chapters 4 and 7 of [228]):

$0 < s' \leq s < t \leq t' \Rightarrow PD_{[s,t]} \subseteq PD_{[s',t']}$;

$PD_{[s,t]}$ is hereditary; and

$PD_{[s,t]}^{-1} = PD_{[\frac{1}{t},\frac{1}{s}]}$.

Because of Theorem 4.2, $A \in PD_{[s,t]}$ then implies $(A/A[\alpha])^{-1} \in PD_{[\frac{1}{t},\frac{1}{s}]}$ or $A/A[\alpha] \in PD_{[s,t]}$. Thus, $PD_{[s,t]}$ is both hereditary and SC–closed, though not generally inverse–closed.

Table for Subsection (d)

Class	SC Closure	Heredity	Inverse Closure	Inverse Class SC Closure	Inverse Class Heredity
TP	*	Y	*	Y	*
DN	N	Y	N	Y	N
CP	N	Y	N	Y	N
$CoPos$	N	Y	N	Y	N
PST	N	N	Y	N	N
ST	N	N	Y	N	N
NN	N	Y	N	Y	N
F	N	Y	N	Y	N
K	Y	Y	Y	Y	Y
$PD_{[s,t]}$	Y	Y	N	Y	Y

* Though in each case the entry is, in general, N, see the discussion about particular index sets and inverse structure.

e) Other Classes.

In this subsection, we collect several classes of interest that have not fit into prior categories.

• Distance matrices

A matrix $A = (a_{ij})$ is called a *distance* (*squared distance*) matrix if there exist points $p_1, p_2, \ldots, p_n$ in Euclidean n-space such that $a_{ij} = d(p_i, p_j)$ ($a_{ij} = [d(p_i, p_j)]^2$) in which $d(p_i, p_j)$ denotes the distance from p_i to p_j. It is clear from the definition that distance (squared distance) matrices are hereditary. We denote the distance (squared distance) matrices by $DIST$

($SQDIST$). Consider the squared distance matrix

$$A = \begin{pmatrix} 0 & 1 & 2 & 2 \\ 1 & 0 & 1 & 5 \\ 2 & 1 & 0 & 10 \\ 2 & 5 & 10 & 0 \end{pmatrix}$$

for the points $p_1 = (1, -1)$, $p_2 = (2, -1)$, $p_3 = (2, -2)$, and $p_4 = (1, 1)$. Since

$$A/A[\{1, 2\}] = \begin{pmatrix} -4 & -2 \\ -2 & -20 \end{pmatrix},$$

one sees that squared distance matrices are not SC–closed and, by Theorem 4.2, are not inverse–closed either. By considering the distance matrix for these four points, one can show distance matrices are neither SC–closed nor inverse–closed as well.

• Sign nonsingular matrices

A real square matrix A is called *sign nonsingular* (SNS) if any matrix with the same sign pattern $(+, -, 0)$ as A is nonsingular, i.e., if $\det(P \circ A) \neq 0$, whenever P is a positive matrix, in which $\circ$ denotes the entry-wise or Hadamard product of matrices. To be consistent with our PN requirement, we suppose that the main diagonal is totally nonzero (which may always be arranged via permutation equivalence) and, for convenience, that the diagonal is positive, which may be assumed without loss of generality via a benign multiplication by a signature matrix (diagonal matrix of ± 1's). We call such SNS matrices "centered" (denoted $CSNS$) and, in particular, "positively centered". As each principal submatrix of a centered SNS matrix is SNS, the centered ones are necessarily PN; in particular, the positively centered ones are P–matrices. Thus, the centered SNS class is hereditary, so that the inverse class is SC–closed. What about SC–closure of centered SNS matrices? This is not the case, as shown by the following example. Let

$$A = \begin{pmatrix} 1 & 1 & 1 \\ -1 & 1 & 2 \\ 0 & -1 & 1 \end{pmatrix}$$

which is SNS, as all nonzero terms in the determinant are positive (and there are some). However,

$$A/A[\{2\}] = \begin{pmatrix} 2 & -1 \\ -1 & 3 \end{pmatrix},$$

which is not SNS, as shown via Hadamard product with

$$P = \begin{pmatrix} 1 & 6 \\ 1 & 1 \end{pmatrix}.$$

As the centered SNS matrices are permutationally similarity invariant, they are, then, not α SC–closed for any α. For $n \geq 3$, it is easily seen by example that centered SNS matrices are not inverse–closed.

• **Scalable matrices**

We call a square real matrix A *scalable*, denoted SCL, if there exist $D, E \in \mathcal{D}_+$ such that DAE has constant, nonzero line (row and column) sums. It is straightforward that the scalable matrices are inverse–closed. However, as the following example shows, the scalable matrices are not hereditary, and, thus, not SC–closed. Let

$$A = \begin{pmatrix} 1 & 0 & 1 \\ 1 & 1 & 0 \\ 0 & 1 & 1 \end{pmatrix}.$$

Clearly, A is scalable (use $D = E = I$), but $A[\{1, 2\}]$ is not.

Table for Subsection (e)

Class	SC Closure	Heredity	Inverse Closure	Inverse Class SC Closure	Inverse Class Heredity
$DIST$	N	Y	N	Y	N
$SQDIST$	N	Y	N	Y	N
$CSNS$	N	Y	N	Y	N
SCL	N	N	Y	N	N

4.3 Singular principal minors

Thus far, we have assumed that each matrix encountered was PN. This is convenient, as it implies not only that the matrix is invertible, so that the formula in Theorem 4.2 may be used, but also that the Schur complement with respect to any principal submatrix may be formed via the standard definition. It need not always be that the PN property is present. This raises two natural questions for a non-PN matrix A:

(1) What if the principal submatrix, $A[\alpha]$, with respect to which the Schur complement is being taken is nonsingular, but other principal minors, including possibly det A, are zero?

(2) What if $A[\alpha]$, with respect to which the Schur complement is being taken, is itself singular?

The question (1) often arises when the class $\mathcal{C}$ is being broadened somewhat to a kind of closure that allows some principal minors to vanish. For example, this is virtually always the case in unrestricted versions of the structured classes in subsection 4.2(a). But it also happens in many other situations, for example if the PD matrices are broadened to the positive semi-definite (PSD) matrices. Fortunately, in such cases, it generally happens that a continuity argument restores any positive conclusions about SC–closure to the broadened class.

An example is the following. Suppose that the P–matrices are broadened to the P_0– matrices, those for which all principal minors are nonnegative. If we form the Schur complement with respect to $A[\alpha]$ in a P_0–matrix A, is $A/A[\alpha]$ P_0 when $A[\alpha]$ is nonsingular? Supposing without loss of generality that $A[\alpha] = A_{11}$, this may be analyzed for

$$A = \begin{pmatrix} A_{11} & A_{12} \\ A_{21} & A_{22} \end{pmatrix}$$

as follows. If we replace A_{22} by $A'_{22} = A_{22} + \epsilon I$, $\epsilon > 0$, to give A', it may be verified that A'_{22} is P and that A' is invertible. By Jacobi's identity, A'^{-1} is P_0 and $(A'/A_{11})^{-1}$, and, thus, A'/A_{11} are P_0. But,

$$A'/A_{11} = (A/A_{11}) + \epsilon I$$

so that

$$\lim_{\epsilon \to 0} A'/A_{11} = A/A_{11}$$

and a limit of P_0–matrices is P_0. Thus, "SC–closure" holds for P_0, as long as $A[\alpha]$ is nonsingular.

Addressing question (2) is necessarily more subtle, as it requires a more general definition of Schur complement to start. Somehow, $A[\alpha]^{-1}$ must be replaced in the definition of Schur complement, and, somehow, what is adopted must mesh with the nonsingular case. There seems to be no universal definition that works well in most cases, though the Moore-Penrose generalized inverse of $A[\alpha]$ (in place of $A[\alpha]^{-1}$) is a natural candidate and has been used. For the assessment of SC–closure of the PSD matrices, we offer here a reasonably simple approach that requires no generalized inverse.

Let A be PSD and α an index. We give a natural definition of $A/A[\alpha]$ for which SC–closure of the PSD matrices may be verified. Heredity in the PSD case is well known (and straightforward), but Corollary 4.3 cannot be used to conclude SC–closure, as A will not be invertible if $A[\alpha]$ is not. Some analog of Theorem 4.4 would be welcome in the general, possibly singular case. When $A[\alpha]$, etc is singular, our definition proceeds in two stages to simulate the important fact, Theorem 4.2, about Schur complements .

First, if $A[\alpha] = 0$, we note that $A[\alpha^c, \alpha]$ and $A[\alpha, \alpha^c]$ must be 0. Then, we take $A/A[\alpha] \equiv A[\alpha^c]$. This seems an obvious candidate in any case, is consistent with using a generalized inverse in place of $A[\alpha]^{-1}$ and would be the case if $A[\alpha]$ were nonsingular and $A[\alpha^c, \alpha]$ (and hence $A[\alpha, \alpha^c]$ since A is symmetric) were 0. Second, if $A[\alpha] \neq 0$ (but is singular), we note that there is a principal submatrix $A[\beta]$ of $A[\alpha]$ that is $\operatorname{rank} A[\alpha]$–by–$\operatorname{rank} A[\alpha]$ and nonsingular (and thus positive definite). There may be several candidates for β, but any one may be used with an unambiguous final result, as discussed later. Now, calculate $A/A[\beta]$ in the usual way, and notice that $A[\alpha]/A[\beta]$, a principal submatrix of $A/A[\beta]$, is 0. By the discussion relative to question (1), $A/A[\beta]$ is PSD and, thus, has 0 rows and columns corresponding to the indices $A[\alpha]/A[\beta]$. Now, to complete the definition of $A/A[\alpha]$, extract the principal submatrix of $A/A[\alpha]$ that is complementary to the 0 block, $A[\alpha]/A[\beta]$. Thus,

$$A/A[\alpha] = (A/A[\beta])[\alpha^c],$$

in case $A[\alpha] \neq 0$. Because $A/A[\beta]$ is PSD and PSD is hereditary, it follows that $A/A[\alpha]$ is PSD in any event, and PSD is SC–closed under the extended definition.

To see that the definition of $A/A[\alpha]$ is independent of the choice of β (subject to the requirement that $\beta \subseteq \alpha$ and $A[\beta]$ is maximal nonsingular in $A[\alpha]$), we will need to assume that, for some $\beta, \gamma \subseteq \alpha \subseteq N$, $A[\beta]$ and $A[\gamma]$ are each maximal nonsingular in $A[\alpha]$ and show that $A/A[\beta] = A/A[\gamma]$. We first consider the case in which $\beta \cap \gamma$ is empty. That is, consider the PSD matrix

$$A = \begin{pmatrix} A_{11} & A_{12} & A_{13} & A_{14} \\ A_{12}^* & A_{22} & A_{23} & A_{24} \\ A_{31}^* & A_{32}^* & A_{33} & A_{34} \\ A_{41}^* & A_{42}^* & A_{43}^* & A_{44} \end{pmatrix} = \begin{pmatrix} B_{11} & B_{12} \\ B_{12}^* & B_{22} \end{pmatrix}$$

in which

$$B_{11} = A[\alpha], \quad B_{22} = A_{44} = A[\alpha^c],$$
$$A_{11} = A[\beta], \quad A_{22} = A[\gamma], \quad A_{33} = A[\alpha - (\beta \cup \gamma)],$$

and

$$\operatorname{rank} B_{11} = \operatorname{rank} A_{11} = \operatorname{rank} A_{22} = |\beta| = |\gamma|$$

(so that A_{11} and A_{22} are nonsingular of the same order). Then, there exist matrices C, D, E such that

$$A_{12} = A_{11}C, \quad A_{13} = A_{11}D, \quad \text{and} \quad A_{14} = A_{11}E$$

or, equivalently,

$$C = A_{11}^{-1}A_{12}, \quad D = A_{11}^{-1}A_{13}, \quad \text{and} \quad E = A_{11}^{-1}A_{14}.$$

It follows from the invertibility of A_{11} that

$$A = \begin{pmatrix} A_{11} & A_{11}C & A_{11}D & A_{11}E \\ C^*A_{11} & C^*A_{11}C & C^*A_{11}D & C^*A_{11}E \\ D^*A_{11} & D^*A_{11}C & D^*A_{11}D & D^*A_{11}E \\ E^*A_{11} & E^*A_{11}C & E^*A_{11}D & A_{44} \end{pmatrix}$$

so that

$$A/A_{11} = \begin{pmatrix} 0 & 0 & 0 \\ 0 & 0 & 0 \\ 0 & 0 & A_{44} - E^*A_{11}E \end{pmatrix}$$

in which

$$\begin{aligned} A_{44} - E^*A_{11}E &= A_{44} - (A_{11}^{-1}A_{14})^*A_{11}(A_{11}^{-1}A_{14}) \\ &= A_{44} - A_{14}^*A_{11}^{-1}A_{14}. \end{aligned}$$

Similarly, for matrices

$$F = A_{22}^{-1}A_{12}^*, \quad G = A_{22}^{-1}A_{23}, \quad \text{and} \quad H = A_{22}^{-1}A_{24},$$

we have

$$A = \begin{pmatrix} F^*A_{22}F & F^*A_{22} & F^*A_{22}G & F^*A_{22}H \\ A_{22}F & A_{22} & A_{22}G & A_{22}H \\ G^*A_{22}F & G^*A_{22} & G^*A_{22}G & G^*A_{22}H \\ H^*A_{22}F & H^*A_{22} & H^*A_{22}G & A_{44} \end{pmatrix}$$

so that

$$A/A_{22} = \begin{pmatrix} 0 & 0 & 0 \\ 0 & 0 & 0 \\ 0 & 0 & A_{44} - H^*A_{22}H \end{pmatrix}$$

in which

$$\begin{aligned} A_{44} - H^*A_{22}H &= A_{44} - (A_{22}^{-1}A_{24})^*A_{22}(A_{22}^{-1}A_{24}) \\ &= A_{44} - A_{24}^*A_{22}^{-1}A_{24}. \end{aligned}$$

Also,

$$A_{24} = C^*A_{11}E = A_{12}^*A_{11}^{-1}A_{14}$$

and

$$A_{22} = C^*A_{11}C = A_{12}^*A_{11}^{-1}A_{12}.$$

Finally, noting that it follows from $A_{22} = C^*A_{11}C$ and $A_{12} = A_{11}C$ that C and, hence, A_{12} have full rank, we have

$$\begin{aligned} A_{24}^*A_{22}^{-1}A_{24} &= (A_{12}^*A_{11}^{-1}A_{14})^*(A_{12}^{-1}A_{11}(A_{12}^*)^{-1})(A_{12}^*A_{11}^{-1}A_{14}) \\ &= A_{14}^*A_{11}^{-1}A_{14}, \end{aligned}$$

establishing the result.

It follows from the quotient formula (Theorem 1.4) that the case in which $\beta \cap \gamma$ is nonempty reduces to the first case upon taking the Schur complement of $A[\beta\cap\gamma]$ in A. Hence, the definition of $A/A[\alpha]$ is independent of the choice of the maximal nonsingular submatrix $A[\beta]$ of $A[\alpha]$.

4.4 Authors' historical notes

Schur complement closure results often arise, sometimes in disguised form, in a variety of mathematical and applied mathematical contexts. (It is often difficult to publish an isolated fact out of context.) A number of the ones mentioned here are surely known, though we know of no derivation of them from a unified point of view (as here). Proper attribution of particular cases is fraught with difficulty. Often such a result is known to a particular community, without having been published or having been published only in a non-transparent way, and then is later published by an unsuspecting author. Or one person who knows the result may not see it as sufficient for publication, and later someone else does. For example, the author Johnson recalls from stimulating discussions, early in his career, that several closure results were known to Velvo Kahan either prior to much later publication by others, or that have not been published.

Chapter 5

Schur Complements and Matrix Inequalities: Operator-Theoretic Approach

5.0 Introduction

The purpose of this chapter is to study the Schur complements for positive semidefinite matrices from the standpoint of order relation and to produce various kinds of matrix inequalities.

For this purpose, we will take an operator theoretic approach. This means that a (complex) matrix is considered as a (continuous) linear operator (= map) on a finite dimensional Hilbert space, say $\mathcal{H}$, with inner product $\langle \cdot, \cdot \rangle$ and norm $\| \cdot \|$. When vectors are represented as (numerical) column vectors, the inner product and the norm are usually defined as

$$\langle x, y \rangle = y^* x \quad \text{and} \quad \|x\| = \sqrt{x^* x}.$$

Here y^* is the complex transpose of y, which is a (numerical) row vector.

As far as possible, we will present the results in basis free form fit to the case of a Hilbert space, so that matrix entries seldom appear on the surface.

In this introduction we prepare standard results concerning linear operators on a Hilbert space and point out the advantage of the finite dimensional situations (see [51, Chapter I]).

Recall that the *adjoint* A^* of a linear operator A on a Hilbert space $\mathcal{H}$ is defined as a linear operator satisfying the condition:

$$\langle Ax, y \rangle = \langle x, A^* y \rangle \qquad (x, y \in \mathcal{H}).$$

When $A = A^*$, it is called *selfadjoint* or, in the case of matrices, *Hermitian*. Selfadjointness is characterized by the requirement that $\langle Ax, x \rangle$ is real for all $x \in \mathcal{H}$. Notice that coincidence of A and B follows from the requirement that $\langle Ax, x\rangle = \langle Bx, x\rangle$ for all x.

The order relation $A \geq B$ (resp. $A > B$) for two linear operators A, B means that both A and B are selfadjoint and $A - B$ is *positive semidefinite* (resp. *positive definite*). In particular, $A \geq 0$ (resp. $A > 0$) means that A is positive semidefinite (resp. positive definite). Here positive semidefiniteness is defined as

$$A \geq 0 \quad \Longleftrightarrow \quad \langle Ax, x\rangle \geq 0 \quad (x \in \mathcal{H}), \tag{5.0.1}$$

and consequently

$$A \geq B \quad \Longleftrightarrow \quad \langle Ax, x\rangle \geq \langle Bx, x\rangle \quad (x \in \mathcal{H}). \tag{5.0.2}$$

The strict order relation $A > 0$ (resp. $A > B$) is defined with $>$ for $x \neq 0$ in (5.0.1) (resp. (5.0.2)) . For the case of matrices positive definiteness is also equivalent to positive semidefiniteness with invertibility.

An important tool for our approach is the existence of a *square root* for any positive semidefinite operator A on a Hilbert space. More exactly there is uniquely a positive semidefinite operator, denoted by $A^{1/2}$, whose square coincides with A.

The order relation $A \geq B$ (≥ 0) is closely related to the range inclusion relations for A and B. We will use $\operatorname{ran}(A)$ and $\ker(A)$ to denote the *range* and the *kernel* of A, respectively;

$$\operatorname{ran}(A) \equiv \{Ax : x \in \mathcal{H}\} \quad \text{and} \quad \ker(A) \equiv \{x : Ax = 0\}.$$

Another important tool is the orthogonal decomposition theorem for a Hilbert space $\mathcal{H}$. More exactly, for any (closed) subspace $\mathcal{M}$ of a Hilbert space $\mathcal{H}$, define its *orthcomplement* $\mathcal{M}^\perp$ as

$$\mathcal{M}^\perp \equiv \{z : \langle y, z\rangle = 0 \ \forall \ y \in \mathcal{M}\}.$$

Then the orthogonal decomposition theorem says that

$$\mathcal{H} = \mathcal{M} + \mathcal{M}^\perp.$$

An immediate consequence is that each vector $x \in \mathcal{H}$ is uniquely written as $x = y + z$ with $y \in \mathcal{M}, z \in \mathcal{M}^\perp$. The correspondence $x \longmapsto y$ defines a linear operator, denoted by $P_{\mathcal{M}}$ and called the *orthoprojection* to the subspace $\mathcal{M}$. It has the following properties:

$$0 \leq P_{\mathcal{M}} \leq I, \ P_{\mathcal{M}}^2 = P_{\mathcal{M}} \quad \text{and} \quad \operatorname{ran}(P_{\mathcal{M}}) = \mathcal{M}, \ \ker(P_{\mathcal{M}}) = \mathcal{M}^\perp,$$

where I denotes the identity operator. Notices that a linear operator P coincides with the orthoprojection $P_{\mathcal{M}}$ if and only if it is selfadjoint, idempotent, that is, $P^2 = P$, and $\operatorname{ran}(P) = \mathcal{M}$.

For a vector x and a (closed) subspace $\mathcal{M}$ the norm $\|P_{\mathcal{M}}x\|$ has the following characterizations:

$$\|P_{\mathcal{M}}x\|^2 = \sup_{0 \neq y \in \mathcal{M}} \frac{|\langle x, y\rangle|^2}{\|y\|^2} = \inf_{y \in \mathcal{M}^\perp} \|x - y\|^2. \tag{5.0.3}$$

In a Hilbert space $\mathcal{H}$ the orthocomplement of $\operatorname{ran}(A)$ coincides with $\ker(A^*)$ while that of $\ker(A)$ coincides not with $\operatorname{ran}(A^*)$ but with its closure $\operatorname{ran}(A^*)^-$:

$$\operatorname{ran}(A)^\perp = \ker(A^*) \quad \text{and} \quad \ker(A)^\perp = \operatorname{ran}(A^*)^-. \tag{5.0.4}$$

For any positive semidefinite operator A on a Hilbert space the identity

$$\ker(A) = \ker(A^{1/2})$$

is always true. Though the inclusion $\operatorname{ran}(A) \subset \operatorname{ran}(A^{1/2})$ is always valid, the equality sign does not occur in general. By (5.0.4) the orthocomplement of $\ker(A)$ coincides not with $\operatorname{ran}(A)$ but with its closure $\operatorname{ran}(A)^-$. Therefore what is derived from the above identity is

$$\operatorname{ran}(A)^- = \operatorname{ran}(A^{1/2})^-,$$

that is, $\operatorname{ran}(A)$ is dense in $\operatorname{ran}(A^{1/2})$. It is easy to see that

$$\operatorname{ran}(A) = \operatorname{ran}(A^{1/2}) \quad \Longleftrightarrow \quad \operatorname{ran}(A) \ \text{ is closed}$$

and that the closedness of $\operatorname{ran}(A)$ is equivalent to that of $\operatorname{ran}(A^{1/2})$.

It is known that for positive semidefinite operators A, B the range inclusion relation

$$\operatorname{ran}(A) \supset \operatorname{ran}(B)$$

is equivalent to the existence of a linear operator C such that $AC = B$. It is further equivalent to the existence of $\gamma > 0$ such that $\gamma A^2 \ \geq \ B^2$.

From now on, let us consider only the case of a finite dimensional Hilbert space $\mathcal{H}$, and use the word "*matrix*" in place of "linear operator". Then since every subspace is closed, we can say that for any positive semidefinite matrix A

$$\operatorname{ran}(A) = \operatorname{ran}(A^{1/2}).$$

Therefore for any positive semidefinite A, B

$$\operatorname{ran}(A) \supset \operatorname{ran}(B) \iff \gamma A \geq B \quad \exists\, \gamma > 0. \tag{5.0.5}$$

Notice that if $P_{\operatorname{ran}(A)}$ is the orthoprojection to the subspace $\operatorname{ran}(A)$ then $P_{\operatorname{ran}(A)}$ and A have the same range. Therefore it follows from (5.0.5) that

$$\gamma P_{\operatorname{ran}(A)} \geq A \geq \frac{1}{\gamma} P_{\operatorname{ran}(A)} \quad \exists\, \gamma > 0. \tag{5.0.6}$$

For positive semidefinite A, B, we use the notation $A \wedge B = 0$ to mean $\operatorname{ran}(A) \cap \operatorname{ran}(B) = \{0\}$. Then it is easy to see from (5.0.5) and (5.0.6) that

$$A \wedge B = 0 \iff \{X \geq 0 :\ X \leq A, B\} = \{0\}. \tag{5.0.7}$$

An advantage of the positive semidefiniteness of A is in the existence of its natural *Moore–Penrose inverse* $A^\dagger$. Since A is bijective on the subspace $\operatorname{ran}(A)$ and vanishes on $\ker(A)$, the Moore–Penrose inverse $A^\dagger$ is defined as $(A|_{\operatorname{ran}(A)})^{-1}$ on the subspace $\operatorname{ran}(A)$ and as 0 on $\ker(A)$. The Moore-Penrose inverse is again positive semidefinite and

$$(A^\dagger)^\dagger = A, \quad A^\dagger A = AA^\dagger = P_{\operatorname{ran}(A)} \quad \text{and} \quad (A^\dagger)^{1/2} = (A^{1/2})^\dagger. \tag{5.0.8}$$

With the help of the Cauchy-Schwarz inequality it is easy to see that when A is positive definite,

$$\langle A^{-1}x, x\rangle = \|(A^{-1})^{1/2}x\|^2 = \sup\left\{\frac{|\langle x, y\rangle|^2}{\langle Ay, y\rangle} : 0 \neq y \in \mathcal{H}\right\}.$$

By definition of the Moore-Penrose inverse, this is extended to the following form for any positive semidefinite matrix A:

$$\langle A^\dagger x, x\rangle = \|(A^\dagger)^{1/2}x\|^2 = \sup\left\{\frac{|\langle x, y\rangle|^2}{\langle Ay, y\rangle} : 0 \neq y \in \operatorname{ran}(A)\right\}. \tag{5.0.9}$$

5.1 Schur complement and orthoprojection

In this section we introduce the Schur complement of a positive semidefinite matrix relative to a subspace. We will show relationship between Schur complement and orthoprojection, and present various characterizations of the Schur complement. We note that some of our results have appeared in the previous chapters in different forms.

Let $\mathcal{H}$ be a finite dimensional Hilbert space, and let $\mathcal{M}$ be a subspace with orthocomplement $\mathcal{M}^\perp$. According to the orthogonal decomposition

$$\mathcal{H} = \mathcal{M} + \mathcal{M}^\perp,$$

every matrix A is written in a block-form

$$A = \begin{pmatrix} A_{11} & A_{12} \\ A_{21} & A_{22} \end{pmatrix},$$

where A_{11} (resp. A_{22}) is considered as a linear operator on $\mathcal{M}$ (resp. $\mathcal{M}^\perp$) while A_{12} (resp. A_{21}) is that from $\mathcal{M}^\perp$ (resp. $\mathcal{M}$) to $\mathcal{M}$ (resp. $\mathcal{M}^\perp$). If A_{22} is invertible (on $\mathcal{M}^\perp$), the *Schur complement* of A relative to A_{22} (or even to the subspace $\mathcal{M}^\perp$), denoted by $A/\mathcal{M}^\perp$, is defined as (see Chapter 1)

$$A/\mathcal{M}^\perp \equiv A_{11} - A_{12}A_{22}^{-1}A_{21}.$$

By definition the Schur complement is a linear operator on $\mathcal{M}$, that is, a matrix of smaller size. But from our standpoint, it is often useful to extend $A/\mathcal{M}^\perp$ to a linear operator on the whole space $\mathcal{H}$ or a matrix of full size, denoted by $[\mathcal{M}]A$, as

$$[\mathcal{M}]A \equiv \begin{pmatrix} A/\mathcal{M}^\perp & 0 \\ 0 & 0 \end{pmatrix} = \begin{pmatrix} A_{11} - A_{12}A_{22}^{-1}A_{21} & 0 \\ 0 & 0 \end{pmatrix}.$$

When A is positive semidefinite, A_{22} is positive semidefinite. Therefore with use of its Moore-Penrose inverse $A_{22}^\dagger$ in place of the inverse, we will define the Schur complement, even when A_{22} is not invertible, as

$$A/\mathcal{M}^\perp \equiv A_{11} - A_{12}A_{22}^\dagger A_{21}, \tag{5.1.10}$$

and the associated linear operator $[\mathcal{M}]A$ on $\mathcal{H}$ or a matrix of full size as

$$[\mathcal{M}]A \equiv \begin{pmatrix} A_{11} - A_{12}A_{22}^\dagger A_{21} & 0 \\ 0 & 0 \end{pmatrix}. \tag{5.1.11}$$

We shall use the term "*Schur complement*" also for $[\mathcal{M}]A$.

Anderson [10] and Anderson-Trapp [13] called $[\mathcal{M}]A$ in (5.1.11) the *shorted operator* in connection with multiports electric network theory.

To get a geometric meaning of the Schur complement (5.1.11), let us introduce a (positive semidefinite) inner product and the associated seminorm, induced by a positive semidefinite matrix A, as

$$\langle x, y\rangle_A \equiv \langle Ax, y\rangle \quad \text{and} \quad ||x||_A \equiv \sqrt{\langle x, x\rangle_A}. \tag{5.1.12}$$

The space $\mathcal{H}$ equipped with this inner product will be denoted by $\mathcal{H}_A$. Notice that when A is not positive definite the space $\mathcal{H}_A$ is merely a *pre-Hilbert* space in the sense that $\|x\|_A = 0$ does not imply $x = 0$. Therefore everything is determined modulo the subspace $\ker(A)$.

A word "A–$\cdots$" will be used with respect to this inner product. For instance

$$x, y \quad A\text{-orthogonal} \quad \Longleftrightarrow \quad \langle x, y\rangle_A = 0.$$

First notice that when A is positive semidefinite then A_{22} is positive semidefinite, $A_{12}^* = A_{21}$ and $\mathrm{ran}(A_{21}) \subset \mathrm{ran}(A_{22})$. Then by (5.0.9) and $\mathrm{ran}(A_{21}) \subset \mathrm{ran}(A_{22})$, for any $x_1 \in \mathcal{M}$ we have, with convention $0/0 = 0$,

$$\begin{aligned}
\langle A_{12}A_{22}^{\dagger}A_{21}x_1, x_1\rangle &= \langle A_{22}^{\dagger}A_{21}x_1, A_{21}x_1\rangle \\
&= \sup\left\{\frac{|\langle A_{21}x_1, y_2\rangle|^2}{\langle A_{22}y_2, y_2\rangle} \;:\; y_2 \in \mathrm{ran}(A_{22})\right\} \\
&= \sup\left\{\frac{|\langle A_{21}x_1, y_2\rangle|^2}{\langle A_{22}y_2, y_2\rangle} \;:\; y_2 \in \mathcal{M}^{\perp}\right\} \\
&= \sup\left\{\frac{|\langle x_1, y_2\rangle_A|^2}{||y_2||_A^2} \;:\; y_2 \in \mathcal{M}^{\perp}\right\}.
\end{aligned}$$

Therefore it follows from (5.0.3) that, denoting by $Q_{\mathcal{M}^{\perp}}$ the A-orthoprojection to the subspace $\mathcal{M}^{\perp}$ in the pre-Hilbert space $\mathcal{H}_A$,

$$\sup\left\{\frac{|\langle x_1, y_2\rangle_A|^2}{||y_2||_A^2} \;;\; y_2 \in \mathcal{M}^{\perp}\right\} = ||Q_{\mathcal{M}^{\perp}}x_1||_A^2 = \langle Q_{\mathcal{M}^{\perp}}x_1, x_1\rangle_A,$$

and hence

$$\begin{aligned}
\langle (A/\mathcal{M}^{\perp})x_1, x_1\rangle &= \langle x_1, x_1\rangle_A - \langle Q_{\mathcal{M}^{\perp}}x_1, x_1\rangle_A \\
&= \langle (I - Q_{\mathcal{M}^{\perp}})x_1, x_1\rangle_A.
\end{aligned}$$

Since, for a vector $x = x_1 + x_2$ with $x_1 \in \mathcal{M},\ x_2 \in \mathcal{M}^{\perp}$,

$$\begin{aligned}
\langle ([\mathcal{M}]A)x, x\rangle &= \langle (A/\mathcal{M}^{\perp})x_1, x_1\rangle \\
&= \langle (I - Q_{\mathcal{M}^{\perp}})x_1, x_1\rangle_A \\
&= \langle (I - Q_{\mathcal{M}^{\perp}})x, x\rangle_A,
\end{aligned}$$

and $I - Q_{\mathcal{M}^{\perp}}$ is the A-orthoprojection to the A-orthocomplement of the subspace $\mathcal{M}^{\perp}$ in the pre-Hilbert space $\mathcal{H}_A$, the following relation is proved.

Theorem 5.1 *For any positive semidefinite matrix A and any subspace $\mathcal{M} \subset \mathcal{H}$ the quantity $\langle ([\mathcal{M}]A)x, x\rangle$ coincides with $||Q_{\tilde{\mathcal{M}}}x||_A^2$, where $Q_{\tilde{\mathcal{M}}}$ is* *the A-orthoprojection to the A-orthocomplement $\tilde{\mathcal{M}}$ of the subspace $\mathcal{M}^\perp$ in the pre-Hilbert space $\mathcal{H}_A$;*

$$\langle([\mathcal{M}]A)x, x\rangle = ||Q_{\tilde{\mathcal{M}}}x||_A^2 = \inf\left\{||x-y||_A^2 \ : \ y \in \mathcal{M}^\perp\right\} \quad (x \in \mathcal{H}),$$

that is,

$$[\mathcal{M}]A = Q_{\tilde{\mathcal{M}}}^* A Q_{\tilde{\mathcal{M}}} = AQ_{\tilde{\mathcal{M}}}.$$

The matrix

$$\begin{pmatrix} A_{12}A_{22}^\dagger A_{21} & A_{12} \\ A_{21} & A_{22} \end{pmatrix} = A - [\mathcal{M}]A$$

should have the properties complementary to those of $[\mathcal{M}]A$.

Theorem 5.2 *For any positive semidefinite matrix A and subspace $\mathcal{M} \subset \mathcal{H}$*

$$\langle(A - [\mathcal{M}]A)x, x\rangle = ||Q_{\mathcal{M}^\perp}x||_A^2,$$

that is,

$$A - [\mathcal{M}]A = Q_{\mathcal{M}^\perp}^* A Q_{\mathcal{M}^\perp} = AQ_{\mathcal{M}^\perp},$$

where $Q_{\mathcal{M}^\perp}$ is the A-orthoprojection to $\mathcal{M}^\perp$ in the pre-Hilbert space $\mathcal{H}_A$.

This is merely a reformulation of Theorem 5.1.

Next we present the explicit forms of the subspace $\tilde{\mathcal{M}}$ and the A-orthoprojections $Q_{\mathcal{M}^\perp}$ and $Q_{\tilde{\mathcal{M}}}$. Since, by the definition,

$$x \in \tilde{\mathcal{M}} \iff Ax \in (\mathcal{M}^\perp)^\perp$$

and $(\mathcal{M}^\perp)^\perp$ coincides with $\mathcal{M}$, we have

$$\tilde{\mathcal{M}} = \{x : Ax \in \mathcal{M}\}.$$

Now we can give the representations

$$Q_{\mathcal{M}^\perp} = \begin{pmatrix} 0 & 0 \\ A_{22}^\dagger A_{21} & I_{\mathcal{M}^\perp} \end{pmatrix} \quad \text{and} \quad Q_{\tilde{\mathcal{M}}} = \begin{pmatrix} I_{\mathcal{M}} & 0 \\ -A_{22}^\dagger A_{21} & 0 \end{pmatrix},$$

where $I_{\mathcal{M}}$ and $I_{\mathcal{M}^\perp}$ are the identity operators on $\mathcal{M}$ and $\mathcal{M}^\perp$, respectively.

In fact, the matrix

$$Q \equiv \begin{pmatrix} 0 & 0 \\ A_{22}^\dagger A_{21} & I_{\mathcal{M}^\perp} \end{pmatrix}$$

is idempotent with $\mathrm{ran}(Q) \subset \mathcal{M}^{\perp}$ and $Qy = y$ ($y \in \mathcal{M}^{\perp}$). It remains to show that Q is A-Hermitian. Since $\mathrm{ran}(A_{21}) \subset \mathrm{ran}(A_{22})$, this is seen by (5.0.8) as follows

$$AQ = \begin{pmatrix} A_{12}A_{22}^{\dagger}A_{21} & A_{12} \\ A_{21} & A_{22} \end{pmatrix} = Q^*A.$$

Recall by (5.0.5) that for any positive semidefinite matrix X and any subspace $\mathcal{M} \subset \mathcal{H}$, the inclusion relation $\mathrm{ran}(X) \subset \mathcal{M}$ is equivalent to the existence $\gamma = \gamma(X) > 0$ such that $X \leq \gamma P_{\mathcal{M}}$, where $P_{\mathcal{M}}$ is the orthoprojection to the subspace $\mathcal{M}$.

The following is the central result of this section.

Theorem 5.3 *For any positive semidefinite matrix A and any subspace $\mathcal{M} \subset \mathcal{H}$ the Schur complement $[\mathcal{M}]A$ satisfies the condition*

$$A \geq [\mathcal{M}]A \geq 0,$$

and more precisely

$$[\mathcal{M}]A = \max\{X : A \geq X \geq 0 \text{ and } X \leq \gamma P_{\mathcal{M}}, \ \exists\, \gamma = \gamma(X) > 0\}.$$

Proof. The inequality in question is immediate from the identity in Theorem 5.1. Now $[\mathcal{M}]A$ satisfies the condition required for X in the right hand side of the above identity. Take B such that $A \geq B \geq 0$ and $\mathrm{ran}(B) \subset \mathcal{M}$. Then since $By = 0$ ($y \in \mathcal{M}^{\perp}$), again by Theorem 5.1,

$$\begin{aligned} \langle([\mathcal{M}]A)x, x\rangle &= \inf\{\|x-y\|_A^2 \ : \ y \in \mathcal{M}^{\perp}\} \\ &\geq \inf\{\|x-y\|_B^2 \ : \ y \in \mathcal{M}^{\perp}\} \\ &= \langle Bx, x\rangle, \end{aligned}$$

so that $[\mathcal{M}]A \geq B$, which proves the maximum property for $[\mathcal{M}]A$. ∎

Theorem 5.4 *For any positive semidefinite matrix $A = \begin{pmatrix} A_{11} & A_{12} \\ A_{21} & A_{22} \end{pmatrix}$ and any subspace $\mathcal{M} \subset \mathcal{H}$,*

$$A_{21}A_{22}^{\dagger}A_{21} = \min\left\{Y : \begin{pmatrix} Y & A_{12} \\ A_{21} & A_{22} \end{pmatrix} \geq 0\right\},$$

where $\begin{pmatrix} Y & A_{12} \\ A_{21} & A_{22} \end{pmatrix}$ is the block-form according to the decomposition $\mathcal{H} = \mathcal{M} + \mathcal{M}^{\perp}$.

This is just a reformulation of Theorem 5.3.

It is obvious from definition that for any subspace $\mathcal{M} \subset \mathcal{H}$ and the identity matrix $I = I_{\mathcal{H}}$

$$[\mathcal{M}]I = P_{\mathcal{M}}.$$

This can be extended to the case of an orthoprojection as follows.

Theorem 5.5 *For any subspaces $\mathcal{M}, \mathcal{N} \subset \mathcal{H}$*

$$[\mathcal{M}]P_{\mathcal{N}} = [\mathcal{N}]P_{\mathcal{M}} = P_{\mathcal{M}\cap\mathcal{N}}.$$

Proof. Since

$$0 \leq P_{\mathcal{M}\cap\mathcal{N}} \leq P_{\mathcal{N}} \quad \text{and} \quad \mathrm{ran}(P_{\mathcal{M}\cap\mathcal{N}}) \subset \mathcal{M},$$

by Theorem 5.3

$$P_{\mathcal{M}\cap\mathcal{N}} \leq [\mathcal{M}]P_{\mathcal{N}}.$$

On the other hand, since

$$[\mathcal{M}]P_{\mathcal{N}} \leq P_{\mathcal{N}} \leq I \quad \text{and} \quad \mathrm{ran}([\mathcal{M}]P_{\mathcal{N}}) \subset \mathcal{M}\cap\mathcal{N},$$

we have

$$[\mathcal{M}]P_{\mathcal{N}} \leq [\mathcal{M}\cap\mathcal{N}]I = P_{\mathcal{M}\cap\mathcal{N}}.$$

These prove one of the desired identities. The other one is proved by interchanging the roles of $\mathcal{M}$ and $\mathcal{N}$. ∎

The range of the Schur complement $[\mathcal{M}]A$ is described in terms of $\mathcal{M}$ and $\mathrm{ran}(A)$ as seen in the following theorem.

Theorem 5.6 *For any positive semidefinite matrix A and subspace $\mathcal{M} \subset \mathcal{H}$*

$$\mathrm{ran}([\mathcal{M}]A) = \mathcal{M} \cap \mathrm{ran}(A).$$

Proof. The left hand side of the above is included in the right hand side by (5.0.5) and Theorem 5.3. To see the converse, take any nonzero vector $a \in \mathcal{M} \cap \mathrm{ran}(A)$. Then since the range of the rank one matrix aa^* is contained in $\mathcal{M} \cap \mathrm{ran}(A)$, again by (5.0.5) and Theorem 5.3, we obtain $[\mathcal{M}]A \geq \gamma aa^*$ for some $\gamma > 0$, which implies $a \in \mathrm{ran}([\mathcal{M}]A)$. Since a is arbitrary, this proves that the right hand side is included in the left hand side, proving the identity. ∎

For any positive semidefinite matrix X and any subspace $\mathcal{M} \subset \mathcal{H}$ the formula (5.0.7) shows that

$$X \wedge P_{\mathcal{M}} = 0 \iff \mathcal{M} \cap \mathrm{ran}(X) = \{0\}.$$

On this basis we can state another characteristic property of the Schur complement $[\mathcal{M}]A$.

Theorem 5.7 *For any positive semidefinite matrix A and subspace $\mathcal{M} \subset \mathcal{H}$*

$$(A - [\mathcal{M}]A) \wedge P_{\mathcal{M}} = 0.$$

Further the decomposition of A as

$$A = [\mathcal{M}]A + (A - [\mathcal{M}]A)$$

is a unique decomposition $A = X + Y$ such that $X, Y \geq 0$ and

$$X \leq \gamma P_{\mathcal{M}} \ \exists\, \gamma > 0 \quad \text{and} \quad Y \wedge P_{\mathcal{M}} = 0.$$

Proof. To see the first assertion, suppose by contradiction that

$$\mathcal{M} \cap \text{ran}(A - [\mathcal{M}]A) \neq \{0\}.$$

Then by (5.0.7) there is nonzero $B \geq 0$ such that

$$\text{ran}(B) \subset \mathcal{M} \quad \text{and} \quad B \leq A - [\mathcal{M}]A.$$

This implies that

$$0 \leq [\mathcal{M}]A + B \leq A \quad \text{and} \quad \text{ran}\,([\mathcal{M}]A + B) \subset \mathcal{M}.$$

Then by Theorem 5.3 we are led to a contradiction.

$$[\mathcal{M}]A + B \leq [\mathcal{M}]A.$$

The uniqueness in the second assertion can be seen as follows. For X, Y in the question, by Theorem 5.3, $X \leq [\mathcal{M}]A$ so that

$$0 \leq [\mathcal{M}]A - X \leq [\mathcal{M}]A.$$

On the other hand, since

$$[\mathcal{M}]A - X \leq A - X = Y,$$

it follows from the assumption that

$$\text{ran}\,([\mathcal{M}]A - X) \subset \mathcal{M} \cap \text{ran}(Y) = \{0\}.$$

This implies $[\mathcal{M}]A = X$. ∎

To close this section, we present an expression of the Schur complement $[\mathcal{M}]A$, corresponding to an identity in Theorem 5.1.

Theorem 5.8 *For any positive semidefinite matrix* A *and subspace* $\mathcal{M} \subset \mathcal{H}$

$$[\mathcal{M}]A = A^{1/2} P_{\mathcal{N}} A^{1/2},$$

where $P_{\mathcal{N}}$ *is the orthoprojection to the subspace* $\mathcal{N} \equiv \{x : A^{1/2}x \in \mathcal{M}\}$.

Proof. By Theorem 5.1,

$$\begin{aligned} \langle([\mathcal{M}]A)x, x\rangle &= \inf\{\|x - y\|_A^2 : y \in \mathcal{M}^\perp\} \\ &= \inf\{\langle A(x-y), x-y\rangle : y \in \mathcal{M}^\perp\} \\ &= \inf\{\|A^{1/2}x - A^{1/2}y\|^2 : y \in \mathcal{M}^\perp\} \\ &= \|P_{\tilde{\mathcal{N}}} A^{1/2}x\|^{1/2}, \end{aligned}$$

where

$$\tilde{\mathcal{N}} = \left\{A^{1/2}y : y \in \mathcal{M}^\perp\right\}^\perp.$$

Since $(\mathcal{M}^\perp)^\perp = \mathcal{M}$, we can see

$$\tilde{\mathcal{N}} = \left\{z : A^{1/2}z \in (\mathcal{M}^\perp)^\perp\right\} = \left\{z : A^{1/2}z \in \mathcal{M}\right\}.$$

Therefore $\tilde{\mathcal{N}}$ coincides with $\mathcal{N} = \{x : A^{1/2}x \in \mathcal{M}\}$ and we can conclude

$$\langle([\mathcal{M}]A)x, x\rangle = \|P_{\mathcal{N}} A^{1/2}x\|^2 = \langle(A^{1/2}P_{\mathcal{N}}A^{1/2})x, x\rangle \qquad (x \in \mathcal{H}),$$

which yields the identity in the assertion. ∎

We end the section by pointing out that the approach to the Schur complement as in Theorem 5.1 is presented here for the first time. But it was implicit in Theorem 5.3 established by M. G. Krein [269] in connection with an extension problem of Hermitian positive semidefinite forms and rediscovered by Anderson [10] and Anderson–Trapp [13] (as well by Ando) in connection with the shorted operator. See also Li–Mathias [280]. Theorem 5.6 was obtained by Anderson–Trapp [13] while Theorem 5.7 was mentioned in Ando [15] under a more general setting. Finally Theorem 5.8 is a restatement of a more general result for a Hilbert space operator by Pekarev [348] and Kosaki [264].

5.2 Properties of the map $A \longmapsto [\mathcal{M}]A$

Each subspace $\mathcal{M} \subset \mathcal{H}$ gives rise to a non-affine map $A \longmapsto [\mathcal{M}]A$ on the class of positive semidefinite matrices. In this section we investigate the properties of this map.

The following properties of this map are easily derived from the maximum property of $[\mathcal{M}]A$ in Theorem 5.3.

Theorem 5.9 *The map $A \longmapsto [\mathcal{M}]A$ has the following properties. Here $A, B, \ldots$ are positive semidefinite matrices and $\mathcal{M}$, $\mathcal{N}$ are subspaces of $\mathcal{H}$.*

(i) $[\mathcal{M}](\lambda A) = \lambda[\mathcal{M}]A \qquad (\lambda \geq 0)$;

(ii) $[\mathcal{M}](A+B) \geq [\mathcal{M}]A + [\mathcal{M}]B$;

(iii) $\mathcal{M} \supset \operatorname{ran}(A) \Longrightarrow [\mathcal{M}]A = A$;

(iv) $\mathcal{N} \supset \mathcal{M} \Longrightarrow [\mathcal{N}]A \geq [\mathcal{M}]A$;

(v) $A_n \downarrow A \Longrightarrow [\mathcal{M}]A_n \downarrow [\mathcal{M}]A$.

Here $A_n \downarrow A$ means that $A_1 \geq A_2 \geq \cdots$ and $\lim_{n\to\infty} A_n = A$.

Our basis free definition of Schur complements makes it possible to generalize the so-called Quotient Formula (see Theorem 1.4) for the case of positive semidefinite matrices.

Theorem 5.10 (Generalized Quotient Formula) *For any subspaces $\mathcal{M}$, $\mathcal{N} \subset \mathcal{H}$, the map $[\mathcal{M}]$ commutes with the map $[\mathcal{N}]$. More precisely*

$$[\mathcal{N}] \circ [\mathcal{M}] = [\mathcal{M}] \circ [\mathcal{N}] = [\mathcal{M} \cap \mathcal{N}],$$

that is,

$$[\mathcal{N}]\,([\mathcal{M}]A) = [\mathcal{M}]\,([\mathcal{N}]A) = [\mathcal{M} \cap \mathcal{N}]A \quad (A \geq 0).$$

Proof. Since by Theorem 5.3

$$[\mathcal{N}]\,([\mathcal{M}]A) \leq [\mathcal{M}]A \leq A$$

and by Theorem 5.6

$$\operatorname{ran}\,([\mathcal{N}]\,([\mathcal{M}]A)) = \mathcal{N} \cap \operatorname{ran}\,([\mathcal{M}]A) = \mathcal{N} \cap (\mathcal{M} \cap \operatorname{ran}(A)) \subset \mathcal{M} \cap \mathcal{N},$$

it follows again from Theorem 5.3

$$[\mathcal{N}]\,([\mathcal{M}]A) \leq [\mathcal{M} \cap \mathcal{N}]A.$$

On the other hand, by Theorem 5.9 (iv)

$$[\mathcal{M}\cap\mathcal{N}]A \leq [\mathcal{M}]A.$$

Since by Theorem 5.3

$$\operatorname{ran}([\mathcal{M}\cap\mathcal{N}]A) \subset \mathcal{M}\cap\mathcal{N} \subset \mathcal{N},$$

it follows again from Theorem 5.3 that

$$[\mathcal{M}\cap\mathcal{N}]A \leq [\mathcal{N}]\big([\mathcal{M}]A\big),$$

which proves

$$[\mathcal{N}]\circ[\mathcal{M}] = [\mathcal{M}\cap\mathcal{N}].$$

Finally interchange of the roles of $\mathcal{M}$ and $\mathcal{N}$ in the above argument yields the other identity. ∎

In our notation the classical quotient formula for Schur complements says that for a general matrix A and subspaces $\mathcal{M}\subset\mathcal{N}\subset\mathcal{H}$

$$A/(\mathcal{N}\cap\mathcal{M}^{\perp})^{\perp} \;=\; (A/\mathcal{N}^{\perp})/(\mathcal{N}\cap\mathcal{M}^{\perp})^{\perp}, \tag{5.2.13}$$

provided that the Schur complements in the expressions are well defined.

When A is positive semidefinite, this formula can be derived from Theorem 5.10. In fact, (5.2.13) means by definition

$$[\mathcal{N}\cap\mathcal{M}^{\perp}]A \;=\; [\mathcal{N}\cap\mathcal{M}^{\perp}]\,([\mathcal{N}]A).$$

Since $(\mathcal{N}\cap\mathcal{M}^{\perp})\cap\mathcal{N} = \mathcal{N}\cap\mathcal{M}^{\perp}$, the identity follows from Theorem 5.10.

Let A be a general matrix with the block-form

$$A \;=\; \begin{pmatrix} A_{11} & A_{12} \\ A_{21} & A_{22} \end{pmatrix}$$

according to the orthogonal decomposition $\mathcal{H} = \mathcal{M} + \mathcal{M}^{\perp}$. If A, A_{11} and A_{22} are all invertible, by the Schur determinant formula (Theorem 1.1)

$$\begin{aligned} \det(A) &= \det(A_{11})\det(A_{22} - A_{21}A_{11}^{-1}A_{12}) \\ &= \det(A_{22})\det(A_{11} - A_{12}A_{22}^{-1}A_{21}), \end{aligned}$$

so that both $A_{11} - A_{12}A_{22}^{-1}A_{21}$ and $A_{22} - A_{21}A_{11}^{-1}A_{12}$ are invertible. Furthermore it is well known that the inverse A^{-1} admits the following block-representation (Theorem 1.2)

$$A^{-1} = \begin{pmatrix} (A_{11} - A_{12}A_{22}^{-1}A_{21})^{-1} & -(A_{11} - A_{12}A_{22}^{-1}A_{21})^{-1}A_{12}A_{22}^{-1} \\ -A_{22}^{-1}A_{21}(A_{11} - A_{12}A_{22}^{-1}A_{21})^{-1} & (A_{22} - A_{21}A_{11}^{-1}A_{12})^{-1} \end{pmatrix}$$

$$= \begin{pmatrix} (A_{11} - A_{12}A_{22}^{-1}A_{21})^{-1} & -A_{11}^{-1}A_{12}(A_{22} - A_{21}A_{11}^{-1}A_{12})^{-1} \\ -(A_{22} - A_{21}A_{11}^{-1}A_{12})^{-1}A_{21}A_{11}^{-1} & (A_{22} - A_{21}A_{11}^{-1}A_{12})^{-1} \end{pmatrix}.$$

If A is positive definite, so are A_{11} and A_{22}; and this formula shows

$$([\mathcal{M}]A)^{\dagger} = P_{\mathcal{M}}A^{-1}P_{\mathcal{M}} \qquad (A > 0). \tag{5.2.14}$$

The identity (5.2.14) is useful to derive inequalities related to positive or negative powers of a positive semidefinite matrix and their Schur complements.

Though integral and fractional powers A^{α} of a positive semidefinite matrix A are defined in a natural way, we confine ourselves here to the observation of the most interesting cases: $\alpha = 2, 1/2$ and -1. In the course of the proof we use rather well known general inequalities ([51, Chap. V])

$$X \geq Y > 0 \implies X^{-1} \leq Y^{-1} \tag{5.2.15}$$

and

$$X \geq Y \geq 0 \implies X^{1/2} \geq Y^{1/2}. \tag{5.2.16}$$

Theorem 5.11 *Let $\mathcal{M}$ be a subspace of $\mathcal{H}$. Then*

(i) $[\mathcal{M}](A^2) \leq ([\mathcal{M}]A)^2 \qquad (A \geq 0)$;

(ii) $[\mathcal{M}](A^{1/2}) \geq ([\mathcal{M}]A)^{1/2} \quad (A \geq 0)$;

(iii) $[\mathcal{M}](A^{-1}) \leq ([\mathcal{M}]A)^{\dagger} \qquad (A > 0)$.

Proof. (i) Since $A_n \equiv A + \frac{1}{n}I$ and A_n^2 are invertible and they converge decreasingly to A and A^2 respectively as $n \to \infty$, by Theorem 5.9 (v),

$$([\mathcal{M}]A)^2 = \lim_{n\to\infty}([\mathcal{M}]A_n)^2 \quad \text{and} \quad [\mathcal{M}](A^2) = \lim_{n\to\infty}([\mathcal{M}](A_n^2).$$

So for the proof of (i) we may assume that A is invertible. By (5.2.14),

$$\begin{aligned}([\mathcal{M}](A^2))^\dagger &= P_{\mathcal{M}}(A^2)^{-1}P_{\mathcal{M}} \\ &= P_{\mathcal{M}}A^{-1}\cdot A^{-1}P_{\mathcal{M}} \\ &\geq (P_{\mathcal{M}}A^{-1}P_{\mathcal{M}})^2 \\ &= \{([\mathcal{M}]A)^\dagger\}^2 \\ &= \{([\mathcal{M}]A)^2\}^\dagger.\end{aligned}$$

Since both $[\mathcal{M}](A^2)$ and $([\mathcal{M}]A)^2$ are invertible on $\mathcal{M}$, we conclude from (5.2.15) that

$$[\mathcal{M}](A^2) = \{([\mathcal{M}](A^2))^\dagger\}^\dagger \leq \{\{([\mathcal{M}]A)^2\}^\dagger\}^\dagger = ([\mathcal{M}]A)^2.$$

(ii) Apply (i) to $A^{1/2}$ in place of A to get

$$[\mathcal{M}]A \leq \{[\mathcal{M}](A^{1/2})\}^2.$$

which implies (ii) by (5.2.16).

(iii) Apply (5.2.14) to A^{-1} in place of A to get

$$\{[\mathcal{M}](A^{-1})\}^\dagger = P_{\mathcal{M}}AP_{\mathcal{M}} \geq [\mathcal{M}]A.$$

Since both $[\mathcal{M}](A^{-1})$ and $[\mathcal{M}]A$ are invertible on $\mathcal{M}$, taking the Moore-Penrose inverses of both side we have (iii) by (5.2.15) as above. ∎

One may ask whether (5.2.14) is true for every positive semidefinite matrix A in the form

$$([\mathcal{M}]A)^\dagger = P_{\mathcal{M}}A^\dagger P_{\mathcal{M}}. \tag{5.2.17}$$

This is, however, not the case in general. For instance, let $A = P_{\mathcal{N}}$ be an orthoprojection which does not commutes with $P_{\mathcal{M}}$. Suppose that (5.2.17) holds for this A. Since by Theorem 5.5

$$([\mathcal{M}]A)^\dagger = (P_{\mathcal{M}\cap\mathcal{N}})^\dagger = P_{\mathcal{M}\cap\mathcal{N}} \quad \text{and} \quad P_{\mathcal{M}}P_{\mathcal{N}}^\dagger P_{\mathcal{M}} = P_{\mathcal{M}}P_{\mathcal{N}}P_{\mathcal{M}}$$

we have

$$P_{\mathcal{M}\cap\mathcal{N}} = P_{\mathcal{M}}P_{\mathcal{N}}P_{\mathcal{M}},$$

which leads to the commutativity of $P_{\mathcal{M}}$ and $P_{\mathcal{N}}$, a contradiction.

Let us present, without proof, the following characterization of a positive semidefinite matrix A and a subspace $\mathcal{M}$ for which (5.2.17) is valid. A proof is based on tedious, repeated uses of Theorem 5.3.

Theorem 5.12 *In order that the relation*

$$([\mathcal{M}]A)^{\dagger} = P_{\mathcal{M}} A^{\dagger} P_{\mathcal{M}}$$

hold for a positive semidefinite matrix A *and a subspace* $\mathcal{M} \subset \mathcal{H}$, *it is necessary and sufficient that*

$$\operatorname{ran}(P_{\mathcal{M}} A) \subset \operatorname{ran}(A).$$

Note that Theorem 5.9 is mentioned in Anderson–Trapp [13] while Theorem 5.10 was established by Ando [17].

5.3 Schur complement and parallel sum

In this section we will show that the Schur complement has intimate connection with an important operation of parallel addition for positive semidefinite matrices. The climax is the recapture of the Schur complement in terms of parallel addition.

For two positive definite matrices A and B, the operation dual to the usual addition $(A, B) \longmapsto A + B$ should be

$$(A, B) \longmapsto \{A^{-1} + B^{-1}\}^{-1}.$$

Following Anderson-Duffin [11] (see also Pekarev-Smulian [349]), we call this operation *parallel addition* and denote the *parallel sum* of A and B by

$$A : B \equiv \{A^{-1} + B^{-1}\}^{-1} \qquad (A, B > 0). \tag{5.3.18}$$

Notice that

$$\begin{aligned} \{A^{-1} + B^{-1}\}^{-1} &= A(A+B)^{-1}B \\ &= B(A+B)^{-1}A \\ &= A - A(A+B)^{-1}A \\ &= B - B(A+B)^{-1}B. \end{aligned}$$

Then considering the block-forms

$$\begin{pmatrix} A & A \\ A & A+B \end{pmatrix} \quad \text{and} \quad \begin{pmatrix} B & B \\ B & A+B \end{pmatrix}$$

as linear operators on the direct sum $\mathcal{H} \oplus \mathcal{H}$, the parallel sum $A : B$ for $A, B > 0$ is written as

$$A : B = \begin{pmatrix} A & A \\ A & A+B \end{pmatrix} / \mathcal{H}_1^{\perp} = \begin{pmatrix} B & B \\ B & A+B \end{pmatrix} / \mathcal{H}_1^{\perp}, \tag{5.3.19}$$

where $\mathcal{H}_1$ is the first summand of $\mathcal{H} \oplus \mathcal{H}$.

The formulas (5.3.18) and (5.3.19) are extended to the case of positive semidefinite matrices. Now the *parallel sum* $A : B$ for positive semidefinite matrices A, B is defined as

$$A : B \equiv A - A(A+B)^{\dagger}A = B - B(A+B)^{\dagger}B \qquad (A, B \geq 0). \quad (5.3.20)$$

Then (5.3.19) is valid. The properties of parallel sum can be derived from those of Schur complements.

Theorem 5.13 *For any positive semidefinite matrices A, B*

$$\langle (A : B)x, x\rangle = \inf\{\langle Ay, y\rangle + \langle Bz, z\rangle \ : \ x = y + z\} \qquad (x \in \mathcal{H}).$$

Proof. By (5.3.20) and Theorem 5.1

$$\begin{aligned}\langle (A : B)x, x\rangle &= \inf\left\{\left\langle \begin{pmatrix} A & A \\ A & A+B \end{pmatrix} \begin{pmatrix} x \\ -z \end{pmatrix}, \begin{pmatrix} x \\ -z \end{pmatrix}\right\rangle \ : \ z \in \mathcal{H}\right\} \\ &= \inf\{\langle A(x-z), x-z\rangle + \langle Bz, z\rangle \ : \ z \in \mathcal{H}\}.\end{aligned}$$

Replacing $x - z$ by y we arrive at the assertion. ∎

Theorem 5.14 *The parallel addition has the following properties. Here A, B, $C, \ldots$ are positive semidefinite matrices.*

(a) $(\alpha A) : (\beta A) = \frac{\alpha\beta}{\alpha+\beta} A \qquad (\alpha, \beta \geq 0)$;

(b) $A : B = B : A$;

(c) $(\lambda A) : (\lambda B) = \lambda(A : B) \qquad (\lambda \geq 0)$;

(d) $(A : B) : C = A : (B : C)$;

(e) $A,\ B \geq A : B \geq 0$;

(f) $A_1 \geq A_2,\ B_1 \geq B_2 \implies A_1 : B_1 \geq A_2 : B_2$.
More precisely, $A_n \downarrow A,\ B_n \downarrow B \implies A_n : B_n \downarrow A : B$;

(g) $(A_1 + A_2) : (B_1 + B_2) \geq A_1 : B_1 + A_2 : B_2$;

(h) $(S^*AS) : (S^*BS) = S^*(A : B)S$ *for all invertible* S.

Proof. (a) to (g) are immediate from Theorem 5.13. Finally (h) is obvious for invertible A, B by (5.3.18) while the general case follows by (f). ∎

Theorem 5.15 *For any positive semidefinite matrices* A, B

$$\operatorname{ran}(A:B) \;=\; \operatorname{ran}(A)\cap\operatorname{ran}(B).$$

Proof. By (5.3.19) and Theorem 5.6

$$\begin{aligned}\operatorname{ran}(A:B) &= \left\{x \,:\, \begin{pmatrix}x\\0\end{pmatrix} = \begin{pmatrix}A & A\\ A & A+B\end{pmatrix}\begin{pmatrix}y\\z\end{pmatrix}\ \exists\, y,z\right\}\\ &= \{x \,:\, x = A(y+z) = -Bz\ \exists\, y,z\}\\ &= \operatorname{ran}(A)\cap\operatorname{ran}(B).\ \blacksquare\end{aligned}$$

Theorem 5.16 *For any positive semidefinite matrices* A, B *and any subspace* $\mathcal{M}\subset\mathcal{H}$

$$[\mathcal{M}](A:B) \;\geq\; [\mathcal{M}]A:[\mathcal{M}]B.$$

Proof. Since by Theorem 5.3

$$A \geq [\mathcal{M}]A \quad\text{and}\quad B\geq[\mathcal{M}]B,$$

we have by Theorem 5.13 (f)

$$A:B \;\geq\; [\mathcal{M}]A:[\mathcal{M}]B.$$

Further since obviously

$$\operatorname{ran}([\mathcal{M}]A:[\mathcal{M}]B) \;\subset\; \mathcal{M}.$$

The assertion follows from the maximum property in Theorem 5.3. ∎

The following result should be compared with Theorem 5.5.

Theorem 5.17 *For any subspaces* $\mathcal{M}$, $\mathcal{N}\subset\mathcal{H}$

$$P_{\mathcal{M}}:P_{\mathcal{N}} \;=\; \tfrac{1}{2}P_{\mathcal{M}\cap\mathcal{N}}.$$

Proof. Since $P_{\mathcal{M}\cap\mathcal{N}}\leq P_{\mathcal{M}},\ P_{\mathcal{N}}$, by Theorem 5.12 (a) and (f)

$$\tfrac{1}{2}P_{\mathcal{M}\cap\mathcal{N}} \;=\; P_{\mathcal{M}\cap\mathcal{N}}:P_{\mathcal{M}\cap\mathcal{N}} \;\leq\; P_{\mathcal{M}}:P_{\mathcal{N}}.$$

On the other hand, since by Theorem 5.14 (f) and Theorem 5.15

$$P_{\mathcal{M}}:P_{\mathcal{N}} \;\leq\; I:I \;=\; \tfrac{1}{2}I \quad\text{and}\quad \operatorname{ran}(P_{\mathcal{M}}:P_{\mathcal{N}}) \;\subset\; \mathcal{M}\cap\mathcal{N}$$

we can conclude from Theorem 5.3 that

$$P_{\mathcal{M}}:P_{\mathcal{N}} \;\leq\; \tfrac{1}{2}[\mathcal{M}\cap\mathcal{N}]I \;=\; \tfrac{1}{2}P_{\mathcal{M}\cap\mathcal{N}}.$$

Putting the above arguments together completes the proof. ∎

We have defined parallel sum $A : B$ in terms of Schur complement. In the converse direction we will show that Schur complement can be recaptured with use of parallel addition.

Consider two positive semidefinite matrices A, B. Since by Theorem 5.14 (f) and (e)

$$A : B \;\leq\; A : (2B) \;\leq \cdots \leq\; A : (nB) \;\leq\; \cdots \;\leq\; A$$

we define the limit of the sequence $A : (nB)$, $n \to \infty$, denoted by $[B]A$, as

$$[B]A \;\equiv\; \lim_{n\to\infty} A : (nB). \tag{5.3.21}$$

Since all nB have the same range $\mathrm{ran}(B)$,

$$[B]A \;\leq\; A \quad \text{and} \quad \mathrm{ran}([B]A) \;\subset\; \mathrm{ran}(B)$$

which implies by Theorem 5.3

$$[B]A \;\leq\; [\mathrm{ran}(B)]A.$$

Let us show that the reversed inequality holds. To this end, let $C \equiv [\mathrm{ran}(B)]A$. Since $\mathrm{ran}(C) \subset \mathrm{ran}(B)$ due to Theorem 5.3, by (5.0.5)

$$C \leq \gamma B \quad \exists\, \gamma > 0,\; C \leq A \quad \text{and} \quad \mathrm{ran}(C + \gamma B) = \mathrm{ran}(B).$$

Then by (5.0.9) and (5.3.20)

$$\langle (C : (nB))x, x\rangle \;=\; \langle Cx, x\rangle - \sup\left\{ \frac{|\langle Cx, y\rangle|^2}{\langle (C + nB)y, y\rangle} \;:\; y \in \mathrm{ran}(B) \right\},$$

it follows by the Cauchy-Schwarz inequality that

$$\begin{aligned} 0 \;&\leq\; \langle Cx, x\rangle - \langle (C : (nB))x, x\rangle \\ &\leq\; \sup\left\{ \frac{\langle Cy, y\rangle \cdot \langle Cx, x\rangle}{(1 + n\gamma^{-1})\langle Cy, y\rangle} \;:\; y \in \mathrm{ran}(B) \right\} \\ &\leq\; \frac{\gamma}{n + \gamma}\langle Cx, x\rangle, \end{aligned}$$

which implies

$$[\mathrm{ran}(B)]A \;=\; C \;=\; \lim_{n\to\infty} C : (nB) \leq \lim_{n\to\infty} A : (nB) \;=\; [B]A.$$

Thus we have proved the following theorem, the central result of the section.

Theorem 5.18 *For any positive semidefinite matrices A, B*

$$[B]A = [\mathrm{ran}(B)]A.$$

Theorem 5.19 *Given two positive semidefinite matrices A, B, let*

$$A_1 \equiv [B]A \quad \text{and} \quad A_2 \equiv A - A_1.$$

Then $A = A_1 + A_2$ is the unique decomposition such that

$$0 \le A_1 \le \gamma B \;\; \exists\, \gamma > 0 \quad \text{and} \quad A_2 \wedge B = 0.$$

This is a reformulation of Theorem 5.7 based on Theorem 5.18.

Theorem 5.20 *Let A, B, and C be positive semidefinite matrices. Then*

(i) $[\alpha B]A = [B]A \quad (\alpha > 0)$;

(ii) $[S^*BS](S^*AS) = S^* \cdot [B]A \cdot S$ *for all invertible S*;

(iii) $0 \le B \le C \implies [B]A \le [C]A$;

(iv) $\mathrm{ran}(A) \subset \mathrm{ran}(B) \implies [B]A = A$;

(v) $[B]A = [A : B]A$.

Proof. (i) follows from Theorem 5.18 because $\mathrm{ran}(\alpha B) = \mathrm{ran}(B)$. (ii) follows from Theorem 5.14 (h) and definition (5.3.21). (iii) and (iv) are obvious. For (v), by Theorem 5.18, Theorem 5.14, and Theorem 5.10

$$\begin{aligned} [A : B]A &= [\mathrm{ran}(A : B)]A \\ &= [\mathrm{ran}(A) \cap \mathrm{ran}(B)]A \\ &= [\mathrm{ran}(B)]\big([\mathrm{ran}(A)]A\big) \\ &= [B]A. \blacksquare \end{aligned}$$

Note that the expression of the parallel sum in terms of the Schur complement was first pointed out by Anderson–Trapp [13]. They established also Theorems 5.13 and 5.15. Theorem 5.17 is due to Anderson–Schreiber [12]. The operation $[B]A$ was introduced by Ando [15], and Theorem 5.18 is a restatement of a more general result for Hilbert space operators by Ando [15].

5.4 Application to the infimum problem

In this section we will present an important application of the map $A \longmapsto [B]A$ to show its usefulness in the investigation of the order structure of the cone of positive semidefinite matrices.

For any pair of positive semidefinite matrices A and B, the set $\{X : X \leq A, B\}$ does not admit the maximum element in the class of Hermitian matrices, except when A and B are *comparable*, i.e., $A \geq B$ or $A \leq B$.

The situation is, however, different if observation is restricted to the cone of positive semidefinite matrices. The purpose of this section is to find a condition so that the set $\{X \geq 0 : X \leq A, B\}$ admits the maximum element. When this is the case, the maximum will be denoted by $A \wedge B$ and referred to as the *infimum* of A and B.

Note that the notation $A \wedge B$ already appeared in Section 5.1 for the case $A \wedge B = 0$. Moreover, it is easy to see that when A, B are orthoprojections, that is, $A = P_{\mathcal{M}}$, $B = P_{\mathcal{N}}$, their infimum always exists, and

$$P_{\mathcal{M}} \wedge P_{\mathcal{N}} = P_{\mathcal{M} \cap \mathcal{N}}.$$

First if $0 \leq X \leq A,\ B$, by Theorem 5.20 (iv)

$$X = [B]X \leq [B]A \quad \text{and} \quad X = [A]X \leq [A]B.$$

Therefore if $[A]B$ and $[B]A$ are comparable, then $A \wedge B$ exists and

$$A \wedge B = \min\{[A]B, [B]A\}.$$

The converse direction for the existence of infimum is the central part of the following theorem. In the investigation of the order structure of the set $\{X \geq 0 : X \leq A, B\}$ we may assume that $A + B = I$. This is seen as follows. Let $\mathcal{M} \equiv \operatorname{ran}(A + B)$. Since $0 \leq X \leq A, B$ implies that the map

$$X \longmapsto \Phi(X) \equiv ((A+B)^{\dagger})^{1/2} \cdot X \cdot ((A+B)^{\dagger})^{1/2}$$

is an affine order-isomorphism from the set $\{X \geq 0 : X \leq A, B\}$ to the set $\{Y \geq 0 : Y \leq \Phi(A), \Phi(B)\}$ in the class of positive semidefinite operators on the Hilbert space $\mathcal{M}$ and $\Phi(A) + \Phi(B) = I_{\mathcal{M}}$, and this isomorphism satisfies, by Theorem 5.20 (ii), the conditions

$$\Phi([A]X) = [\Phi(A)]\Phi(X) \quad \text{and} \quad \Phi([B]X) = [\Phi(B)]\Phi(X).$$

Theorem 5.21 *Positive semidefinite matrices A and B admit the infimum $A \wedge B$ if and only if $[A : B]A$ and $[A : B]B$ are comparable. In this case*

$$A \wedge B = \min\{[A : B]A, [A : B]B\}.$$

Proof. Notice first that by Theorem 5.20 (v)

$$[A]B = [A:B]B \quad \text{and} \quad [B]A = [A:B]A. \tag{5.4.22}$$

Now let A, B be positive semidefinite matrices of order n for which the infimum $A \wedge B$ exists. As explained above, we may assume that

$$A + B = I. \tag{5.4.23}$$

We may further assume that A (and so B) is not an orthoprojection. Otherwise under (5.4.23)

$$A \wedge B = 0 \quad \text{and} \quad [A:B]A = [A:B]B = 0.$$

By the spectral theorem (see [51, p. 5]) there is a unitary matrix U for which U^*AU (and hence $U^*BU = I - U^*AU$) is diagonal. By Theorem 5.20 (ii) we may further assume that both A and B are diagonal and of the form

$$A = \operatorname{diag}(\lambda_1, \lambda_2, \cdots, \lambda_n), \quad \text{where } 0 \leq \lambda_1 \leq \lambda_2 \leq \cdots \leq \lambda_n \leq 1. \tag{5.4.24}$$

and by (5.4.23)

$$B = \operatorname{diag}(1 - \lambda_1, 1 - \lambda_2, \cdots, 1 - \lambda_n). \tag{5.4.25}$$

Since A is not an orthoprojection by assumption, with $\lambda_0 \equiv 0$ and $\lambda_{n+1} \equiv 1$, there are $1 \leq p \leq q \leq n$ such that

$$\lambda_{p-1} = 0 < \lambda_p \leq \lambda_q < \lambda_{q+1} = 1. \tag{5.4.26}$$

Let

$$C \equiv \operatorname{diag}\big(\min(\lambda_j, 1 - \lambda_j)\big)_{j=1}^n. \tag{5.4.27}$$

Clearly $0 \leq C \leq A, B$ so that by assumption $0 \leq C \leq A \wedge B$. Take any D such that $C \leq D \leq A, B$. Then since

$$0 \leq D - C \leq A - C \quad \text{and} \quad 0 \leq D - C \leq B - C,$$

we have by (5.0.5)

$$\operatorname{ran}(D - C) \subset \operatorname{ran}(A - C) \quad \text{and} \quad \operatorname{ran}(D - C) \subset \operatorname{ran}(B - C).$$

On the other hand, since by (5.4.24), (5.4.25) and (5.4.27)

$$A - C = \operatorname{diag}\big(\max(2\lambda_j - 1, 0)\big)_{j=1}^n$$

and

$$B - C = \operatorname{diag}\left(\max(1-2\lambda_j, 0)\right)_{j=1}^{n},$$

we can infer

$$\operatorname{ran}(A-C) \cap \operatorname{ran}(B-C) \;=\; \{0\},$$

so that $\operatorname{ran}(D-C) = \{0\}$ hence $D = C$. This shows that C is a *maximal* element in the set $\{X \geq 0 : X \leq A, B\}$, thus it must coincide with $A \wedge B$.

It follows, by (5.4.24) and (5.4.25), from definition (5.3.21), that

$$[B]A \;=\; \operatorname{diag}(0, \ldots, 0, \lambda_p, \ldots, \lambda_q, 0, \ldots, 0)$$

and

$$[A]B \;=\; \operatorname{diag}(0, \ldots, 0, 1-\lambda_p, \ldots, 1-\lambda_q, \ldots, 0).$$

Suppose, by contradiction, that $[B]A$ and $[A]B$ are not comparable, i.e.,

$$[B]A \;\not\geq\; [A]B \quad \text{and} \quad [B]A \;\not\leq\; [A]B,$$

which is equivalent by (5.4.26) to saying that $p < q$ and

$$1 - 2\lambda_p \;>\; 0 \quad \text{and} \quad 2\lambda_q - 1 \;>\; 0.$$

Choose $\epsilon > 0$ and $\delta > 0$ such that

$$\epsilon < \delta < 2\epsilon < \gamma - \epsilon, \text{ where } \gamma \equiv \min\{\lambda_p, 1-2\lambda_p, 1-\lambda_q, 2\lambda_q - 1\}, \quad (5.4.28)$$

and consider the Hermitian matrix $F \equiv (f_{ij})_{i,j=1}^{n}$ with entries

$$f_{pp} = f_{qq} = \epsilon, \quad f_{pq} = f_{qp} = \delta \quad \text{and} \quad f_{ij} = 0 \quad \text{for other } i, j,$$

First, F is not positive semidefinite because by (5.4.28)

$$\det F[p,q] = \epsilon^2 - \delta^2 < 0,$$

where $F[p,q]$ is the 2×2 principal submatrix of F indexed by p, q.

Next $C - F$ is positive semidefinite. To see this, it suffices to show the positive semidefiniteness of the principal submatrix

$$(C-F)[p,q] = \begin{pmatrix} \lambda_p - \epsilon & -\delta \\ -\delta & 1 - \lambda_q - \epsilon \end{pmatrix}.$$

We can see from (5.4.28) that

$$\begin{pmatrix} \lambda_p - \epsilon & -\delta \\ -\delta & 1 - \lambda_q - \epsilon \end{pmatrix} \geq \begin{pmatrix} \gamma - \epsilon & -\delta \\ -\delta & \gamma - \epsilon \end{pmatrix} \geq 0.$$

Finally, let us show that

$$C - F \le A,\ B \quad \text{or equivalently} \quad A - C + F,\ B - C + F \ge 0.$$

As above, it is sufficient to show the positive semidefiniteness

$$(A - C + F)[p,q] \ge 0 \quad \text{and} \quad (B - C + F)[p,q] \ge 0.$$

Since

$$\max(2\lambda_p - 1, 0) = 0 \quad \text{and} \quad \max(2\lambda_q - 1, 0) = 2\lambda_q - 1$$

we can see from (5.4.27) that

$$(A - C + F)[p,q] = \begin{pmatrix} \epsilon & \delta \\ \delta & 2\lambda_q - 1 + \epsilon \end{pmatrix}$$

and

$$\det(A - C + F)[p,q] = \epsilon(2\lambda_q - 1 + \epsilon) - \delta^2 \ge \epsilon(\gamma + \epsilon) - \delta^2.$$

By (5.4.28)

$$\epsilon\gamma > (\delta - \epsilon)(\delta + \epsilon)$$

which implies the positivity of the determinant, so $(A - C + F)[p,q] \ge 0$, hence $A - C + F \ge 0$. That $B - C + F \ge 0$ is proved in a similar way.

Now we are in the position that

$$0 \le C - F \le A, B.$$

Since $C = A \wedge B$, we have $C - F \le C$. This contradicts the fact that F is not positive semidefinite. Therefore $[A]B$ and $[B]A$ have to be comparable. Thus $[A : B]A$ and $[A : B]B$ are comparable by (5.4.22). ∎

Notice that if A, B are orthoprojection, then $[A]B = [B]A$ by Theorem 5.5 so that $A \wedge B = [A]B = [B]A$ by Theorem 5.21.

There are positive semidefinite matrices B for which the infimum $A \wedge B$ exists for all A in a wide class of positive semidefinite matrices.

Theorem 5.22 *Let B be a matrix such that $0 \le B \le I$ and $\operatorname{rank}(B) \ge 2$. Then in order that the infimum $A \wedge B$ exist for all $0 \le A \le I$, it is necessary and sufficient that B is an orthoprojection. In this case $A \wedge B = [B]A$.*

Proof. First we suppose that B is an orthoprojection, $B \equiv P_{\mathcal{M}}$ for some subspace $\mathcal{M}$. By Theorem 5.20 we have

$$[P_{\mathcal{M}}]A \ \le \ [P_{\mathcal{M}}]I \ = \ P_{\mathcal{M}},$$

and

$$[P_{\mathcal{M}}]A \;=\; [[P_{\mathcal{M}}]A]\,([P_{\mathcal{M}}]A) \;\le\; [[P_{\mathcal{M}}]A]P_{\mathcal{M}} \;\le [A]P_{\mathcal{M}},$$

which shows that $[P_{\mathcal{M}}]A$ and $[A]P_{\mathcal{M}}$ are comparable and

$$\min\{[P_{\mathcal{M}}]A, [A]P_{\mathcal{M}}\} \;=\; [P_{\mathcal{M}}]A,$$

hence by Theorem 5.21 $A \wedge B = [B]A$.

Suppose now that B is not an orthoprojection. Since B is of rank ≥ 2, there are mutually annihilating orthoprojection $P_1,\ P_2$, commuting with B, and $0 < \epsilon < \gamma < 1$ such that

$$BP_1 \;\geq\; \gamma P_1 \quad \text{and} \quad \gamma P_2 \geq BP_2 \geq \epsilon P_2. \tag{5.4.29}$$

Let

$$A \;\equiv\; \frac{\gamma+\epsilon}{2}P_1 + \frac{1+\gamma}{2}P_2. \tag{5.4.30}$$

Then $0 \leq A \leq I$, and A and $BP_1 + BP_2$ are not comparable. By (5.4.29) and (5.4.30)

$$\frac{1+\gamma}{2\epsilon}B \;\geq\; \gamma P_1 + \frac{1+\gamma}{2}P_2 \geq A,$$

which implies by (5.0.5)

$$\operatorname{ran}(B) \supset \operatorname{ran}(A)$$

so that, by Theorem 5.20 and (5.4.22),

$$[B]A = [A : B]A = [A]A = A.$$

On the other hand, by (5.4.30)

$$P_1 + P_2 \;\geq\; A \;\geq\; \epsilon(P_1 + P_2)$$

and since $P_1 + P_2$ is an orthoprojection commuting with B, we have by definition (5.3.20)

$$[A]B \;=\; [P_1 + P_2]B \;=\; BP_1 + BP_2.$$

Therefore $[A]B$ and $[B]A$ are not comparable. By Theorem 5.21 the infimum $A \wedge B$ does not exists. ∎

Theorem 5.23 *Let B be a positive semidefinite matrix. Then in order that the infimum $A \wedge B$ exist for all $A \geq 0$, it is necessary and sufficient that B is of rank less than or equal to 1.*

Proof. Suppose that B is of rank ≤ 1. By Theorem 5.18 and Theorem 5.3

$$\operatorname{ran}([A]B), \ \operatorname{ran}([B]A) \ \subset \ \operatorname{ran}(B)$$

and since B is of rank ≤ 1, both $[A]B$ and $[B]A$ are nonnegative scalar multiplies of B so that they are comparable. Then the existence of $A \wedge B$ follows from Theorem 5.21.

Suppose now that B is of rank ≥ 2. Via multiplication by a positive scalar we may assume that $B \leq \frac{1}{2}I$. Then since B is not an orthoprojection, as in the proof of Theorem 5.22 there is $0 \leq A \leq I$ such that the infimum $A \wedge B$ does not exists. ∎

We conclude the section by pointing out that the problem of infimum in the cone of positive semidefinite operators on a Hilbert space has been discussed by a group of mathematical physicists. The results of this section for the case of matrices are obtained by Moreland–Gudder [324]. The proofs presented here, however, are quite different from theirs and adopted from those of the corresponding results for Hilbert space operators by Ando [20].

Chapter 6

Schur complements in statistics and probability

6.0 Basic results on Schur complements

In this chapter we survey the use of the Schur complement in statistics and probability, building upon the surveys by Ouellette [345] and Styan [432] published, respectively, in 1981 and 1985. We will use Roman boldface capital letters for matrices and Roman boldface lower case letters for vectors. We use a prime to denote transpose and all our row vectors are primed. Scalars will be denoted by lower case lightface italic letters. In Section 6.0 our matrices are all complex, but in all subsequent sections unless stated explicitly to the contrary, our matrices are real.

We will use several basic results on Schur components as studied in Chapter 0. These results include the important Aitken block-diagonalization formula as well as the Haynsworth inertia additivity formula, and the Banachiewicz and Duncan matrix inversion formulas. We also look at the inversion formulas due to Bartlett, to Sherman–Morrison, and to Woodbury. We introduce the Albert nonnegative (positive) definiteness conditions, as well as the generalized quotient property and the notion of generalized Schur complement.

In Section 6.1 we consider several matrix inequalities which are useful in statistics and probability. In 6.2 we study correlation and in 6.3, the general linear model and multiple linear regression. In Section 6.4 we look at experimental design, where the Schur complement plays a crucial role as the so-called C-matrix. We end the chapter with an analysis-of-covariance model associated with Broyden's mark-scaling algorithm [97], published in 1983.

6.0.1 The Aitken block-diagonalization formula

Let the partitioned (block) matrix

$$\mathbf{M} = \begin{pmatrix} \mathbf{P} & \mathbf{Q} \\ \mathbf{R} & \mathbf{S} \end{pmatrix}, \tag{6.0.1}$$

where the square matrix $\mathbf{P}$ is nonsingular (invertible); the matrix $\mathbf{M}$ need not be square. Then

$$\mathbf{M}/\mathbf{P} = \mathbf{S} - \mathbf{R}\mathbf{P}^{-1}\mathbf{Q} \tag{6.0.2}$$

is the Schur complement of $\mathbf{P}$ in the partitioned matrix $\mathbf{M}$.

We find the *Aitken block-diagonalization formula*

$$\begin{pmatrix} \mathbf{I} & \mathbf{0} \\ -\mathbf{R}\mathbf{P}^{-1} & \mathbf{I} \end{pmatrix} \begin{pmatrix} \mathbf{P} & \mathbf{Q} \\ \mathbf{R} & \mathbf{S} \end{pmatrix} \begin{pmatrix} \mathbf{I} & -\mathbf{P}^{-1}\mathbf{Q} \\ \mathbf{0} & \mathbf{I} \end{pmatrix} = \begin{pmatrix} \mathbf{P} & \mathbf{0} \\ \mathbf{0} & \mathbf{M}/\mathbf{P} \end{pmatrix}, \tag{6.0.3}$$

introduced in (0.9.1), to be very useful. In (6.0.3) neither $\mathbf{M}/\mathbf{P}$ nor $\mathbf{S}$ need be square. The two triangular matrices in (6.0.3) each have determinant equal to 1 and so, they are both nonsingular. It then follows at once that rank is additive on the Schur complement

$$\operatorname{rank}(\mathbf{M}) = \operatorname{rank}(\mathbf{P}) + \operatorname{rank}(\mathbf{M}/\mathbf{P}) = \operatorname{rank}(\mathbf{P}) + \operatorname{rank}(\mathbf{S} - \mathbf{R}\mathbf{P}^{-1}\mathbf{Q}), \tag{6.0.4}$$

which we refer to as the *Guttman rank additivity formula*, see also (0.9.2). Since $\mathbf{P}$ is nonsingular, we see that the nullity of $\mathbf{M}$ and the nullity of the Schur complement $\mathbf{M}/\mathbf{P}$ are the same

$$\nu(\mathbf{M}) = \nu(\mathbf{M}/\mathbf{P}) = \nu(\mathbf{S} - \mathbf{R}\mathbf{P}^{-1}\mathbf{Q}). \tag{6.0.5}$$

When $\mathbf{M}$ is square, taking determinants of (6.0.3) shows that determinant is multiplicative on the Schur complement as established by Schur [404] in 1917.

$$\det(\mathbf{M}) = \det(\mathbf{P}) \cdot \det(\mathbf{M}/\mathbf{P}) = \det(\mathbf{P}) \cdot \det(\mathbf{S} - \mathbf{R}\mathbf{P}^{-1}\mathbf{Q}). \tag{6.0.6}$$

We refer to (6.0.6) as the *Schur determinant formula*, see also (0.3.2).

6.0.2 The Banachiewicz, Duncan, Sherman–Morrison and Woodbury matrix inversion formulas

Let us consider again the square complex nonsingular partitioned matrix $\mathbf{M} = \begin{pmatrix} \mathbf{P} & \mathbf{Q} \\ \mathbf{R} & \mathbf{S} \end{pmatrix}$ as in (6.0.1) above, with $\mathbf{P}$ nonsingular, and therefore the

Schur complement $\mathbf{M/P} = \mathbf{S} - \mathbf{RP}^{-1}\mathbf{Q}$ also nonsingular. Then from the Aitken block-diagonalization formula (6.0.3) above, we obtain the *Banachiewicz inversion formula*, see also (0.7.2),

$$\mathbf{M}^{-1} = \begin{pmatrix} \mathbf{P}^{-1} + \mathbf{P}^{-1}\mathbf{Q}(\mathbf{M/P})^{-1}\mathbf{RP}^{-1} & -\mathbf{P}^{-1}\mathbf{Q}(\mathbf{M/P})^{-1} \\ -(\mathbf{M/P})^{-1}\mathbf{RP}^{-1} & (\mathbf{M/P})^{-1} \end{pmatrix}. \tag{6.0.7}$$

When $\mathbf{S}$ is nonsingular, then the Schur complement $\mathbf{M/S} = \mathbf{P} - \mathbf{QS}^{-1}\mathbf{R}$ is also nonsingular and

$$\mathbf{M}^{-1} = \begin{pmatrix} (\mathbf{M/S})^{-1} & -(\mathbf{M/S})^{-1}\mathbf{QS}^{-1} \\ -\mathbf{S}^{-1}\mathbf{R}(\mathbf{M/S})^{-1} & \mathbf{S}^{-1} + \mathbf{S}^{-1}\mathbf{R}(\mathbf{M/S})^{-1}\mathbf{QS}^{-1} \end{pmatrix}, \tag{6.0.8}$$

with $\mathbf{P}$ not necessarily nonsingular. When, however, both $\mathbf{S}$ and $\mathbf{P}$ are nonsingular, then we may equate the top left-hand corners in (6.0.7) and (6.0.8) to yield $(\mathbf{M/S})^{-1} = \mathbf{P}^{-1} + \mathbf{P}^{-1}\mathbf{Q}(\mathbf{M/P})^{-1}\mathbf{RP}^{-1}$ or explicitly, as in (0.8.3),

$$(\mathbf{P} - \mathbf{QS}^{-1}\mathbf{R})^{-1} = \mathbf{P}^{-1} + \mathbf{P}^{-1}\mathbf{Q}(\mathbf{S} - \mathbf{RP}^{-1}\mathbf{Q})^{-1}\mathbf{RP}^{-1}, \tag{6.0.9}$$

which we refer to as the *Duncan inversion formula*; we believe (6.0.9) was first established by Duncan [151]; see also Guttman [197]. Grewal & Andrews [189, p. 366] call (6.0.9) the *Hemes inversion formula* following a reference to H. Hemes by Bodewig [64, p. 218]; see also [190, p. 309] and our Chapter 0.

The formula (6.0.9) was also obtained six years later in 1950 by Woodbury [461], who established that

$$(\mathbf{P} + \mathbf{QTR})^{-1} = \mathbf{P}^{-1} - \mathbf{P}^{-1}\mathbf{QT}(\mathbf{T} + \mathbf{TRP}^{-1}\mathbf{QT})^{-1}\mathbf{TRP}^{-1}, \tag{6.0.10}$$

which we refer to as the *Woodbury inversion formula* and which follows easily from (6.0.9) by substituting $\mathbf{S} = -\mathbf{T}^{-1}$. At first glance (6.0.10) seems not to require that $\mathbf{T}$ be nonsingular. But $\mathbf{T}$ is a factor of $\mathbf{T} + \mathbf{TRP}^{-1}\mathbf{QT}$, which is nonsingular, and hence so is $\mathbf{T}$; see also Henderson & Searle [219], where many special cases and variations of the inversion formulas (6.0.9) and (6.0.10) are presented.

Hager [200] focuses on the *inverse matrix modification formula*

$$(\mathbf{P} - \mathbf{QR})^{-1} = \mathbf{P}^{-1} + \mathbf{P}^{-1}\mathbf{Q}(\mathbf{I} - \mathbf{RP}^{-1}\mathbf{Q})^{-1}\mathbf{RP}^{-1}, \tag{6.0.11}$$

and observes that the matrix $\mathbf{I} - \mathbf{RP}^{-1}\mathbf{Q}$ is often called the *capacitance matrix*, see also [356]. The inverse matrix modification formula (6.0.11) is the special case of our Duncan inversion formula (6.0.9) with $\mathbf{S} = \mathbf{I}$ and the special case of our Woodbury inversion formula (6.0.10) with $\mathbf{T} = -\mathbf{I}$. Hager [200] notes moreover that his inverse matrix modification formula (6.0.11) is frequently called the *Woodbury formula.*

When

$$\mathbf{M} = \begin{pmatrix} \mathbf{P} & \mathbf{q} \\ \mathbf{r}^* & s \end{pmatrix},$$

with $\mathbf{q}$ and $\mathbf{r}$ column vectors and s a nonzero scalar, then the Schur complement $\mathbf{M}/\mathbf{S}$ becomes

$$\mathbf{M}/s = \mathbf{P} - \frac{1}{s}\mathbf{q}\mathbf{r}^*,$$

and the Duncan inversion formula (6.0.9) becomes

$$(\mathbf{P} - \frac{1}{s}\mathbf{q}\mathbf{r}^*)^{-1} = \mathbf{P}^{-1} + \frac{1}{s - \mathbf{r}^*\mathbf{P}^{-1}\mathbf{q}}\mathbf{P}^{-1}\mathbf{q}\mathbf{r}^*\mathbf{P}^{-1}, \qquad (6.0.12)$$

where the scalar Schur complement $\mathbf{M}/\mathbf{P} = s - \mathbf{r}^*\mathbf{P}^{-1}\mathbf{q} \neq 0$.

The special case of (6.0.12) with $s = -1$,

$$(\mathbf{P} + \mathbf{q}\mathbf{r}^*)^{-1} = \mathbf{P}^{-1} - \frac{1}{1 + \mathbf{r}^*\mathbf{P}^{-1}\mathbf{q}}\mathbf{P}^{-1}\mathbf{q}\mathbf{r}^*\mathbf{P}^{-1}, \qquad (6.0.13)$$

was apparently first established explicitly by Bartlett [36] in 1951 and so we refer to (6.0.13) as the *Bartlett inversion formula.* We note that (6.0.13) is the special case of the inverse matrix modification formula (6.0.11) with $\mathbf{Q} = -\mathbf{q}$ and $\mathbf{R} = \mathbf{r}$.

The special case of the Bartlett inversion formula (6.0.13) with $\mathbf{q} = \mathbf{e}_i$, the column vector with 1 in its ith position and 0 elsewhere, is

$$(\mathbf{P} + \mathbf{e}_i\mathbf{r}^*)^{-1} = \mathbf{P}^{-1} - \frac{1}{1 + \mathbf{r}^*\mathbf{P}^{-1}\mathbf{e}_i}\mathbf{P}^{-1}\mathbf{e}_i\mathbf{r}^*\mathbf{P}^{-1}, \qquad (6.0.14)$$

which shows how the inverse changes when the row vector $\mathbf{r}^*$ is added to the ith row; the column vector $\mathbf{P}^{-1}\mathbf{e}_i$ is the ith column of $\mathbf{P}^{-1}$. The special case of (6.0.14) with $\mathbf{r} = k\mathbf{e}_j$,

$$\begin{aligned}(\mathbf{P} + k\mathbf{e}_i\mathbf{e}_j')^{-1} &= \mathbf{P}^{-1} - \frac{1}{1 + k\mathbf{e}_j'\mathbf{P}^{-1}\mathbf{e}_i}\mathbf{P}^{-1}\mathbf{e}_i\mathbf{e}_j'\mathbf{P}^{-1} \\ &= \mathbf{P}^{-1} - \frac{1}{1 + kp^{(ji)}}\boldsymbol{u}_i\mathbf{v}_j' \qquad (6.0.15)\end{aligned}$$

shows how the inverse changes when k is added to the (i, j)th element. In (6.0.15) the column vector $\boldsymbol{u}_i = \mathbf{P}^{-1}\mathbf{e}_i$ is the ith column of $\mathbf{P}^{-1}$, the row vector $\mathbf{v}_j'$ is the jth row of $\mathbf{P}^{-1}$, and $p^{ji} = \mathbf{e}_j'\mathbf{P}^{-1}\mathbf{e}_i$ is the (j,i)th element of $\mathbf{P}^{-1}$. The matrix $\mathbf{P}$ remains nonsingular after the (i,j) element has been changed from p_{ij} to $p_{ij} + k$ if and only if $k \neq -p^{ji}$.

We believe that the formula (6.0.15) was first explicitly given in 1949 by Sherman & Morrison [416] and first published in 1950 [418]. The formula (6.0.14) was apparently first given in 1949 at a meeting in Colorado by both Sherman & Morrison [416] and by Woodbury [325, p. 192], whose formula was presented on his behalf at this meeting by the economist Oskar Morgenstern (1902–1976).

We refer to (6.0.15) as the *Sherman–Morrison inversion formula* and (6.0.14) as the *Sherman–Morrison–Woodbury inversion formula.*

For the inverse matrix modification formula (6.0.11), the term *Sherman–Morrison–Woodbury formula* is used by Golub & Van Loan [185, p. 50] in their well-known *Matrix Computations* book; see also [190, p. 309], [258, pp. 52–53], and [313, p. 124], while Duda, Hart & Stork [149] in their *Pattern Recognition* book, call our Bartlett inversion formula (6.0.13) the *Sherman–Morrison–Woodbury matrix identity.* In their *Global Positioning Systems* book, Grewal, Weill & Andrews [190, p. 309] call our Bartlett inversion formula (6.0.13) the *Sherman–Morrison formula*; see also [258, p. 52]. Meyer [313, p. 124] in his recent *Matrix Analysis and Applied Linear Algebra* book calls (6.0.13) the *Sherman–Morrison rank-one update formula.*

6.0.3 The Haynsworth inertia additivity formula

For the partitioned complex Hermitian matrix

$$\mathbf{A} = \begin{pmatrix} \mathbf{A}_{11} & \mathbf{A}_{12} \\ \mathbf{A}_{12}^* & \mathbf{A}_{22} \end{pmatrix},$$

we define the *inertia triple*

$$\operatorname{In}\mathbf{A} = \{\pi,\ \gamma,\ \zeta\},$$

where the nonnegative integers $\pi = \pi(\mathbf{A})$, $\gamma = \gamma(\mathbf{A})$, and $\zeta = \zeta(\mathbf{A})$ give the numbers, respectively, of positive, negative and zero eigenvalues of $\mathbf{A}$. Here $\mathbf{A}_{11}$ is nonsingular and $\mathbf{A}_{12}^*$ is the conjugate transpose of $\mathbf{A}_{12}$. And so, since $\mathbf{A}$ is Hermitian,

$$\operatorname{rank}(\mathbf{A}) = \pi(\mathbf{A}) + \gamma(\mathbf{A}) \quad \text{and} \quad \nu(\mathbf{A}) = \zeta(\mathbf{A}),$$

where $\nu(\cdot)$ denotes nullity.

Then Haynsworth [210, 211] proved the *Haynsworth inertia additivity formula* (0.10.1)

$$\begin{aligned}\mathrm{In}(\mathbf{A}) &= \mathrm{In}(\mathbf{A}_{11}) + \mathrm{In}(\mathbf{A}/\mathbf{A}_{11}) \\ &= \mathrm{In}(\mathbf{A}_{11}) + \mathrm{In}(\mathbf{A}_{22} - \mathbf{A}_{12}^{*}\mathbf{A}_{11}^{-1}\mathbf{A}_{12}) \end{aligned} \tag{6.0.16}$$

from which it follows at once that $\mathbf{A}$ is nonnegative definite if and only if $\mathbf{A}_{11}$ is positive definite and the Schur complement $\mathbf{A}/\mathbf{A}_{11}$ is nonnegative definite. And then $\mathbf{A}$ and $\mathbf{A}/\mathbf{A}_{11}$ have the same nullity, $\nu(\mathbf{A}) = \nu(\mathbf{A}/\mathbf{A}_{11})$.

6.0.4 The generalized Schur complement and the Albert nonnegative definiteness conditions

Let us now suppose that $\mathbf{A}$ is nonnegative definite and that $\mathbf{A}_{11}$ is possibly singular. Then we define the *generalized Schur complement*

$$\mathbf{A}/\mathbf{A}_{11} = \mathbf{A}_{22} - \mathbf{A}_{12}^{*}\mathbf{A}_{11}^{-}\mathbf{A}_{12}, \tag{6.0.17}$$

where the superscript $^{-}$ denotes generalized inverse so that $\mathbf{G}^{-}$ is a generalized inverse of the (possible rectangular) matrix $\mathbf{G}$ whenever $\mathbf{G} = \mathbf{G}\mathbf{G}^{-}\mathbf{G}$. For more details on generalized inverses, see, e.g., Ben-Israel & Greville [45], and for generalized Schur complements, see, e.g., Ouellette [345].

To show that our generalized Schur complement as defined in (6.0.17) does not depend on the choice of generalized inverse, we write the nonnegative definite matrix

$$\mathbf{A} = \mathbf{B}\mathbf{B}' = \begin{pmatrix}\mathbf{B}_1 \\ \mathbf{B}_2\end{pmatrix}\begin{pmatrix}\mathbf{B}_1^{*} & \mathbf{B}_2^{*}\end{pmatrix} = \begin{pmatrix}\mathbf{B}_1\mathbf{B}_1^{*} & \mathbf{B}_1\mathbf{B}_2^{*} \\ \mathbf{B}_2\mathbf{B}_1^{*} & \mathbf{B}_2\mathbf{B}_2^{*}\end{pmatrix}, \tag{6.0.18}$$

and so we see that the generalized Schur complement

$$\begin{aligned}\mathbf{A}/\mathbf{A}_{11} &= \mathbf{A}_{22} - \mathbf{A}_{12}^{*}\mathbf{A}_{11}^{-}\mathbf{A}_{12} \\ &= \mathbf{B}_2\mathbf{B}_2^{*} - \mathbf{B}_2\mathbf{B}_1^{*}(\mathbf{B}_1\mathbf{B}_1^{*})^{-}\mathbf{B}_1\mathbf{B}_2^{*}\end{aligned} \tag{6.0.19}$$

does not depend on the choice of generalized inverse since the orthogonal projector $\mathbf{B}_1^{*}(\mathbf{B}_1\mathbf{B}_1^{*})^{-}\mathbf{B}_1$ does not depend on the choice of generalized inverse. This leads to the *generalized Aitken block-diagonalization formula*

$$\begin{pmatrix}\mathbf{I} & \mathbf{0} \\ -\mathbf{A}_{12}^{*}\mathbf{A}_{11}^{-} & \mathbf{I}\end{pmatrix}\begin{pmatrix}\mathbf{A}_{11} & \mathbf{A}_{12} \\ \mathbf{A}_{12}^{*} & \mathbf{A}_{22}\end{pmatrix}\begin{pmatrix}\mathbf{I} & -\mathbf{A}_{11}^{-}\mathbf{A}_{12} \\ \mathbf{0} & \mathbf{I}\end{pmatrix} = \begin{pmatrix}\mathbf{A}_{11} & \mathbf{0} \\ \mathbf{0} & (\mathbf{A}/\mathbf{A}_{11})\end{pmatrix}. \tag{6.0.20}$$

And so when $\mathbf{A}$ is nonnegative definite and Hermitian, we see that inertia is additive on the generalized Schur complement

$$\begin{aligned} \mathrm{In}(\mathbf{A}) &= \mathrm{In}(\mathbf{A}_{11}) + \mathrm{In}(\mathbf{A}/\mathbf{A}_{11}), \\ &= \mathrm{In}(\mathbf{A}_{11}) + \mathrm{In}(\mathbf{A}_{22} - \mathbf{A}_{12}^*\mathbf{A}_{11}^-\mathbf{A}_{12}), \end{aligned} \tag{6.0.21}$$

and hence so is rank

$$\begin{aligned} \mathrm{rank}(\mathbf{A}) &= \mathrm{rank}(\mathbf{A}_{11}) + \mathrm{rank}(\mathbf{A}/\mathbf{A}_{11}) \\ &= \mathrm{rank}(\mathbf{A}_{11}) + \mathrm{rank}(\mathbf{A}_{22} - \mathbf{A}_{12}^*\mathbf{A}_{11}^-\mathbf{A}_{12}), \end{aligned} \tag{6.0.22}$$

and nullity

$$\nu(\mathbf{A}) = \nu(\mathbf{A}_{11}) + \nu(\mathbf{A}/\mathbf{A}_{11}) = \nu(\mathbf{A}_{11}) + \nu(\mathbf{A}_{22} - \mathbf{A}_{12}^*\mathbf{A}_{11}^-\mathbf{A}_{12}). \tag{6.0.23}$$

The three additivity formulas (6.0.21), (6.0.22) and (6.0.23) use the nonnegative definiteness of $\mathbf{A}$ to ensure that the generalized Schur complement $\mathbf{A}_{22} - \mathbf{A}_{12}^*\mathbf{A}_{11}^-\mathbf{A}_{12}$ does not depend on the choice of generalized inverse $\mathbf{A}_{11}^-$. And as Albert [6] noted, see also Baksalary & Kala [24], the generalized Schur complement $\mathbf{A}_{22} - \mathbf{A}_{12}^*\mathbf{A}_{11}^-\mathbf{A}_{12}$ does not depend on the choice of generalized inverse $\mathbf{A}_{11}^-$ if and only if $\mathrm{rank}(\mathbf{A}_{11} : \mathbf{A}_{12}) = \mathrm{rank}(\mathbf{A}_{11})$, or equivalently if and only if $\mathcal{C}(\mathbf{A}_{12}) \subset \mathcal{C}(\mathbf{A}_{11})$, where $\mathcal{C}(\cdot)$ denotes column space (range). And so as shown by Albert [6], see also Pukelsheim [357, p. 75], the following three statements are equivalent:

$$(\mathrm{a}_0)\ \mathbf{A} \geq_{\mathsf{L}} \mathbf{0},$$

$$(\mathrm{b}_0)\ \mathbf{A}_{11} \geq_{\mathsf{L}} \mathbf{0} \quad \text{and} \quad \mathbf{A}_{22} - \mathbf{A}_{12}^*\mathbf{A}_{11}^-\mathbf{A}_{12} \geq_{\mathsf{L}} \mathbf{0} \quad \text{and} \quad \mathcal{C}(\mathbf{A}_{12}) \subset \mathcal{C}(\mathbf{A}_{11}),$$

$$(\mathrm{c}_0)\ \mathbf{A}_{22} \geq_{\mathsf{L}} \mathbf{0} \quad \text{and} \quad \mathbf{A}_{11} - \mathbf{A}_{12}\mathbf{A}_{22}^-\mathbf{A}_{12}^* \geq_{\mathsf{L}} \mathbf{0} \quad \text{and} \quad \mathcal{C}(\mathbf{A}_{12}^*) \subset \mathcal{C}(\mathbf{A}_{22}),$$

where $\geq_{\mathsf{L}}$ denotes the Löwner partial ordering so that $\mathbf{A} \geq_{\mathsf{L}} \mathbf{0}$ means that $\mathbf{A}$ is nonnegative definite. We refer to the set of statements $(\mathrm{a}_0), (\mathrm{b}_0), (\mathrm{c}_0)$ as the *Albert nonnegative definiteness conditions.*

While we believe that Albert [6] was the first to establish (in 1969) these nonnegative definiteness conditions, we are well aware of Stigler's Law of Eponymy [428, ch. 14], which "in its simplest form" states that "no scientific discovery is named after its original discoverer" [428, p. 277].

When $\mathbf{A}_{11} >_{\mathsf{L}} \mathbf{0}$, i.e., $\mathbf{A}_{11}$ is positive definite, it follows at once that $\mathrm{rank}(\mathbf{A}_{11} : \mathbf{A}_{12}) = \mathrm{rank}(\mathbf{A}_{11})$ and so $\mathcal{C}(\mathbf{A}_{12}) \subset \mathcal{C}(\mathbf{A}_{11})$. And, of course, the Schur complement $\mathbf{A}_{11} - \mathbf{A}_{12}\mathbf{A}_{22}^{-1}\mathbf{A}_{12}^*$ is unique! Hence the following three statements are equivalent:

$$(\mathrm{a}_1)\ \mathbf{A} >_{\mathsf{L}} \mathbf{0},$$

$$(\mathrm{b}_1)\ \mathbf{A}_{11} >_{\mathsf{L}} \mathbf{0} \quad \text{and} \quad \mathbf{A}_{22} - \mathbf{A}_{12}^{*}\mathbf{A}_{11}^{-1}\mathbf{A}_{12} >_{\mathsf{L}} \mathbf{0},$$

$$(\mathrm{c}_1)\ \mathbf{A}_{22} >_{\mathsf{L}} \mathbf{0} \quad \text{and} \quad \mathbf{A}_{11} - \mathbf{A}_{12}\mathbf{A}_{22}^{-1}\mathbf{A}_{12}^{*} >_{\mathsf{L}} \mathbf{0},$$

see also the inertia formula (6.0.16) above. We refer to the set of statements $(\mathrm{a}_1), (\mathrm{b}_1), (\mathrm{c}_1)$ as the *Albert positive definiteness conditions.*

An interesting special case is when $\mathbf{A}_{12} = \mathbf{I}$ and

$$\mathbf{A} = \begin{pmatrix} \mathbf{A}_{11} & \mathbf{I} \\ \mathbf{I} & \mathbf{A}_{22} \end{pmatrix} \geq_{\mathsf{L}} \mathbf{0}. \tag{6.0.24}$$

Then both $\mathbf{A}_{11}$ and $\mathbf{A}_{22}$ are positive definite. To show this, we note from the conditions in (b_0) and (c_0) above, $\mathbf{A} \geq_{\mathsf{L}} \mathbf{0}$ implies $\mathcal{C}(\mathbf{I}) \subset \mathcal{C}(\mathbf{A}_{11})$ and $\mathcal{C}(\mathbf{I}) \subset \mathcal{C}(\mathbf{A}_{22})$ and so both $\mathbf{A}_{11}$ and $\mathbf{A}_{22}$ are positive definite. Hence

$$\begin{pmatrix} \mathbf{A}_{11} & \mathbf{I} \\ \mathbf{I} & \mathbf{A}_{22} \end{pmatrix} \geq_{\mathsf{L}} \mathbf{0} \quad \Leftrightarrow \quad \mathbf{A}_{11} \geq_{\mathsf{L}} \mathbf{A}_{22}^{-1} \quad \Leftrightarrow \quad \mathbf{A}_{22} \geq_{\mathsf{L}} \mathbf{A}_{11}^{-1}$$

and

$$\begin{pmatrix} \mathbf{A}_{11} & \mathbf{I} \\ \mathbf{I} & \mathbf{A}_{22} \end{pmatrix} >_{\mathsf{L}} \mathbf{0} \quad \Leftrightarrow \quad \mathbf{A}_{11} >_{\mathsf{L}} \mathbf{A}_{22}^{-1} \quad \Leftrightarrow \quad \mathbf{A}_{22} >_{\mathsf{L}} \mathbf{A}_{11}^{-1}.$$

6.0.5 The quotient property and the generalized quotient property

Let us consider the partitioned matrix

$$\mathbf{M} = \begin{pmatrix} \mathbf{P} & \mathbf{Q} \\ \mathbf{R} & \mathbf{S} \end{pmatrix} = \begin{pmatrix} \mathbf{P}_{11} & \mathbf{P}_{12} & \vdots & \mathbf{Q}_1 \\ \mathbf{P}_{21} & \mathbf{P}_{22} & \vdots & \mathbf{Q}_2 \\ \cdots & \cdots & \vdots & \cdots \\ \mathbf{R}_1 & \mathbf{R}_2 & \vdots & \mathbf{S} \end{pmatrix} \tag{6.0.25}$$

where the matrices $\mathbf{P}$ and $\mathbf{P}_{11}$ are both nonsingular (invertible); the matrix $\mathbf{M}$ need not be square. Then Haynsworth [210] proved

$$\mathbf{M}/\mathbf{P} = (\mathbf{M}/\mathbf{P}_{11})/(\mathbf{P}/\mathbf{P}_{11}), \tag{6.0.26}$$

which we refer to as the *quotient property*; see also Crabtree & Haynsworth [131] and Ostrowski [342].

The property (6.0.26) has also been called the *Crabtree–Haynsworth quotient property*.

On the right-hand side of (6.0.26), we see that the "denominator" matrix $\mathbf{P}_{11}$ "cancels" as if the expressions were scalar fractions. When $\mathbf{M}$ is square and nonsingular, then we may take determinants throughout (6.0.26) and apply the Schur determinant formula (0.3.2) to obtain

$$\det(\mathbf{M}/\mathbf{P}) = \frac{\det(\mathbf{M}/\mathbf{P}_{11})}{\det(\mathbf{P}/\mathbf{P}_{11})}.$$

This quotient property may be generalized as follows. Suppose now that the matrix $\mathbf{M}$ is Hermitian and nonnegative definite and partitioned essentially as in (6.0.25), with $\mathbf{P}_{21} = \mathbf{P}_{12}^*$ and $\mathbf{R} = \mathbf{Q}^*$

$$\mathbf{M} = \begin{pmatrix} \mathbf{P} & \mathbf{Q} \\ \mathbf{Q}^* & \mathbf{S} \end{pmatrix} = \begin{pmatrix} \mathbf{P}_{11} & \mathbf{P}_{12} & \vdots & \mathbf{Q}_1 \\ \mathbf{P}_{12}^* & \mathbf{P}_{22} & \vdots & \mathbf{Q}_2 \\ \cdots & \cdots & \vdots & \cdots \\ \mathbf{Q}_1^* & \mathbf{Q}_2^* & \vdots & \mathbf{S} \end{pmatrix}. \tag{6.0.27}$$

Then the generalized Schur complements $(\mathbf{M}/\mathbf{P}), (\mathbf{M}/\mathbf{P}_{11})$ and $(\mathbf{P}/\mathbf{P}_{11})$ are each unique and well defined. And so (6.0.26) still holds and we have the *generalized quotient property*

$$\mathbf{M}/\mathbf{P} = (\mathbf{M}/\mathbf{P}_{11})/(\mathbf{P}/\mathbf{P}_{11}), \tag{6.0.28}$$

see, e.g., [345, Th. 4.8]. We will use the quotient property (6.0.26) and the generalized quotient property (6.0.28) to prove an interesting property of conditioning in the multivariate normal distribution, see Section 6.2 below.

6.1 Some matrix inequalities in statistics and probability

In this section we study certain matrix inequalities which are useful in statistics and probability and in which the Schur complement plays a role. We begin with the Cramér–Rao Inequality, which provides a lower bound for the variance of an unbiased estimator. In addition we look at the Groves–Rothenberg inequality and matrix convexity, as well as a multivariate Cauchy–Schwarz inequality.

6.1.1 The Cramér–Rao Inequality and the unbiased estimation of a vector-valued parameter

According to *Kendall's Advanced Theory of Statistics* [430, Section 17.15], the inequality, popularly known as the *Cramér–Rao Inequality*, is "the fundamental inequality for the variance of an estimator" and was first given implicitly in 1942 by Aitken & Silverstone [5]. As noted by C. R. Rao [376, p. 400], the "Cramér–Rao bound has acquired the status of a technical term with its listing in the *McGraw-Hill Dictionary of Scientific and Technical Terms* [304] and has been applied in physics [447].

It seems that the Cramér–Rao Inequality was so named by Neyman & Scott [333] but was mentioned by C. R. Rao in the course on estimation he gave in 1943 [376, p. 400]. It "is named after two of its several discoverers": Harald Cramér (1893–1985) and C. Radhakrishna Rao (b. 1920), see Cramér [132, Section 32.3] and Rao [367]. The inequality was given earlier in 1943 by Maurice René Fréchet (1878–1973) in [172] and is also known as the *Fréchet–Cramér–Rao Inequality* or as *Fréchet's Inequality*, see Sverdrup [434, ch. XIII, Section 2.1, pp. 72–81].

The lower bound is sometimes called the *amount of information* or the *Fisher information* in the sample [430, Section 17.15] and Savage [397, p. 238] proposed that the inequality be called the *information inequality*. For a vector-valued parameter the lower bound is a covariance matrix known as the *Fisher information matrix*, see, e.g., Palmgren [346], or just the "information matrix", see, e.g., Rao [372, Section 5a.3, p. 326], who also considers the situation when this lower bound matrix is singular; see also Rao [376, p. 398]. The inequality is in the Löwner partial ordering.

Let $L_{\boldsymbol\theta}(\mathbf{y})$ denote the likelihood function of the unknown parameter vector $\boldsymbol\theta$ corresponding to the observable random vector $\mathbf{y}$. Let $\mathbf{g}(\boldsymbol\theta)$ be a vector-valued function of $\boldsymbol\theta$ and let

$$\mathbf{F}_{\boldsymbol\theta} = \mathrm{E}\left(\frac{\partial \log L_{\boldsymbol\theta}(\mathbf{y})}{\partial \boldsymbol\theta}\frac{\partial \log L_{\boldsymbol\theta}(\mathbf{y})}{\partial \boldsymbol\theta'}\right) \quad \text{and} \quad \mathbf{G}_{\boldsymbol\theta} = \frac{\partial \mathbf{g}(\boldsymbol\theta)}{\partial \boldsymbol\theta'}. \tag{6.1.1}$$

Let $\mathbf{t}(\mathbf{y})$ denote an unbiased estimator of $\mathbf{g}(\boldsymbol\theta)$; then, under certain regularity conditions, see, e.g., Rao [372, Section 5a.3, p. 326], the covariance matrix

$$\operatorname{cov}\big(\mathbf{t}(\mathbf{y})\big) \geq_{\mathsf{L}} \mathbf{G}_{\boldsymbol\theta}\mathbf{F}_{\boldsymbol\theta}^{-}\mathbf{G}_{\boldsymbol\theta}' \tag{6.1.2}$$

in the Löwner partial ordering. The lower bound matrix $\mathbf{G}_{\boldsymbol\theta}\mathbf{F}_{\boldsymbol\theta}^{-}\mathbf{G}_{\boldsymbol\theta}'$ does not depend on the choice of generalized inverse $\mathbf{F}_{\boldsymbol\theta}^{-}$, see (6.0.17) above. We refer to (6.1.2) as the *generalized multiparameter Cramér–Rao Inequality*.

When as unbiased estimator $\mathbf{t}_0$, say, exists such that its covariance matrix equals the lower bound matrix $\mathbf{G}_{\boldsymbol{\theta}}\mathbf{F}_{\boldsymbol{\theta}}^{-}\mathbf{G}_{\boldsymbol{\theta}}'$ then $\mathbf{t}_0$ is known as the *Markov estimator* [434, ch. XIII, Section 2, pp. 72–86] or *minimum variance unbiased estimator* of $\mathbf{g}(\boldsymbol{\theta})$.

To prove the generalized multiparameter Cramér–Rao Inequality (6.1.2), it suffices to show that the joint covariance matrix of $\mathbf{t}(\mathbf{y})$ and the *score vector* $\partial \log L_{\boldsymbol{\theta}}/\partial\boldsymbol{\theta}$ is

$$\boldsymbol{\Psi} = \operatorname{cov}\begin{pmatrix} \mathbf{t}(\mathbf{y}) \\ \partial \log L_{\boldsymbol{\theta}}(\mathbf{y})/\partial\boldsymbol{\theta} \end{pmatrix} = \begin{pmatrix} \operatorname{cov}\big(\mathbf{t}(\mathbf{y})\big) & \mathbf{G}_{\boldsymbol{\theta}} \\ \mathbf{G}_{\boldsymbol{\theta}}' & \mathbf{F}_{\boldsymbol{\theta}} \end{pmatrix}. \tag{6.1.3}$$

Since a covariance matrix is always nonnegative definite, it follows at once that $\mathbf{G}_{\boldsymbol{\theta}}\mathbf{F}_{\boldsymbol{\theta}}^{-}\mathbf{G}_{\boldsymbol{\theta}}'$ does not depend on the choice of generalized inverse $\mathbf{F}_{\boldsymbol{\theta}}^{-}$ and that the generalized Schur complement

$$\boldsymbol{\Psi}/\mathbf{F}_{\boldsymbol{\theta}} = \operatorname{cov}\big(\mathbf{t}(\mathbf{y})\big) - \mathbf{G}_{\boldsymbol{\theta}}\mathbf{F}_{\boldsymbol{\theta}}^{-}\mathbf{G}_{\boldsymbol{\theta}}' \geq_{\mathsf{L}} \mathbf{0}, \tag{6.1.4}$$

which leads directly to the inequality (6.1.2).

To prove (6.1.3), we first note that the covariance matrix

$$\operatorname{cov}\left(\frac{\partial \log L_{\boldsymbol{\theta}}(\mathbf{y})}{\partial\boldsymbol{\theta}}\right) = \mathrm{E}\left(\frac{\partial \log L_{\boldsymbol{\theta}}(\mathbf{y})}{\partial\boldsymbol{\theta}}\frac{\partial \log L_{\boldsymbol{\theta}}(\mathbf{y})}{\partial\boldsymbol{\theta}'}\right) = \mathbf{F}_{\boldsymbol{\theta}}, \tag{6.1.5}$$

since

$$\begin{aligned}\mathrm{E}\left(\frac{\partial \log L_{\boldsymbol{\theta}}(\mathbf{y})}{\partial\boldsymbol{\theta}}\right) &= \int \frac{\partial \log L_{\boldsymbol{\theta}}(\mathbf{y})}{\partial\boldsymbol{\theta}} L_{\boldsymbol{\theta}}(\mathbf{y})\, d\mathbf{y} \\ &= \int \frac{\partial L_{\boldsymbol{\theta}}(\mathbf{y})}{\partial\boldsymbol{\theta}}\, d\mathbf{y} = \frac{\partial}{\partial\boldsymbol{\theta}} \int L_{\boldsymbol{\theta}}(\mathbf{y})\, d\mathbf{y} = \mathbf{0}\end{aligned}$$

as $\int L_{\boldsymbol{\theta}}(\mathbf{y})\, d\mathbf{y} = 1$. The third equality depends on the regularity condition that we may interchange the integral sign with the differential $\partial/\partial\boldsymbol{\theta}$.

Furthermore, the cross-covariance matrix

$$\begin{aligned}\operatorname{cov}\left(\mathbf{t}(\mathbf{y}), \frac{\partial \log L_{\boldsymbol{\theta}}(\mathbf{y})}{\partial\boldsymbol{\theta}}\right) &= \mathrm{E}\left(\mathbf{t}(\mathbf{y})\frac{\partial \log L_{\boldsymbol{\theta}}(\mathbf{y})}{\partial\boldsymbol{\theta}'}\right) \\ &= \int \mathbf{t}(\mathbf{y})\frac{\partial \log L_{\boldsymbol{\theta}}(\mathbf{y})}{\partial\boldsymbol{\theta}'} L_{\boldsymbol{\theta}}(\mathbf{y})\, d\mathbf{y} \\ &= \int \mathbf{t}(\mathbf{y})\frac{\partial L_{\boldsymbol{\theta}}(\mathbf{y})}{\partial\boldsymbol{\theta}'}\, d\mathbf{y} = \frac{\partial}{\partial\boldsymbol{\theta}'}\int \mathbf{t}(\mathbf{y}) L_{\boldsymbol{\theta}}(\mathbf{y})\, d\mathbf{y} \\ &= \frac{\partial}{\partial\boldsymbol{\theta}'}\mathrm{E}\big(\mathbf{t}(\mathbf{y})\big) = \frac{\partial \mathbf{g}(\boldsymbol{\theta})}{\partial\boldsymbol{\theta}'} = \mathbf{G}_{\boldsymbol{\theta}}.\end{aligned} \tag{6.1.6}$$

An important special case of the inequality (6.1.2) is when $g(\boldsymbol{\theta}) = \boldsymbol{\theta}$ and so $\mathrm{E}\big(\mathbf{t}(\mathbf{y})\big) = \boldsymbol{\theta}$. Here $\mathbf{G}_{\boldsymbol{\theta}} = \partial g(\boldsymbol{\theta})/\partial\boldsymbol{\theta}' = \partial(\boldsymbol{\theta})/\partial\boldsymbol{\theta}' = \mathbf{I}$, the identity matrix and the joint covariance matrix

$$\operatorname{cov}\begin{pmatrix} \mathbf{t}(\mathbf{y}) \\ \partial \log L_{\boldsymbol{\theta}}(\mathbf{y})/\partial\boldsymbol{\theta} \end{pmatrix} = \begin{pmatrix} \operatorname{cov}\big(\mathbf{t}(\mathbf{y})\big) & \mathbf{I} \\ \mathbf{I} & \mathbf{F}_{\boldsymbol{\theta}} \end{pmatrix}. \tag{6.1.7}$$

It follows that $\mathbf{F}_{\boldsymbol{\theta}}$ is positive definite and the Cramér–Rao Inequality (6.1.2) becomes the *reduced multiparameter Cramér–Rao Inequality*

$$\operatorname{cov}\big(\mathbf{t}(\mathbf{y})\big) \geq_{\mathsf{L}} \mathbf{F}_{\boldsymbol{\theta}}^{-1}. \tag{6.1.8}$$

A simple example in which the reduced multiparameter Cramér–Rao Inequality (6.1.8) yields the minimum variance unbiased estimator is in the usual general linear model with normality, see, e.g., [345, p. 290] Let $\mathbf{y}$ follow the multivariate normal distribution $\mathrm{N}(\mathbf{X}\boldsymbol{\gamma}, \sigma^2\mathbf{I})$, where $\mathbf{X}$ has full column rank; the vector $\boldsymbol{\gamma}$ is unknown and is to be estimated based on a single realization of $\mathbf{y}$. The log-likelihood

$$\log L = -\frac{n}{2}\log 2\pi - n\log\sigma - \frac{1}{2\sigma^2}(\mathbf{y} - \mathbf{X}\boldsymbol{\gamma})'(\mathbf{y} - \mathbf{X}\boldsymbol{\gamma}) \tag{6.1.9}$$

and so the score vector

$$\frac{\partial \log L}{\partial \boldsymbol{\gamma}} = \frac{1}{\sigma^2}\mathbf{X}'(\mathbf{y} - \mathbf{X}\boldsymbol{\gamma}) = \frac{1}{\sigma^2}(\mathbf{X}'\mathbf{y} - \mathbf{X}'\mathbf{X}\boldsymbol{\gamma}), \tag{6.1.10}$$

which has covariance matrix $\operatorname{cov}(\partial \log L/\partial\boldsymbol{\gamma}) = (1/\sigma^2)\mathbf{X}'\mathbf{X}$. The lower bound matrix is, therefore, $\mathbf{F}_{\boldsymbol{\gamma}}^{-1} = \sigma^2(\mathbf{X}'\mathbf{X})^{-1}$, the covariance matrix of the maximum likelihood estimator $\hat{\boldsymbol{\gamma}} = (\mathbf{X}'\mathbf{X})^{-1}\mathbf{X}'\mathbf{y}$, which we may obtain by equating the score vector (6.1.10) to $\mathbf{0}$. Hence $\hat{\boldsymbol{\gamma}}$ is the minimum variance unbiased (or Markov) estimator of $\boldsymbol{\gamma}$.

6.1.2 The Cramér–Rao Inequality and Schur products

In this subsection we will show how the reduced multiparameter Cramér–Rao Inequality (6.1.8) yields interesting matrix inequalities involving Schur products.

Let $\mathbf{A}$ denote a $p \times p$ real symmetric positive definite matrix. Then using the reduced multiparameter Cramér–Rao inequality (6.1.8), Styan [431] proved that

$$\mathbf{A} * \mathbf{A} \geq_{\mathsf{L}} 2(\mathbf{A} * \mathbf{I})(\mathbf{A}^{-1} * \mathbf{A} + \mathbf{I})^{-1}(\mathbf{A} * \mathbf{I}) \tag{6.1.11}$$

and

$$\mathbf{A}^{-1} * \mathbf{A} + \mathbf{I} \geq_{\mathsf{L}} 2(\mathbf{A} * \mathbf{I})(\mathbf{A} * \mathbf{A})^{-1}(\mathbf{A} * \mathbf{I}). \tag{6.1.12}$$

Here $*$ denotes the Schur product (or Hadamard product) of two matrices multiplied together elementwise. Also used here is the theorem established in 1911 by Issai Schur [403, Th. VII] stating that the Schur product of two positive definite matrices is positive definite, see also [431, Th. 3.1].

Ando [16, Th. 20] proved the inequality (6.1.11) in a different (nonstatistical) way and also showed that if both $\mathbf{A}$ and $\mathbf{B}$ are $p \times p$ and positive definite then

$$\mathbf{A} * \mathbf{B} \geq_{\mathsf{L}} \big((\mathbf{A} + \mathbf{B}) * \mathbf{I}\big)\big(\mathbf{A}^{-1} * \mathbf{B} + \mathbf{B}^{-1} * \mathbf{A} + 2\mathbf{I}\big)^{-1}\big((\mathbf{A} + \mathbf{B}) * \mathbf{I}\big),$$

which becomes (6.1.11) when $\mathbf{A} = \mathbf{B}$. Ando [16] also established other extensions of (6.1.11).

To prove (6.1.11), Styan [431] considered the problem of estimating the unknown positive standard deviations $\sigma_1, ..., \sigma_p$ in a p-variate normal distribution with zero mean vector and known positive definite correlation matrix $\mathbf{R}$. A random sample of $p \times 1$ vectors $\mathbf{x}_1, ..., \mathbf{x}_n$, with $n > p$, is available and we will let the sample covariance matrix

$$\mathbf{S} = \frac{1}{n} \sum_{\alpha=1}^{n} \mathbf{x}_\alpha \mathbf{x}_\alpha' . \tag{6.1.13}$$

With $n > p$ the matrix $\mathbf{S}$ is positive definite with probability 1, see, e.g., Das Gupta [138]. Let $l = -(2/n)\log L - p \log 2\pi$, where L is the joint likelihood of $\mathbf{x}_1, ..., \mathbf{x}_n$ and log is the natural logarithm. Then, as Olkin & Siotani [340] and Styan [431] have shown,

$$\frac{\partial l}{\partial \boldsymbol{\sigma}^{(-1)}} = 2\big((\mathbf{R}^{-1} * \mathbf{S})\boldsymbol{\sigma}^{(-1)} - \boldsymbol{\sigma}\big), \tag{6.1.14}$$

where the $p \times 1$ vector $\boldsymbol{\sigma} = \{\sigma_i\}$, with $\boldsymbol{\sigma}^{(-1)} = \{1/\sigma_i\}$. From (6.1.14), it follows at once that the likelihood equations may be written as

$$(\mathbf{R}^{-1} * \mathbf{S})\hat{\boldsymbol{\sigma}}^{(-1)} = \hat{\boldsymbol{\sigma}}. \tag{6.1.15}$$

The maximum likelihood estimator $\hat{\boldsymbol{\sigma}}$ in (6.1.15) is unique since the Hessian matrix

$$\frac{\partial^2 l}{\partial \boldsymbol{\sigma}^{(-1)} \partial \boldsymbol{\sigma}^{(-1)\prime}} = 2(\mathbf{R}^{-1} * \mathbf{S} + \boldsymbol{\Delta}^2) \tag{6.1.16}$$

is positive definite (with probability 1). Here $\boldsymbol{\Delta}$ is the diagonal matrix of the σ_i $(i = 1, \ldots, p)$.

Let $\mathbf{s}$ denote the $p \times 1$ vector formed from the diagonal elements of the sample covariance matrix defined in (6.1.13) and so $\mathrm{E}(\mathbf{s}) = \boldsymbol{\sigma}^{(2)}$, the $p \times 1$ vector of the σ_i^2 $(i = 1, \ldots, p)$. To prove (6.1.11), we will use (6.1.8) with the parameter vector $\boldsymbol{\theta} = \boldsymbol{\sigma}^{(2)}$, the unbiased estimator $\mathbf{t}(\mathbf{y}) = \mathbf{s}$, and the Fisher information matrix

$$\mathbf{F} = \mathrm{cov}(\partial \log L / \partial \boldsymbol{\sigma}^{(2)}).$$

We will evaluate

$$\mathrm{cov}\begin{pmatrix} \mathbf{s} \\ \partial \log L / \partial \boldsymbol{\sigma}^{(-1)} \end{pmatrix} = \begin{pmatrix} \mathrm{cov}(\mathbf{s}) & \mathbf{I} \\ \mathbf{I} & \mathbf{F} \end{pmatrix} \tag{6.1.17}$$

from which, using (6.1.8) the inequality (6.1.11) follows directly.

We begin by proving that the covariance matrix

$$\mathrm{cov}(\mathbf{s}) = \frac{2}{n}(\boldsymbol{\Sigma} * \boldsymbol{\Sigma}) = \frac{2}{n}\boldsymbol{\Delta}^2(\mathbf{R} * \mathbf{R})\boldsymbol{\Delta}^2, \tag{6.1.18}$$

where the covariance matrix $\mathrm{cov}(\mathbf{x}_\alpha) = \boldsymbol{\Sigma} = \boldsymbol{\Delta}\mathbf{R}\boldsymbol{\Delta}$. Let s_i and s_j denote, respectively, the ith and jth element of $\mathbf{s}$. Then for $i, j = 1, \ldots, p$,

$$\begin{aligned} \mathrm{cov}(s_i,\, s_j) &= \frac{1}{n^2}\mathrm{cov}\left(\sum_{\alpha=1}^{n} x_{\alpha i}^2,\, \sum_{\beta=1}^{n} x_{\beta j}^2\right) = \frac{1}{n}\mathrm{cov}(X_i^2,\, X_j^2) \\ &= \frac{1}{2n}\left(\mathrm{var}(X_i^2 + X_j^2) - \mathrm{var}(X_i^2) - \mathrm{var}(X_j^2)\right) \\ &= \frac{1}{n}\left(\mathrm{tr}(\mathbf{V}^2) - \sigma_{ii}^2 - \sigma_{jj}^2\right), \end{aligned} \tag{6.1.19}$$

see, e.g., Searle [410, p. 57]; here X_i and X_j are bivariate normal with zero means and covariance matrix

$$\mathbf{V} = \begin{pmatrix} \sigma_{ii} & \sigma_{ij} \\ \sigma_{ij} & \sigma_{jj} \end{pmatrix}. \tag{6.1.20}$$

Since $\mathrm{tr}(\mathbf{V}^2) = \sigma_{ii}^2 + \sigma_{jj}^2 + 2\sigma_{ij}^2$ it follows at once that for $i, j = \cdots = p$, the covariance $\mathrm{cov}(s_i,\, s_j) = (2/n)\sigma_{ij}^2$, which is equivalent to (6.1.18).

Moreover the Fisher information matrix

$$\mathbf{F}_{\boldsymbol{\theta}} = \mathrm{cov}\left(\frac{\partial \log L}{\partial \boldsymbol{\sigma}^{(2)}}\right) = \mathrm{cov}\left(-\frac{n}{2}\frac{\partial l}{\partial \boldsymbol{\sigma}^{(2)}}\right) = \frac{n^2}{4}\mathrm{cov}\left(\frac{\partial \boldsymbol{\sigma}^{(-1)}}{\partial \boldsymbol{\sigma}^{(2)\prime}}\frac{\partial l}{\partial \boldsymbol{\sigma}^{(-1)}}\right),$$

where $l = -(2/n)\log L - p\log 2\pi$. Clearly

$$\frac{\partial \boldsymbol{\sigma}^{(-1)}}{\partial \boldsymbol{\sigma}^{(2)\prime}} = -\frac{1}{2}\boldsymbol{\Delta}^{-3}$$

and

$$\operatorname{cov}\left(\frac{\partial l}{\partial \boldsymbol{\sigma}^{(-1)}}\right) = \frac{2}{n}\mathrm{E}\frac{\partial^2 l}{\partial \boldsymbol{\sigma}^{(-1)}\partial \boldsymbol{\sigma}^{(-1)\prime}} = \frac{4}{n}\boldsymbol{\Delta}(\mathbf{R}^{-1} * \mathbf{R} + \mathbf{I})\boldsymbol{\Delta}. \tag{6.1.21}$$

The first equality in (6.1.21) holds since here

$$\mathrm{E}\left(\frac{\partial \log L}{\partial \boldsymbol{\sigma}^{(-1)}}\frac{\partial \log L}{\partial \boldsymbol{\sigma}^{(-1)\prime}}\right) = -\mathrm{E}\left(\frac{\partial^2 \log L}{\partial \boldsymbol{\sigma}^{(-1)}\partial \boldsymbol{\sigma}^{(-1)\prime}}\right),$$

see, e.g., [430, §17.14], while for the second equality in (6.1.21), we use (6.1.16) and $\mathrm{E}(\mathbf{S}) = \boldsymbol{\Sigma}$. Therefore

$$\begin{aligned}\mathbf{F}_{\boldsymbol{\theta}} =& \frac{-1}{2}\boldsymbol{\Delta}^{-3} \cdot \frac{4}{n}\boldsymbol{\Delta}(\mathbf{R}^{-1} * \mathbf{R} + \mathbf{I})\boldsymbol{\Delta} \cdot \frac{-1}{2}\boldsymbol{\Delta}^{-3} \\ =& \frac{1}{n}\boldsymbol{\Delta}^{-2}(\mathbf{R}^{-1} * \mathbf{R} + \mathbf{I})\boldsymbol{\Delta}^{-2},\end{aligned} \tag{6.1.22}$$

which completes our proof of (6.1.11).

6.1.3 The Cramér–Rao Inequality and BLUP

The Cramér–Rao Inequality (6.1.2) can be nicely generalized to cover the case of predicting values of a random vector. Let $\mathbf{y}$ be an observable random vector and $\mathbf{y}_f$ be an unobservable (observable in future) random vector with joint density $f_{\boldsymbol{\theta}}(\mathbf{y}, \mathbf{y}_f)$, where $\boldsymbol{\theta}$ is an unknown parameter vector.

Let us consider the problem of predicting $\mathbf{y}_f$ based on $\mathbf{y}$. Let the statistic $\mathbf{t}(\mathbf{y})$ be an unbiased predictor of $\mathbf{y}_f$ so that

$$\mathrm{E}_{\boldsymbol{\theta}}\big(\mathbf{t}(\mathbf{y})\big) = \mathrm{E}_{\boldsymbol{\theta}}(\mathbf{y}_f) \qquad \forall\, \boldsymbol{\theta}, \tag{6.1.23}$$

and let now

$$\mathbf{F}_{\boldsymbol{\theta}} = \mathrm{E}_{\boldsymbol{\theta}}\left(\frac{\partial \log L_{\boldsymbol{\theta}}(\mathbf{y})}{\partial \boldsymbol{\theta}}\frac{\partial \log L_{\boldsymbol{\theta}}(\mathbf{y})}{\partial \boldsymbol{\theta}'}\right) \quad \text{and} \quad \mathbf{G}_{\boldsymbol{\theta}} = \frac{\partial \mathrm{E}_{\boldsymbol{\theta}}(\mathbf{y}_f|\mathbf{y})}{\partial \boldsymbol{\theta}'}. \tag{6.1.24}$$

Then, under certain regularity conditions, the covariance matrix of the prediction error

$$\operatorname{cov}_{\boldsymbol{\theta}}\big(\mathbf{t}(\mathbf{y}) - \mathbf{y}_f\big) \geq_{\mathsf{L}} \mathrm{E}_{\boldsymbol{\theta}}\big(\operatorname{cov}_{\boldsymbol{\theta}}(\mathbf{y}_f|\mathbf{y})\big) + \mathbf{G}_{\boldsymbol{\theta}}\mathbf{F}_{\boldsymbol{\theta}}^{-}\mathbf{G}_{\boldsymbol{\theta}}' \tag{6.1.25}$$

in the Löwner partial ordering.

The proof of (6.1.25) follows from the fact that

$$\operatorname{cov}_{\boldsymbol{\theta}}\big(\mathbf{t}(\mathbf{y}) - \mathbf{y}_f\big) = \operatorname{cov}_{\boldsymbol{\theta}}\big(\mathbf{t}(\mathbf{y}) - \mathrm{E}_{\boldsymbol{\theta}}(\mathbf{y}_f|\mathbf{y})\big) + \mathrm{E}_{\boldsymbol{\theta}}\big(\operatorname{cov}_{\boldsymbol{\theta}}(\mathbf{y}_f|\mathbf{y})\big),$$

which was first considered by Yatracos [466]; see also Nayak [329, 330]. The unbiased predictor the covariance matrix of which attains the lower bound matrix (6.1.25) is called the *uniformly minimum mean squared error unbiased predictor* (UMMSEUP).

Consider the linear model under normality. Then the joint distribution of $\mathbf{y}$ and $\mathbf{y}_f$ is

$$\begin{pmatrix} \mathbf{y} \\ \mathbf{y}_f \end{pmatrix} \sim N\left(\begin{pmatrix} \mathbf{X}\boldsymbol{\beta} \\ \mathbf{X}_f\boldsymbol{\beta} \end{pmatrix}, \quad \sigma^2 \begin{pmatrix} \mathbf{V} & \mathbf{W} \\ \mathbf{W}' & \mathbf{V}_f \end{pmatrix} \right), \tag{6.1.26}$$

where $\mathbf{X}$ and $\mathbf{X}_f$ are known model (design) matrices, $\boldsymbol{\beta}$ is a vector of unknown parameters, $\mathbf{V}$ and $\mathbf{V}_f$ are known positive definite matrices, $\mathbf{W}$ is a known cross-covariance matrix and σ^2 is an unknown positive constant.

Let $\mathbf{Ty}$ be a linear predictor of $\mathbf{y}_f$. The linear predictor $\mathbf{Ty}$ is unbiased, as in (6.1.23), if and only if the matrix equality

$$\mathbf{TX} = \mathbf{X}_f \tag{6.1.27}$$

holds. Furthermore, under normality, the Cramér–Rao lower bound for the covariance matrix of the prediction error is

$$\sigma^2(\mathbf{V}_f - \mathbf{W}'\mathbf{V}^{-1}\mathbf{W} + (\mathbf{X}_f - \mathbf{W}'\mathbf{V}^{-1}\mathbf{X})(\mathbf{X}'\mathbf{V}^{-1}\mathbf{X})^{-1}(\mathbf{X}_f - \mathbf{W}'\mathbf{V}^{-1}\mathbf{X})') \tag{6.1.28}$$

and hence the matrix $\mathbf{T}$, which satisfies (6.1.27), also satisfies the matrix inequality

$$\begin{aligned} &\sigma^2(\mathbf{TVT}' - \mathbf{TW} - \mathbf{W}'\mathbf{T}') \geq_{\mathsf{L}} \\ &\quad \sigma^2((\mathbf{X}_f - \mathbf{W}'\mathbf{V}^{-1}\mathbf{X})(\mathbf{X}'\mathbf{V}^{-1}\mathbf{X})^{-1}(\mathbf{X}_f - \mathbf{W}'\mathbf{V}^{-1}\mathbf{X})' - \mathbf{W}'\mathbf{V}^{-1}\mathbf{W}). \end{aligned} \tag{6.1.29}$$

Goldberger [184] showed that the *best linear unbiased predictor* (BLUP) of $\mathbf{y}_f$ is

$$\mathrm{BLUP}(\mathbf{y}_f) = \mathbf{X}_f\tilde{\boldsymbol{\beta}} + \mathbf{W}'\mathbf{V}^{-1}(\mathbf{y} - \mathbf{X}\tilde{\boldsymbol{\beta}}), \tag{6.1.30}$$

where $\tilde{\boldsymbol{\beta}} = (\mathbf{X}'\mathbf{V}^{-1}\mathbf{X})^{-1}\mathbf{X}'\mathbf{V}^{-1}\mathbf{y}$. And so for BLUP, the matrix $\mathbf{T}$ as introduced in (6.1.27) has the form

$$\mathbf{T} = \mathbf{X}_f(\mathbf{X}'\mathbf{V}^{-1}\mathbf{X})^{-1}\mathbf{X}'\mathbf{V}^{-1} + \mathbf{W}'\mathbf{V}^{-1}(\mathbf{I} - \mathbf{X}(\mathbf{X}'\mathbf{V}^{-1}\mathbf{X})^{-1}\mathbf{X}'\mathbf{V}^{-1}). \tag{6.1.31}$$

We note that the matrix product $\mathbf{W}'\mathbf{V}^{-1}(\mathbf{I} - \mathbf{X}(\mathbf{X}'\mathbf{V}^{-1}\mathbf{X})^{-1}\mathbf{X}'\mathbf{V}^{-1})$ in (6.1.31) is the Schur complement of $\mathbf{X}'\mathbf{V}^{-1}\mathbf{X}$ in

$$\begin{pmatrix} \mathbf{W}'\mathbf{V}^{-1} & \mathbf{W}'\mathbf{V}^{-1}\mathbf{X} \\ \mathbf{X}'\mathbf{V}^{-1} & \mathbf{X}'\mathbf{V}^{-1}\mathbf{X} \end{pmatrix}.$$

The matrix $\mathbf{T}$ given by (6.1.31) satisfies the unbiasedness condition (6.1.27) and the covariance matrix of the linear predictor $\mathbf{Ty}$ of $\mathbf{y}_f$ is the lower bound matrix in the matrix inequality (6.1.29). Thus, under normality, the best linear unbiased predictor (BLUP) of $\mathbf{y}_f$ is also the uniformly minimum mean squared error unbiased predictor (UMMSEUP).

6.1.4 The Groves–Rothenberg inequality and matrix convexity

As Groves & Rothenberg [196] observed, "it is well known that if X is any positive [scalar] random variable then the expectation

$$\mathrm{E}\left(\frac{1}{X}\right) \geq \frac{1}{\mathrm{E}(X)},$$

provided that the expectations exist". A quick proof uses the Cauchy–Schwarz inequality $\mathrm{E}(U^2)\mathrm{E}(V^2) \geq \mathrm{E}^2(UV)$, with $U = \sqrt{X}$ and $V = 1/\sqrt{X}$.

We now consider the $n \times n$ random matrix $\mathbf{A}$, which we assume to be real and symmetric, and positive definite with probability 1. Then, as Groves & Rothenberg (1969) proved,

$$\mathrm{E}(\mathbf{A}^{-1}) \geq_{\mathsf{L}} \left(\mathrm{E}(\mathbf{A})\right)^{-1} \tag{6.1.32}$$

provided the expectations exist. To prove (6.1.32), Groves & Rothenberg [196] used a convexity argument. As Styan [432, pp. 43–44] showed, we may prove (6.1.32) very quickly using a Schur complement argument.

The Schur complement of $\mathbf{A}$ in the $2n \times 2n$ random matrix

$$\mathbf{M} = \begin{pmatrix} \mathbf{A} & \mathbf{I} \\ \mathbf{I} & \mathbf{A}^{-1} \end{pmatrix} \tag{6.1.33}$$

is $\mathbf{0}$ and so using the Haynsworth inertia additivity formula (6.0.13), we see that $\mathbf{M}$ has inertia triple $\{n, 0, n\}$ and hence is nonnegative definite with probability 1. It follows that the expectation matrix

$$\mathrm{E}(\mathbf{M}) = \begin{pmatrix} \mathrm{E}(\mathbf{A}) & \mathbf{I} \\ \mathbf{I} & \mathrm{E}(\mathbf{A}^{-1}) \end{pmatrix} \tag{6.1.34}$$

is nonnegative definite and hence the Schur complement

$$\mathrm{E}(\mathbf{M})/\mathrm{E}(\mathbf{A}) = \mathrm{E}(\mathbf{A}^{-1}) - \big(\mathrm{E}(\mathbf{A})\big)^{-1} \geq_{\mathsf{L}} \mathbf{0},$$

which establishes (6.1.32).

Moore [322, 323] showed that the matrix-inverse function is "matrix convex" on the class of all real symmetric positive definite matrices in that

$$\lambda\mathbf{A}^{-1} + (1-\lambda)\mathbf{B}^{-1} \geq_{\mathsf{L}} \big(\lambda\mathbf{A} + (1-\lambda)\mathbf{B}\big)^{-1}, \tag{6.1.35}$$

see also W. N. Anderson & Trapp [14], Lieb [283], Marshall & Olkin [301, pp. 469–471], Styan [432, p. 43].

To prove (6.1.35), we note that

$$\begin{aligned} \mathbf{G}_\lambda &= \begin{pmatrix} \lambda\mathbf{A} + (1-\lambda)\mathbf{B} & \mathbf{I} \\ \mathbf{I} & \lambda\mathbf{A}^{-1} + (1-\lambda)\mathbf{B}^{-1} \end{pmatrix} \\ &= \lambda \begin{pmatrix} \mathbf{A} & \mathbf{I} \\ \mathbf{I} & \mathbf{A}^{-1} \end{pmatrix} + (1-\lambda) \begin{pmatrix} \mathbf{B} & \mathbf{I} \\ \mathbf{I} & \mathbf{B}^{-1} \end{pmatrix} \geq_{\mathsf{L}} \mathbf{0}, \end{aligned} \tag{6.1.36}$$

and so the Schur complement

$$\mathbf{G}_\lambda/(\lambda\mathbf{A} + (1-\lambda)\mathbf{B}) \; = \; \lambda\mathbf{A}^{-1} + (1-\lambda)\mathbf{B}^{-1} - \big(\lambda\mathbf{A} + (1-\lambda)\mathbf{B}\big)^{-1} \geq_{\mathsf{L}} \mathbf{0},$$

from which (6.1.35) follows at once.

The inequality (6.1.35) is a special case of

$$\sum_{j=1}^{k} \lambda_j \mathbf{X}_j' \mathbf{A}_j^{-1} \mathbf{X}_j \geq_{\mathsf{L}} \sum_{j=1}^{k} \lambda_j \mathbf{X}_j' \mathbf{X}_j \Big(\sum_{j=1}^{k} \lambda_j \mathbf{X}_j' \mathbf{A}_j \mathbf{X}_j \Big)^{-1} \sum_{j=1}^{k} \lambda_j \mathbf{X}_j' \mathbf{X}_j \,, \tag{6.1.37}$$

where the matrices $\mathbf{X}_j$ are all $n \times p$ with full column rank $p \leq n$ and the $\mathbf{A}_j$ are all $n \times n$ positive definite; the scalars λ_j are all positive with $\lambda_1 + \cdots + \lambda_k = 1$. When $k = 2$, $\lambda_1 = \lambda$, $\mathbf{A}_1 = \mathbf{A}$, $\mathbf{A}_2 = \mathbf{B}$, $\mathbf{X}_1 = \mathbf{X}_2 = \mathbf{I}$ then (6.1.37) becomes (6.1.35). The inequality (6.1.37) is easily proved with Schur complements similarly to our proof of (6.1.35).

The inequality (6.1.37) was apparently first established in 1959 by Kiefer [259, Lemma 3.2]; see also Drury *et al.* [147, p. 456], Gaffke & Krafft [178, Lemma 2.1], Lieb [283], Lieb & Ruskai [284], Nakamoto & Takahashi [327, Th. 5], Rao [368, Lemma 2c].

6.1.5 A multivariate Cauchy–Schwarz Inequality

Let us consider again the $2n \times 2n$ nonnegative definite matrix

$$\mathbf{M} = \begin{pmatrix} \mathbf{A} & \mathbf{I} \\ \mathbf{I} & \mathbf{A}^{-1} \end{pmatrix}, \tag{6.1.38}$$

where the matrix $\mathbf{A}$ is positive definite. Let $\mathbf{X}$ denote an $n \times p$ matrix and let the $n \times q$ matrix $\mathbf{Y}$ have full column rank q. Then

$$\begin{pmatrix} \mathbf{X}' & \mathbf{0} \\ \mathbf{0} & \mathbf{Y}' \end{pmatrix} \begin{pmatrix} \mathbf{A} & \mathbf{I} \\ \mathbf{I} & \mathbf{A}^{-1} \end{pmatrix} \begin{pmatrix} \mathbf{X} & \mathbf{0} \\ \mathbf{0} & \mathbf{Y} \end{pmatrix} = \begin{pmatrix} \mathbf{X}'\mathbf{A}\mathbf{X} & \mathbf{X}'\mathbf{Y} \\ \mathbf{Y}'\mathbf{X} & \mathbf{Y}'\mathbf{A}^{-1}\mathbf{Y} \end{pmatrix} \geq_{\mathsf{L}} \mathbf{0} \tag{6.1.39}$$

and $\mathbf{Y}'\mathbf{A}^{-1}\mathbf{Y}$ is positive definite. And so the Schur complement

$$\mathbf{S} = \mathbf{X}'\mathbf{A}\mathbf{X} - \mathbf{X}'\mathbf{Y}(\mathbf{Y}'\mathbf{A}^{-1}\mathbf{Y})^{-1}\mathbf{Y}'\mathbf{X} \geq_{\mathsf{L}} \mathbf{0} \tag{6.1.40}$$

and hence

$$\mathbf{X}'\mathbf{A}\mathbf{X} \geq_{\mathsf{L}} \mathbf{X}'\mathbf{Y}(\mathbf{Y}'\mathbf{A}^{-1}\mathbf{Y})^{-1}\mathbf{Y}'\mathbf{X}, \tag{6.1.41}$$

which we may call a *multivariate Cauchy–Schwarz Inequality*. Equality holds in (6.1.41) if and only if $\mathbf{S} = \mathbf{0}$ and this occurs if and only if

$$\operatorname{rank} \begin{pmatrix} \mathbf{X}'\mathbf{A}\mathbf{X} & \mathbf{X}'\mathbf{Y} \\ \mathbf{Y}'\mathbf{X} & \mathbf{Y}'\mathbf{A}^{-1}\mathbf{Y} \end{pmatrix} = \operatorname{rank}(\mathbf{Y}'\mathbf{A}^{-1}\mathbf{Y}) = q \tag{6.1.42}$$

However, from (6.1.39),

$$\begin{aligned} \operatorname{rank} \begin{pmatrix} \mathbf{X}'\mathbf{A}\mathbf{X} & \mathbf{X}'\mathbf{Y} \\ \mathbf{Y}'\mathbf{X} & \mathbf{Y}'\mathbf{A}^{-1}\mathbf{Y} \end{pmatrix} &= \operatorname{rank} \begin{pmatrix} \mathbf{A} & \mathbf{I} \\ \mathbf{I} & \mathbf{A}^{-1} \end{pmatrix} \begin{pmatrix} \mathbf{X} & \mathbf{0} \\ \mathbf{0} & \mathbf{Y} \end{pmatrix} \\ &= \operatorname{rank} \begin{pmatrix} \mathbf{A}\mathbf{X} & \mathbf{Y} \\ \mathbf{X} & \mathbf{A}^{-1}\mathbf{Y} \end{pmatrix} = \operatorname{rank}(\mathbf{A}\mathbf{X} : \mathbf{Y}) = q \end{aligned}$$

if and only if $\mathbf{Y} = \mathbf{A}\mathbf{X}\mathbf{F}$ for some conformable nonnull matrix $\mathbf{F}$.

When $p = q = 1$, then $\mathbf{X} = \mathbf{x}$ and $\mathbf{Y} = \mathbf{y}$ are column vectors and the inequality (6.1.41) becomes the (univariate) Cauchy–Schwarz Inequality

$$\mathbf{x}'\mathbf{A}\mathbf{x} \geq \frac{(\mathbf{x}'\mathbf{y})^2}{\mathbf{y}'\mathbf{A}^{-1}\mathbf{y}}, \tag{6.1.43}$$

with equality (for nonnull $\mathbf{x}$ and $\mathbf{y}$) if and only if $\mathbf{y} = f\mathbf{A}\mathbf{x}$ for some nonzero scalar f. When $\mathbf{A} = \mathbf{I}$, the identity matrix, then (6.1.43) becomes the familiar Cauchy–Schwarz inequality

$$\mathbf{x}'\mathbf{x} \cdot \mathbf{y}'\mathbf{y} \geq (\mathbf{x}'\mathbf{y})^2, \tag{6.1.44}$$

with equality (for nonnull $\mathbf{x}$ and $\mathbf{y}$) if and only if $\mathbf{y} = f\mathbf{x}$ for some nonzero scalar f. For more about the Cauchy–Schwarz Inequality and its history, see [456].

6.2 Correlation

As observed by Rodriguez [384] in the *Encyclopedia of Statistical Sciences*, "Correlation methods for determining the strength of the linear relationship between two or more variables are among the most widely applied statistical techniques. Theoretically, the concept of correlation has been the starting point or building block in the development of a number of areas of statistical research." In this section we show how the Schur complement plays a role in partial, conditional, multiple, and canonical correlations. In addition we show how the generalized quotient property of Schur complements plays a key role in the analysis of the conditional multivariate normal distribution.

6.2.1 When is a so-called "correlation matrix" really a correlation matrix?

We define a *correlation matrix* $\mathbf{R}$ to be a square $p \times p$ symmetric nonnegative definite matrix with all diagonal elements equal to 1. This definition is natural in the sense that given such a matrix $\mathbf{R}$, we can always construct an $n \times p$ data matrix whose associated correlation matrix is $\mathbf{R}$.

Consider p variables $u_1, \ldots, u_p$ whose observed centered values are the columns of $\mathbf{U} = (\mathbf{u}_1 : \cdots : \mathbf{u}_p)$, and assume that each variable has a nonzero variance, i.e., $\mathbf{u}_i \neq \mathbf{0}$ for each i. Let each column of $\mathbf{U}$ have unit length. Now since the correlation coefficient r_{ij} is the cosine between centered vectors $\mathbf{u}_i$ and $\mathbf{u}_j$, the correlation matrix $\mathbf{R}$ is simply $\mathbf{U}'\mathbf{U}$, and thereby nonnegative definite. Note that $\mathbf{U}'\mathbf{U}$ is not necessarily a correlation matrix of u_i-variables if $\mathbf{U}$ is not centered (even though the columns of $\mathbf{U}$ have unit length). It is also clear that orthogonality and uncorrelatedness are equivalent concepts when the data are centered.

Interpreted as cosines, the off-diagonal elements r_{ij} of such a correlation matrix then satisfy the inequality $r_{ij}^2 \leq 1$ for all $i \neq j$; $i, j = 1, 2, \ldots, p$. To go the other way, suppose that the square $p \times p$ symmetric matrix $\mathbf{R}$ has all its diagonal elements equal to 1 and all its off-diagonal elements $r_{ij}^2 \leq 1$; $i \neq j \, (i, j = 1, 2, \ldots, n)$. Then when is $\mathbf{R}$ a correlation matrix? That is, when is it nonnegative definite?

The special case when $p = 3$ is of particular interest. Indeed Sobel [424] posed the following problem in *The IMS Bulletin* Problems Corner:

> *Problem 6.2.1.* Let X, Y, and Z be random variables. If the correlations $\rho(X,Y)$ and $\rho(Y,Z)$ are known, what are the sharp lower and upper bounds for $\rho(X,Z)$?

This problem was apparently first solved by Priest [355] and Stanley & Wang [426]. See also Baksalary [22], Elffers [153, 154, 155, 156], Glass & Collins [182], Good [187], Olkin [339], and Rousseeuw & Molberghs [386]. For an approach using spherical trigonometry, see Good [186], Kendall & Stuart [256, Sections 27.28–27.19], and Stuart, Ord & Arnold [430, Sections 28.15–28.16]; see also Elffers [153, 154, 155] and Yanai [462].

Let us consider the 3×3 symmetric matrix

$$\mathbf{R} = \begin{pmatrix} 1 & r_{12} & r_{13} \\ r_{12} & 1 & r_{23} \\ r_{13} & r_{23} & 1 \end{pmatrix}, \quad \text{where all } r_{ij}^2 \leq 1 \ \ (i,j = 1,2,3). \tag{6.2.1}$$

Using the Albert nonnegative definite conditions given in Section 6.0.4 above, it is now easy to prove the following theorem.

THEOREM 6.2.1. *The following statements about* (6.2.1) *are equivalent:*

(i) $\mathbf{R}$ *is a correlation matrix,*

(ii) $\det(\mathbf{R}) = 1 - r_{12}^2 - r_{13}^2 - r_{23}^2 + 2r_{12}r_{13}r_{23} \geq 0$,

(iii) $(r_{12} - r_{13}r_{23})^2 \leq (1 - r_{13}^2)(1 - r_{23}^2)$, *or equivalently,*

$$r_{13}r_{23} - \sqrt{(1 - r_{13}^2)(1 - r_{23}^2)} \leq r_{12} \leq r_{13}r_{23} + \sqrt{(1 - r_{13}^2)(1 - r_{23}^2)}, \tag{6.2.2}$$

(iv) $(r_{23} - r_{12}r_{31})^2 \leq (1 - r_{12}^2)(1 - r_{31}^2)$,

(v) $(r_{13} - r_{12}r_{32})^2 \leq (1 - r_{12}^2)(1 - r_{32}^2)$,

(vi) (a) $\mathbf{r}_2 \in \mathcal{C}(\mathbf{R}_1)$ *and* (b) $\mathbf{R}/\mathbf{R}_1 = 1 - \mathbf{r}_2'\mathbf{R}_1^-\mathbf{r}_2 \geq 0$, *where*

$$\mathbf{R} = \begin{pmatrix} 1 & r_{12} & r_{13} \\ r_{12} & 1 & r_{23} \\ r_{13} & r_{23} & 1 \end{pmatrix} = \begin{pmatrix} \mathbf{R}_1 & \mathbf{r}_2 \\ \mathbf{r}_2' & 1 \end{pmatrix}, \tag{6.2.3}$$

(vii)

$$\mathbf{R}_1 - \mathbf{r}_2\mathbf{r}_2' = \begin{pmatrix} 1 - r_{13}^2 & r_{12} - r_{13}r_{23} \\ r_{12} - r_{13}r_{23} & 1 - r_{23}^2 \end{pmatrix} \geq_{\mathsf{L}} \mathbf{0}. \tag{6.2.4}$$

The proof of Theorem 6.2.1 relies on the fact that (i) is equivalent to the nonnegative definiteness of $\mathbf{R}$. Let r_{33} denote the (3,3) element of $\mathbf{R}$. Using the Albert conditions, and since $r_{33} = 1$ is certainly nonnegative, and since the column space $\mathcal{C}(\mathbf{r}_2') \subset \mathbb{R}$, (i) holds if and only if the Schur complement

$$\mathbf{R}/r_{33} = \mathbf{R}_1 - \mathbf{r}_2\mathbf{r}_2' \geq_{\mathsf{L}} \mathbf{0}, \tag{6.2.5}$$

and thus (vii) is proved. Similarly, $\mathbf{R} \geq_{\mathsf{L}} \mathbf{0}$ if and only if

$$\mathbf{r}_2 \in \mathcal{C}(\mathbf{R}_1) \text{ and } \mathbf{R}/\mathbf{R}_1 = 1 - \mathbf{r}_2'\mathbf{R}_1^-\mathbf{r}_2 \geq 0 \text{ and } \mathbf{R}_1 \geq_{\mathsf{L}} \mathbf{0}. \tag{6.2.6}$$

In (6.2.6), the last condition is always true since $r_{12}^2 \leq 1$, and hence (vi) is obtained. Conditions (ii)–(v) follow from (vii) at once.

The quantity

$$r_{12\cdot3}^2 = \frac{(r_{12} - r_{13}r_{23})^2}{(1 - r_{13}^2)(1 - r_{23}^2)} \tag{6.2.7}$$

is defined only with both $r_{13}^2 \neq 1$ and $r_{23}^2 \neq 1$ and then is the formula for the partial correlation coefficient (squared) between variables (say) x_1 and x_2 when x_3 is held constant.

The quadratic form $\mathbf{r}_2'\mathbf{R}_1^-\mathbf{r}_2 = 1 - (\mathbf{R}/\mathbf{R}_1)$ represents the multiple correlation coefficient (squared) when x_3 is regressed on the first two variables x_1 and x_2: $R_{3\cdot12}^2 = \mathbf{r}_2'\mathbf{R}_1^-\mathbf{r}_2$. Matrix $\mathbf{R}_1$ is singular if and only if $r_{12}^2 = 1$. If $r_{12} = 1$, then (a) of (vi) forces $r_{13} = r_{23}$. Choosing

$$\mathbf{R}_1^- = \mathbf{R}_1^+ = \tfrac{1}{4}\begin{pmatrix} 1 & 1 \\ 1 & 1 \end{pmatrix}, \tag{6.2.8}$$

we get $\mathbf{r}_2'\mathbf{R}_1^-\mathbf{r}_2 = r_{13}^2$. Similarly, if $r_{12} = -1$, then (a) of (vi) implies that $r_{13} = -r_{23}$, and again $\mathbf{r}_2'\mathbf{R}_1^-\mathbf{r}_2 = r_{13}^2$. If $r_{12}^2 \neq 1$, then we have

$$R_{3\cdot12}^2 = \frac{r_{13}^2 + r_{23}^2 - 2r_{12}r_{13}r_{23}}{1 - r_{12}^2}, \tag{6.2.9}$$

Of course, if $r_{12} = 0$, then $R_{3\cdot12}^2 = r_{13}^2 + r_{23}^2$, but is interesting to observe that both

$$R_{3\cdot12}^2 < r_{13}^2 + r_{23}^2 \quad \text{and} \quad R_{3\cdot12}^2 > r_{13}^2 + r_{23}^2 \tag{6.2.10}$$

can occur. As Shieh [419] points out, "...the second inequality in (6.2.10) may seem surprising and counterintuitive at first ... but it occurs more often than one may think" and he calls such an occurrence an "enhancement-synergism", while Currie & Korabinski [135] call it "enhancement" and Hamilton [202] suggests "synergism".

Using the Albert conditions, we see immediately that the first part of (6.2.10) is equivalent to

$$\mathbf{R}_1 - \frac{1}{\mathbf{r}_2'\mathbf{r}_2}\mathbf{r}_2\mathbf{r}_2' >_{\mathsf{L}} \mathbf{0}, \tag{6.2.11}$$

which further is equivalent to

$$r_{12}\left(r_{12} - \frac{2r_{13}r_{23}}{r_{13}^2 + r_{23}^2}\right) > 0, \tag{6.2.12}$$

which condition was shown by Hamilton [202] by other means; see also [50, 134, 175, 321].

Let us now take two simple examples. First, consider the intraclass correlation structure:

$$\mathbf{A} = \begin{pmatrix} 1 & r & r \\ r & 1 & r \\ r & r & 1 \end{pmatrix}. \tag{6.2.13}$$

The determinant $\det(\mathbf{A}) = (1-r)^2(1+2r)$, and hence $\mathbf{A}$ in (6.2.13) is indeed a correlation matrix if and only if

$$-\frac{1}{2} \leq r \leq 1. \tag{6.2.14}$$

In general, the $p \times p$ intraclass correlation matrix must satisfy

$$-\frac{1}{p-1} \leq r \leq 1, \tag{6.2.15}$$

which condition can be expressed also as

$$R^2_{p\cdot 12\ldots p-1} = \mathbf{r}_2'\mathbf{R}_1^-\mathbf{r}_2 = \frac{(p-1)r^2}{1+(p-2)r} \leq 1. \tag{6.2.16}$$

As another example, consider

$$\mathbf{B} = \begin{pmatrix} 1 & a & r \\ a & 1 & r \\ r & r & 1 \end{pmatrix}, \tag{6.2.17}$$

where a is a given real number, $a^2 \leq 1$. What are the possible values for r such that $\mathbf{B}$ is a correlation matrix? It is now easy to confirm that the answer is

$$r^2 \leq \frac{1+a}{2}. \tag{6.2.18}$$

Olkin [339] extended the above results (i)–(vii) to the case where three sets of variables are available. For a general mathematical (and essentially nonstatistical) treatment of the null space of the correlation matrix, see Barretta & Pierce [34].

6.2.2 The conditional multivariate normal distribution and the generalized quotient property

As Cottle [128] observed, the "multivariate normal distribution provides a magnificent example of how the Schur complement and the quotient property (6.0.26) arise naturally". Let the $p \times 1$ random vector

$$\mathbf{x} = \begin{pmatrix} \mathbf{x}_1 \\ \mathbf{x}_2 \end{pmatrix}, \tag{6.2.19}$$

where $\mathbf{x}_1$ is $p_1 \times 1$ and $\mathbf{x}_2$ is $p_2 \times 1$, with $p_1 + p_2 = p$. We suppose that $\mathbf{x}$ follows a multivariate normal distribution with mean vector

$$\boldsymbol{\mu} = \begin{pmatrix} \boldsymbol{\mu}_1 \\ \boldsymbol{\mu}_2 \end{pmatrix} \tag{6.2.20}$$

and nonnegative definite, possibly singular, covariance matrix

$$\boldsymbol{\Sigma} = \begin{pmatrix} \boldsymbol{\Sigma}_{11} & \boldsymbol{\Sigma}_{12} \\ \boldsymbol{\Sigma}'_{12} & \boldsymbol{\Sigma}_{22} \end{pmatrix}. \tag{6.2.21}$$

Then the conditional distribution of $\mathbf{x}_1$ given $\mathbf{x}_2$ is multivariate normal with mean vector $\boldsymbol{\nu}_1 = \boldsymbol{\mu}_1 + \boldsymbol{\Sigma}_{12}\boldsymbol{\Sigma}_{22}^{-}(\mathbf{x}_2 - \boldsymbol{\mu}_2)$ and covariance matrix the generalized Schur complement of $\boldsymbol{\Sigma}_{22}$ in $\boldsymbol{\Sigma}$,

$$\boldsymbol{\Sigma}/\boldsymbol{\Sigma}_{22} = \boldsymbol{\Sigma}_{11} - \boldsymbol{\Sigma}_{12}\boldsymbol{\Sigma}_{22}^{-}\boldsymbol{\Sigma}'_{12} = \boldsymbol{\Sigma}_{11\cdot 2}, \tag{6.2.22}$$

see, e.g., Anderson [9, Th. 2.5.1, p. 35], Ouellette [345, Section 6.1].

To prove this result, we note first that the distribution of

$$\begin{pmatrix} \mathbf{x}_1 - \boldsymbol{\Sigma}_{12}\boldsymbol{\Sigma}_{22}^{-}\mathbf{x}_2 \\ \mathbf{x}_2 \end{pmatrix} = \begin{pmatrix} \mathbf{I} & -\boldsymbol{\Sigma}_{12}\boldsymbol{\Sigma}_{22}^{-} \\ \mathbf{0} & \mathbf{I} \end{pmatrix} \begin{pmatrix} \mathbf{x}_1 \\ \mathbf{x}_2 \end{pmatrix} \tag{6.2.23}$$

is multivariate normal with mean vector

$$\begin{pmatrix} \mathbf{I} & -\boldsymbol{\Sigma}_{12}\boldsymbol{\Sigma}_{22}^{-} \\ \mathbf{0} & \mathbf{I} \end{pmatrix} \begin{pmatrix} \boldsymbol{\mu}_1 \\ \boldsymbol{\mu}_2 \end{pmatrix} = \begin{pmatrix} \boldsymbol{\mu}_1 - \boldsymbol{\Sigma}_{12}\boldsymbol{\Sigma}_{22}^{-}\boldsymbol{\mu}_2 \\ \boldsymbol{\mu}_2 \end{pmatrix} \tag{6.2.24}$$

and covariance matrix

$$\begin{pmatrix} \mathbf{I} & -\boldsymbol{\Sigma}_{12}\boldsymbol{\Sigma}_{22}^{-} \\ \mathbf{0} & \mathbf{I} \end{pmatrix} \begin{pmatrix} \boldsymbol{\Sigma}_{11} & \boldsymbol{\Sigma}_{12} \\ \boldsymbol{\Sigma}_{12}' & \boldsymbol{\Sigma}_{22} \end{pmatrix} \begin{pmatrix} \mathbf{I} & \mathbf{0} \\ -(\boldsymbol{\Sigma}_{22}^{-})'\boldsymbol{\Sigma}_{12}' & \mathbf{I} \end{pmatrix} = \begin{pmatrix} (\boldsymbol{\Sigma}/\boldsymbol{\Sigma}_{22}) & \mathbf{0} \\ \mathbf{0} & \boldsymbol{\Sigma}_{22} \end{pmatrix},$$

see also the generalized Aitken block-diagonalization formula (6.0.20). Hence $\mathbf{x}_1 - \boldsymbol{\Sigma}_{12}\boldsymbol{\Sigma}_{22}^{-}\mathbf{x}_2$ is distributed independently of $\mathbf{x}_2$, and so its conditional distribution given $\mathbf{x}_2$ is the same as its unconditional distribution. Thus $\mathbf{x}_1$ given $\mathbf{x}_2$ is multivariate normal with mean vector $\boldsymbol{\nu}_1 = \boldsymbol{\mu}_1 + \boldsymbol{\Sigma}_{12}\boldsymbol{\Sigma}_{22}^{-}(\mathbf{x}_2 - \boldsymbol{\mu}_2)$ and covariance matrix (6.2.22).

Cottle [128, p. 195] gives an interesting interpretation of the quotient property for the multivariate normal distribution, see also [345, Section 6.1]. Let

$$\mathbf{x} = \begin{pmatrix} \mathbf{x}_1 \\ \mathbf{x}_2 \\ \mathbf{x}_3 \end{pmatrix} \quad \text{and} \quad \boldsymbol{\Sigma} = \begin{pmatrix} \boldsymbol{\Sigma}_{11} & \boldsymbol{\Sigma}_{12} & \vdots & \boldsymbol{\Sigma}_{13} \\ \boldsymbol{\Sigma}_{12}' & \boldsymbol{\Sigma}_{22} & \vdots & \boldsymbol{\Sigma}_{23} \\ \cdots & \cdots & \vdots & \cdots \\ \boldsymbol{\Sigma}_{13}' & \boldsymbol{\Sigma}_{23}' & \vdots & \boldsymbol{\Sigma}_{33} \end{pmatrix} = \begin{pmatrix} \boldsymbol{\Sigma}_{1\&2} & \boldsymbol{\Sigma}_{1\&2;\,3} \\ \boldsymbol{\Sigma}_{1\&2;\,3}' & \boldsymbol{\Sigma}_{33} \end{pmatrix},$$

where $\boldsymbol{\Sigma}$ is the covariance matrix of the random vector $\mathbf{x}$. Then the conditional distribution of $\mathbf{x}_3$ given $\mathbf{x}_1$ and $\mathbf{x}_2$ is the conditional distribution of $\mathbf{x}_2$ and $\mathbf{x}_3$ given $\mathbf{x}_1$ conditional on the distribution of $\mathbf{x}_2$ given $\mathbf{x}_1$, in other words, we may condition sequentially.

To see this, it suffices to apply the generalized quotient property (6.0.28) to the covariance matrix $\boldsymbol{\Sigma}$ and so

$$\boldsymbol{\Sigma}/\boldsymbol{\Sigma}_{1\&2} = (\boldsymbol{\Sigma}/\boldsymbol{\Sigma}_{11})/(\boldsymbol{\Sigma}_{1\&2}/\boldsymbol{\Sigma}_{11}).$$

6.2.3 Partial and conditional correlation coefficients

Let $\mathbf{z}$ be a partitioned random vector with covariance matrix

$$\operatorname{cov}(\mathbf{z}) = \operatorname{cov}\begin{pmatrix} \mathbf{z}_1 \\ \mathbf{z}_2 \end{pmatrix} = \begin{pmatrix} \boldsymbol{\Sigma}_{11} & \boldsymbol{\Sigma}_{12} \\ \boldsymbol{\Sigma}_{21} & \boldsymbol{\Sigma}_{22} \end{pmatrix} = \boldsymbol{\Sigma}. \tag{6.2.25}$$

We now establish a very powerful result:

$$\operatorname{cov}(\mathbf{z}_2 - \mathbf{F}\mathbf{z}_1) \geq_{\mathsf{L}} \operatorname{cov}(\mathbf{z}_2 - \boldsymbol{\Sigma}_{21}\boldsymbol{\Sigma}_{11}^{-}\mathbf{z}_1) \text{ for all conformable } \mathbf{F}, \tag{6.2.26}$$

where the inequality refers the Löwner partial ordering. The minimal covariance matrix is the Schur complement

$$\mathrm{cov}(\mathbf{z}_2 - \mathbf{\Sigma}_{21}\mathbf{\Sigma}_{11}^{-}\mathbf{z}_1) = \mathbf{\Sigma}_{22} - \mathbf{\Sigma}_{21}\mathbf{\Sigma}_{11}^{-}\mathbf{\Sigma}_{12} = \mathbf{\Sigma}_{22\cdot 1} = \mathbf{\Sigma}/\mathbf{\Sigma}_{11}. \tag{6.2.27}$$

We note that the covariance matrix in (6.2.27) is invariant for any choice of $\mathbf{\Sigma}_{11}^{-}$, but this is not necessarily so with $\mathbf{\Sigma}_{21}\mathbf{\Sigma}_{11}^{-}\mathbf{z}_1$. The covariance matrix $\mathbf{\Sigma}/\mathbf{\Sigma}_{11}$ in (6.2.27) is the covariance of the conditional normal distribution discussed in Section 6.2.2. The correlations associated with $\mathbf{\Sigma}/\mathbf{\Sigma}_{11}$ are known as *partial correlation coefficients* when the underlying distribution is not necessarily multivariate normal. For more about the connection between conditional and partial correlation coefficients, see Lewis & Styan [278].

The proof of (6.2.26) is very simple with the help of the generalized Aitken block-diagonalization formula (6.0.3). Let us denote

$$\mathbf{u} = \begin{pmatrix} \mathbf{u}_1 \\ \mathbf{u}_2 \end{pmatrix} = \begin{pmatrix} \mathbf{I} & \mathbf{0} \\ -\mathbf{\Sigma}_{21}\mathbf{\Sigma}_{11}^{-} & \mathbf{I} \end{pmatrix} \begin{pmatrix} \mathbf{z}_1 \\ \mathbf{z}_2 \end{pmatrix} = \begin{pmatrix} \mathbf{z}_1 \\ \mathbf{z}_2 - \mathbf{\Sigma}_{21}\mathbf{\Sigma}_{11}^{-}\mathbf{z}_1 \end{pmatrix}. \tag{6.2.28}$$

Then the covariance matrix

$$\mathrm{cov}(\mathbf{u}) = \begin{pmatrix} \mathbf{\Sigma}_{11} & \mathbf{0} \\ \mathbf{0} & \mathbf{\Sigma}_{22} - \mathbf{\Sigma}_{21}\mathbf{\Sigma}_{11}^{-}\mathbf{\Sigma}_{12} \end{pmatrix}, \tag{6.2.29}$$

indicating that $\mathbf{u}_1$ and $\mathbf{u}_2$ are uncorrelated. We can, therefore, write

$$\begin{aligned} \mathrm{cov}(\mathbf{z}_2 - \mathbf{F}\mathbf{z}_1) &= \mathrm{cov}[(\mathbf{z}_2 - \mathbf{\Sigma}_{21}\mathbf{\Sigma}_{11}^{-}\mathbf{z}_1) + (\mathbf{\Sigma}_{21}\mathbf{\Sigma}_{11}^{-} - \mathbf{F})\mathbf{z}_1] \\ &:= \mathrm{cov}(\mathbf{v}_1 + \mathbf{v}_2), \end{aligned} \tag{6.2.30}$$

where the random vectors $\mathbf{v}_1$ and $\mathbf{v}_2$ are uncorrelated and hence,

$$\begin{aligned} \mathrm{cov}(\mathbf{z}_2 - \mathbf{F}\mathbf{z}_1) &= \mathrm{cov}(\mathbf{z}_2 - \mathbf{\Sigma}_{21}\mathbf{\Sigma}_{11}^{-}\mathbf{z}_1) + \mathrm{cov}[(\mathbf{\Sigma}_{21}\mathbf{\Sigma}_{11}^{-} - \mathbf{F})\mathbf{z}_1] \\ &\geq_{\mathsf{L}} \mathrm{cov}(\mathbf{z}_2 - \mathbf{\Sigma}_{21}\mathbf{\Sigma}_{11}^{-}\mathbf{z}_1), \end{aligned} \tag{6.2.31}$$

which proves (6.2.27).

In particular, if the random vector $\mathbf{z}$ is partitioned so that

$$\mathrm{cov}(\mathbf{z}) = \mathrm{cov}\begin{pmatrix} \mathbf{x} \\ y \end{pmatrix} = \begin{pmatrix} \mathbf{\Sigma}_{11} & \boldsymbol{\sigma}_2 \\ \boldsymbol{\sigma}_2' & \sigma_y^2 \end{pmatrix} = \mathbf{\Sigma} \in \mathbb{R}^{(p+1)\times(p+1)}, \tag{6.2.32}$$

where y is a scalar, then

$$\min_{\mathbf{f}} \mathrm{var}(y - \mathbf{f}'\mathbf{x}) = \mathrm{var}(y - \boldsymbol{\sigma}_2'\mathbf{\Sigma}_{11}^{-}\mathbf{x}) = \sigma_y^2 - \boldsymbol{\sigma}_2'\mathbf{\Sigma}_{11}^{-}\boldsymbol{\sigma}_2 = \sigma_{yy\cdot\mathbf{x}}. \tag{6.2.33}$$

Similarly we can find the vector $\hat{\mathbf{g}} \in \mathbb{R}^p$ which gives the maximal value for the correlation between the random variables y and $\mathbf{g}'\mathbf{x}$, i.e.,

$$\max_{\mathbf{b}} \operatorname{corr}(y, \mathbf{g}'\mathbf{x}) = \operatorname{corr}(y, \hat{\mathbf{g}}'\mathbf{x}). \tag{6.2.34}$$

In view of the Cauchy–Schwarz inequality, we have

$$\operatorname{corr}^2(y, \mathbf{g}'\mathbf{x}) = \frac{(\boldsymbol{\sigma}_2'\mathbf{g})^2}{\sigma_y^2 \cdot \mathbf{g}'\boldsymbol{\Sigma}_{11}\mathbf{g}} \leq \frac{\boldsymbol{\sigma}_2'\boldsymbol{\Sigma}_{11}^{-}\boldsymbol{\sigma}_2 \cdot \mathbf{g}'\boldsymbol{\Sigma}_{11}\mathbf{g}}{\sigma_y^2 \cdot \mathbf{g}'\boldsymbol{\Sigma}_{11}\mathbf{g}} = \frac{\boldsymbol{\sigma}_2'\boldsymbol{\Sigma}_{11}^{-}\boldsymbol{\sigma}_2}{\sigma_y^2}, \tag{6.2.35}$$

with equality if and only if $\boldsymbol{\Sigma}_{11}\mathbf{g}$ is a scalar multiple of $\boldsymbol{\sigma}_2$, i.e., $\boldsymbol{\Sigma}_{11}\mathbf{g} = \lambda\boldsymbol{\sigma}_2$, for some $\lambda \in \mathbb{R}$. From this, it follows that a solution to (6.2.34) is $\hat{\mathbf{g}} = \boldsymbol{\Sigma}_{11}^{-}\boldsymbol{\sigma}_2$ (which could, of course, be multiplied with any nonzero scalar) and

$$\max_{\mathbf{g}} \operatorname{corr}(y, \mathbf{g}'\mathbf{x}) = \frac{\sqrt{\boldsymbol{\sigma}_2'\boldsymbol{\Sigma}_{11}^{-}\boldsymbol{\sigma}_2}}{\sigma_y} := \rho_{y\cdot 1\ldots p} := \rho_{y\cdot\mathbf{x}} := \mathcal{R}, \tag{6.2.36}$$

the *population multiple correlation coefficient* [9, Section 2.5.2]. We note that since

$$1 - \mathcal{R}^2 = \frac{\sigma_y^2 - \boldsymbol{\sigma}_2'\boldsymbol{\Sigma}_{11}^{-}\boldsymbol{\sigma}_2}{\sigma_y^2}, \tag{6.2.37}$$

it follows that $\sigma_{yy\cdot\mathbf{x}} = \sigma_y^2(1 - \mathcal{R}^2) \leq \sigma_y^2$. If $\boldsymbol{\Sigma}$ is positive definite, we have

$$1 - \mathcal{R}^2 = \frac{\sigma_y^2 - \boldsymbol{\sigma}_2'\boldsymbol{\Sigma}_{11}^{-1}\boldsymbol{\sigma}_2}{\sigma_y^2} = \frac{1}{\sigma^{yy}\sigma_y^2}, \tag{6.2.38}$$

where σ^{yy} is the last diagonal element of $\boldsymbol{\Sigma}^{-1}$. Using the Schur determinant formula, we immediately obtain, since $\boldsymbol{\Sigma}_{11}$ is now positive definite,

$$\det \boldsymbol{\Sigma} = (\sigma_y^2 - \boldsymbol{\sigma}_2'\boldsymbol{\Sigma}_{11}^{-1}\boldsymbol{\sigma}_2) \det \boldsymbol{\Sigma}_{11} \leq \sigma_y^2 \det \boldsymbol{\Sigma}_{11} \tag{6.2.39}$$

with equality if and only if $\boldsymbol{\sigma}_2 = \mathbf{0}$. Hence

$$\det \boldsymbol{\Sigma} \leq \sigma_y^2\sigma_1^2\sigma_2^2 \cdots \sigma_p^2$$

with equality if and only if $\boldsymbol{\Sigma}$ is diagonal, see, e.g., Mirsky [314, Th. 13.5.2, p. 417], Zhang [468, Th. 6.11, p. 176]. This result is the special case of *Hadamard's determinant theorem* or *Hadamard's inequality*

$$|\det(\mathbf{A})|^2 \leq \prod_{h=1}^{n} (|a_{1h}|^2 + \cdots + |a_{nr}|^2) \tag{6.2.40}$$

which holds for any complex nonsingular $n \times n$ matrix $\mathbf{A} = \{a_{ij}\}$. The inequality (6.2.40) was first established by Jacques Hadamard (1865–1963) in 1893 [199]; see also Mirsky [314, Th. 13.5.3, p. 418], Zhang [468, p. 176].

A statistical proof goes as follows. Write $\mathbf{\Sigma} = \mathbf{\Delta R \Delta}$, where $\mathbf{R}$ is the associated correlation matrix and $\mathbf{\Delta}$ is the diagonal matrix of standard deviations. Then (6.2.39) becomes

$$\det(\mathbf{R}) \leq 1 \tag{6.2.41}$$

with equality if and only if $\mathbf{R} = \mathbf{I}$, the identity matrix. To prove (6.2.41), we note that

$$\det(\mathbf{R}) = \prod \mathrm{ch}(\mathbf{R}) \leq \left(\frac{1}{p+1}\,\mathrm{tr}(\mathbf{R})\right)^{p+1} = 1 \tag{6.2.42}$$

using the arithmetic mean–geometric mean inequality on the eigenvalues $\mathrm{ch}(\mathbf{R})$ of the correlation matrix $\mathbf{R}$. Equality in (6.2.42) holds if and only if all the eigenvalues are equal and this occurs if and only if $\mathbf{R} = \mathbf{I}$.

Consider partitioned random vector $\mathbf{z}$ as in (6.2.32). The *best linear predictor* (BLP) of y is defined as

$$\mathrm{BLP}(y) = \hat{\mathbf{a}}'\mathbf{x} + \hat{b}, \tag{6.2.43}$$

if it minimizes the mean squared error:

$$\min_{\mathbf{a},\, b} \mathrm{E}[y - (\mathbf{a}'\mathbf{x} + b)]^2 = \mathrm{E}[y - (\hat{\mathbf{a}}'\mathbf{x} + \hat{b})]^2. \tag{6.2.44}$$

It is easy to see that

$$\mathrm{E}[y - (\mathbf{a}'\mathbf{x} + b)]^2 = \mathrm{var}(y - \mathbf{a}'\mathbf{x}) + (\mu_2 - \mathbf{a}'\boldsymbol{\mu}_1 - b)^2, \tag{6.2.45}$$

where $\mu_2 = \mathrm{E}(y)$ and $\boldsymbol{\mu}_1 = \mathrm{E}(\mathbf{x})$. Hence (6.2.26) gives immediately the formula for $\mathrm{BLP}(y)$:

$$\mathrm{BLP}(y) = \mu_2 + \boldsymbol{\sigma}_2'\mathbf{\Sigma}_{11}^{-}(\mathbf{x} - \boldsymbol{\mu}_1); \tag{6.2.46}$$

see [9, p. 37], [122, Th. 6.3.2]. Clearly, if a random vector $\mathbf{z}$ is partitioned so that

$$\mathrm{cov}(\mathbf{z}) = \mathrm{cov}\begin{pmatrix}\mathbf{x}\\ \mathbf{y}\end{pmatrix} = \begin{pmatrix}\mathbf{\Sigma}_{11} & \mathbf{\Sigma}_{12}\\ \mathbf{\Sigma}_{21} & \mathbf{\Sigma}_{22}\end{pmatrix}, \quad \mathrm{E}(\mathbf{z}) = \begin{pmatrix}\boldsymbol{\mu}_1\\ \boldsymbol{\mu}_2\end{pmatrix}, \tag{6.2.47}$$

then the random vector $\boldsymbol{\mu}_2 + \mathbf{\Sigma}_{21}\mathbf{\Sigma}_{11}^{-}(\mathbf{x} - \boldsymbol{\mu}_1)$ is the best linear predictor of $\mathbf{y}$ in the sense that it minimizes, in the Löwner sense, the matrix

$$\mathrm{E}\big((\mathbf{y} - \mathbf{Ax} - \mathbf{b})(\mathbf{y} - \mathbf{Ax} - \mathbf{b})'\big). \tag{6.2.48}$$

We complete this section by noting the close connection between the results above and an important result concerning *best prediction*. Namely, if $m(\mathbf{x}) = \mathrm{E}(y \mid \mathbf{x})$ denotes the conditional expectation of y, where $m(\mathbf{x})$ is now considered as a random variable, then it can be shown, see, e.g., [373, p. 264], [122, Th. 6.3.1], that

$$\mathrm{E}[y - m(\mathbf{x})]^2 \leq \mathrm{E}[y - f(\mathbf{x})]^2 \tag{6.2.49}$$

for any other predictor $f(\mathbf{x})$. When the underlying distribution is multivariate normal then

$$m(\mathbf{x}) = \mathrm{E}(y \mid \mathbf{x}) = \mu_2 - \boldsymbol{\sigma}_2' \boldsymbol{\Sigma}_{11}^{-} (\mathbf{x} - \boldsymbol{\mu}_1), \tag{6.2.50}$$

and hence (6.2.49) means that then $m(\mathbf{x})$ is not only the best linear predictor but also the best predictor.

6.3 The general linear model and multiple linear regression

In this section, we consider the general linear model

$$\mathbf{y} = \mathbf{X}\boldsymbol{\beta} + \boldsymbol{\varepsilon}, \tag{6.3.1}$$

where

$$\mathrm{E}(\mathbf{y}) = \mathbf{X}\boldsymbol{\beta}, \quad \mathrm{E}(\boldsymbol{\varepsilon}) = \mathbf{0}, \quad \mathrm{cov}(\mathbf{y}) = \mathrm{cov}(\boldsymbol{\varepsilon}) = \sigma^2 \mathbf{V}. \tag{6.3.2}$$

We suppose that $\mathbf{y}$ is an $n \times 1$ observable random vector, that $\boldsymbol{\varepsilon}$ is an $n \times 1$ random error vector, that $\mathbf{X}$ is a known $n \times p$ model (design) matrix with rank $r \leq p$, that $\boldsymbol{\beta}$ is a $p \times 1$ vector of unknown parameters, that $\mathbf{V}$ is a known $n \times n$ nonnegative definite matrix, and that σ^2 is an unknown positive constant. The other notation is

$$\mathcal{M} = \{\mathbf{y}, \mathbf{X}\boldsymbol{\beta}, \sigma^2 \mathbf{V}\}. \tag{6.3.3}$$

We use the notation

$$\mathbf{H} = \mathbf{P}_{\mathbf{X}}, \ \mathbf{M} = \mathbf{I} - \mathbf{H}, \tag{6.3.4}$$

thereby obtaining the *ordinary least squares estimator* (OLSE) of $\mathbf{X}\boldsymbol{\beta}$ as

$$\mathrm{OLSE}(\mathbf{X}\boldsymbol{\beta}) = \mathbf{X}\hat{\boldsymbol{\beta}} = \hat{\mathbf{y}} = \mathbf{H}\mathbf{y} = \mathbf{P}_{\mathbf{X}}\mathbf{y}, \tag{6.3.5}$$

the corresponding vector of residuals being $\mathbf{y} - \mathbf{H}\mathbf{y} = \mathbf{M}\mathbf{y}$. Matrix $\mathbf{H}$ is known as the *hat matrix*, see, e.g., [114, 115, 223]; matrix $\mathbf{M} = \mathbf{I} - \mathbf{H}$ is the *residual matrix*. Clearly we have

$$\mathrm{E}(\mathbf{H}\mathbf{y}) = \mathbf{H}\mathbf{X}\boldsymbol{\beta} = \mathbf{X}\boldsymbol{\beta}, \quad \mathrm{cov}(\mathbf{H}\mathbf{y}) = \sigma^2 \mathbf{H}\mathbf{V}\mathbf{H}. \tag{6.3.6}$$

In (6.3.5) the vector $\hat{\boldsymbol{\beta}} = (\mathbf{X}'\mathbf{X})^{-}\mathbf{X}'\mathbf{y}$, which becomes $\hat{\boldsymbol{\beta}} = (\mathbf{X}'\mathbf{X})^{-1}\mathbf{X}'\mathbf{y}$ when $\mathbf{X}$ has full column rank.

When $\mathbf{X}$ does not have full column rank, then the vector $\hat{\boldsymbol{\beta}} = (\mathbf{X}'\mathbf{X})^{-}\mathbf{X}'\mathbf{y}$ is not unique and is not a *proper* estimator: it is merely a *solution* to the normal equations; as Searle [410, p. 169] observes "this point cannot be overemphasized".

A linear estimator $\mathbf{Gy}$ is the best linear unbiased estimator (BLUE) of $\mathbf{X}\boldsymbol{\beta}$ if it has the smallest covariance matrix (in the Löwner sense) among all unbiased linear estimators:

$$\operatorname{cov}(\mathbf{Gy}) \leq \operatorname{cov}(\mathbf{By}) \text{ for all } \mathbf{By} \text{ such that } \mathrm{E}(\mathbf{By}) = \mathbf{X}\boldsymbol{\beta}. \tag{6.3.7}$$

Since $\operatorname{cov}(\mathbf{Gy}) = \sigma^2\mathbf{GVG}'$ and the unbiasedness of $\mathbf{Gy}$ means that $\mathbf{GX} = \mathbf{X}$, we can rewrite (6.3.7) as

$$\mathbf{GVG}' \leq \mathbf{BVB}' \text{ for all } \mathbf{B} \text{ such that } \mathbf{BX} = \mathbf{X}. \tag{6.3.8}$$

We may recall here the fundamental "BLUE equation", see, e.g., [371, p. 282], [148, p. 55], that $\mathbf{G}$ has to satisfy for $\mathbf{Gy}$ to be the BLUE of $\mathbf{X}\boldsymbol{\beta}$ under a general linear model $\{\mathbf{y}, \mathbf{X}\boldsymbol{\beta}, \sigma^2\mathbf{V}\}$:

$$\mathbf{G}(\mathbf{X} : \mathbf{VM}) = (\mathbf{X} : \mathbf{0}). \tag{6.3.9}$$

For recent proofs of the BLUE equations, see [23, 193, 364].

In Section 6.3.2, we will consider the situation when $\mathbf{V}$ is positive definite, in which case the BLUE of $\mathbf{X}\boldsymbol{\beta}$ is

$$\mathrm{BLUE}(\mathbf{X}\boldsymbol{\beta}) = \mathbf{X}(\mathbf{X}'\mathbf{V}^{-1}\mathbf{X})^{-}\mathbf{X}'\mathbf{V}^{-1}\mathbf{y} = \mathbf{P}_{\mathbf{X};\mathbf{V}^{-1}}\mathbf{y}. \tag{6.3.10}$$

Above $\mathbf{P}_{\mathbf{X};\mathbf{V}^{-1}}$ denotes the orthogonal projector onto $\mathcal{C}(\mathbf{X})$ when the inner product between the vectors $\mathbf{a}$ and $\mathbf{b}$ is defined as $\mathbf{a}'\mathbf{V}^{-1}\mathbf{b}$. When $\mathbf{V}$ is singular, we have to use general representations for the $\mathrm{BLUE}(\mathbf{X}\boldsymbol{\beta})$, see, e.g., [7, 371]:

$$\mathrm{BLUE}(\mathbf{X}\boldsymbol{\beta}) = \mathbf{Hy} - \mathbf{HVM}(\mathbf{MVM})^{-}\mathbf{My} := \mathbf{G}_1\mathbf{y}, \tag{6.3.11a}$$

$$\mathrm{BLUE}(\mathbf{X}\boldsymbol{\beta}) = \mathbf{y} - \mathbf{VM}(\mathbf{MVM})^{-}\mathbf{My} := \mathbf{G}_2\mathbf{y}, \tag{6.3.11b}$$

$$\mathrm{BLUE}(\mathbf{X}\boldsymbol{\beta}) = \mathbf{X}(\mathbf{X}'\mathbf{W}^{-}\mathbf{X})^{-}\mathbf{X}'\mathbf{W}^{-}\mathbf{y} := \mathbf{G}_3\mathbf{y}, \tag{6.3.11c}$$

where $\mathbf{W} = \mathbf{V} + \mathbf{XUX}'$ and $\mathcal{C}(\mathbf{W}) = \mathcal{C}(\mathbf{X} : \mathbf{V})$. When $\mathbf{V}$ is nonsingular, the matrix $\mathbf{G}$ such that $\mathbf{Gy}$ is the BLUE of $\mathbf{X}\boldsymbol{\beta}$ is unique, but when $\mathbf{V}$ is singular this may not be so. However, the numerical value of $\mathrm{BLUE}(\mathbf{X}\boldsymbol{\beta})$ is unique with probability 1. The matrices $\mathbf{X}$ and $\mathbf{V}$ can be of arbitrary rank but the model must be *consistent* in that $\mathbf{y} \in \mathcal{C}(\mathbf{X} : \mathbf{V})$ with probability

1. The consistency condition means, for example, that whenever we have some statements where the random vector $\mathbf{y}$ is involved, these statements need to hold only for those values of $\mathbf{y}$ which belong to $\mathcal{C}(\mathbf{X} : \mathbf{V})$.

We will use the notation $\mathrm{BLUE}(\mathbf{X}\boldsymbol{\beta}) = \mathbf{X}\tilde{\boldsymbol{\beta}}$, but it is important to realize that in this notation vector $\tilde{\boldsymbol{\beta}}$ is unique (with probability 1) only if $\mathbf{X}$ has full column rank. In this case, the vector $\boldsymbol{\beta}$ is *estimable*, i.e., it has an unbiased linear estimator, and one general expression for $\tilde{\boldsymbol{\beta}}$ is

$$\begin{aligned}\tilde{\boldsymbol{\beta}} &= (\mathbf{X}'\mathbf{X})^{-1}\mathbf{X}'\mathbf{y} - (\mathbf{X}'\mathbf{X})^{-1}\mathbf{X}'\mathbf{V}\mathbf{Z}(\mathbf{Z}'\mathbf{V}\mathbf{Z})^{-}\mathbf{Z}'\mathbf{y} \\ &= \mathbf{X}^{+}\mathbf{y} - \mathbf{X}^{+}\mathbf{V}\mathbf{Z}(\mathbf{Z}'\mathbf{V}\mathbf{Z})^{-}\mathbf{Z}'\mathbf{y},\end{aligned} \tag{6.3.12}$$

where $\mathbf{Z}$ is satisfies $\mathcal{C}(\mathbf{Z}) = \mathcal{C}(\mathbf{M})$. We can also define $\tilde{\boldsymbol{\beta}}$ as a solution to the *generalized normal equations* [25, 317, 370] $\mathbf{X}'\mathbf{W}^{-}\mathbf{X}\tilde{\boldsymbol{\beta}} = \mathbf{X}'\mathbf{W}^{-}\mathbf{y}$, where $\mathbf{W}$ is defined as above. The quantity $\mathrm{SSE}(V) = (\mathbf{y} - \mathbf{X}\tilde{\boldsymbol{\beta}})'\mathbf{W}^{-}(\mathbf{y} - \mathbf{X}\tilde{\boldsymbol{\beta}})$, is needed for finding the quadratic unbiased estimator of σ^2; see, e.g., [369].

6.3.1 A generalized Gauß–Markov Theorem

Let us consider the full rank general linear or Gauß–Markov model

$$\mathbf{y} = \mathbf{X}\boldsymbol{\beta} + \boldsymbol{\varepsilon}, \tag{6.3.13}$$

where $\mathbf{X}$ is a known $n \times p$ model or design matrix with full column rank $p < n$, and $\mathrm{cov}(\mathbf{y}) = \mathbf{V}$ is an $n \times n$ positive definite matrix. We are interested in estimating the unknown parameter vector $\boldsymbol{\beta}$ based on a single realization of $\mathbf{y}$. The linear estimator $\mathbf{A}'\mathbf{y}$ of $\boldsymbol{\beta}$ has covariance matrix

$$\sigma^2\mathbf{A}'\mathbf{V}\mathbf{A} \tag{6.3.14}$$

and is an unbiased estimator of $\boldsymbol{\beta}$ if and only if $\mathbf{A}'$ is a left-inverse of $\mathbf{X}$, i.e., $\mathbf{A}'\mathbf{X} = \mathbf{I}$. The ordinary least squares (OLS) estimator

$$\hat{\boldsymbol{\beta}} = (\mathbf{X}'\mathbf{X})^{-1}\mathbf{X}'\mathbf{y} \tag{6.3.15}$$

is, therefore, an unbiased estimator of $\boldsymbol{\beta}$, and has covariance matrix

$$\sigma^2(\mathbf{X}'\mathbf{X})^{-1}\mathbf{X}'\mathbf{V}\mathbf{X}(\mathbf{X}'\mathbf{X})^{-1}. \tag{6.3.16}$$

The generalized least squares (GLS) or Aitken estimator [2, 222, 412],

$$\tilde{\boldsymbol{\beta}} = (\mathbf{X}'\mathbf{V}^{-1}\mathbf{X})^{-1}\mathbf{X}'\mathbf{V}^{-1}\mathbf{y} \tag{6.3.17}$$

is also unbiased for $\boldsymbol{\beta}$, and has covariance matrix $\sigma^2(\mathbf{X}'\mathbf{V}^{-1}\mathbf{X})^{-1}$.

The well-known Gauß–Markov Theorem [337] states that when $\mathbf{V} = \mathbf{I}$ then the ordinary least squares (OLSE) estimator $\hat{\boldsymbol{\beta}}$ is the "best" linear unbiased estimator (BLUE) of $\boldsymbol{\beta}$, where "best" is taken in the Löwner partial ordering of the covariance matrices, i.e.,

$$\sigma^2 \mathbf{A}'\mathbf{A} \geq_{\mathsf{L}} \sigma^2 (\mathbf{X}'\mathbf{X})^{-1} \quad : \quad \mathbf{A}'\mathbf{X} = \mathbf{I}. \tag{6.3.18}$$

A generalized Gauß–Markov Theorem says that when $\mathbf{V}$ is not necessarily equal to $\mathbf{I}$ then the GLS estimator $\tilde{\boldsymbol{\beta}}$ is the BLUE of $\boldsymbol{\beta}$ in that

$$\sigma^2 \mathbf{A}'\mathbf{V}\mathbf{A} \geq_{\mathsf{L}} \sigma^2 (\mathbf{X}'\mathbf{V}^{-1}\mathbf{X})^{-1} \quad : \quad \mathbf{A}'\mathbf{X} = \mathbf{I}. \tag{6.3.19}$$

When $\mathbf{V} = \mathbf{I}$ then (6.3.19) becomes (6.3.18).

We may prove (6.3.19) very quickly using Schur complements. We begin with the $2p \times 2p$ matrix

$$\begin{aligned} \mathbf{W}_1 &= \begin{pmatrix} \mathbf{A}' & \mathbf{0} \\ \mathbf{0} & \mathbf{X}' \end{pmatrix} \begin{pmatrix} \mathbf{V} & \mathbf{I} \\ \mathbf{I} & \mathbf{V}^{-1} \end{pmatrix} \begin{pmatrix} \mathbf{A} & \mathbf{0} \\ \mathbf{0} & \mathbf{X} \end{pmatrix} \\ &= \begin{pmatrix} \mathbf{A}'\mathbf{V}\mathbf{A} & \mathbf{A}'\mathbf{X} \\ \mathbf{X}'\mathbf{A} & \mathbf{X}'\mathbf{V}^{-1}\mathbf{X} \end{pmatrix} = \begin{pmatrix} \mathbf{A}'\mathbf{V}\mathbf{A} & \mathbf{I} \\ \mathbf{I} & \mathbf{X}'\mathbf{V}^{-1}\mathbf{X} \end{pmatrix}, \end{aligned} \tag{6.3.20}$$

since $\mathbf{A}'\mathbf{X} = \mathbf{I}$. The $2n \times 2n$ matrix

$$\mathbf{W}_2 = \begin{pmatrix} \mathbf{V} & \mathbf{I} \\ \mathbf{I} & \mathbf{V}^{-1} \end{pmatrix} \tag{6.3.21}$$

is nonnegative definite since $\mathbf{V}$ is positive definite and the Schur complement $\mathbf{W}_2/\mathbf{V} = \mathbf{0}$. Hence $\mathbf{W}_1$ is nonnegative definite and, therefore, so is the Schur complement

$$\mathbf{W}_1/\mathbf{X}'\mathbf{V}^{-1}\mathbf{X} = \mathbf{A}'\mathbf{V}\mathbf{A} - (\mathbf{X}'\mathbf{V}^{-1}\mathbf{X})^{-1} \tag{6.3.22}$$

and the inequality (6.3.19) is established.

It follows at once from (6.3.22) that $\operatorname{cov}(\mathbf{A}'\mathbf{y}) = \operatorname{cov}(\tilde{\boldsymbol{\beta}})$ if and only if the Schur complement

$$\mathbf{W}_1/\mathbf{X}'\mathbf{V}^{-1}\mathbf{X} = \mathbf{0}$$

and hence if and only if

$$\operatorname{rank}(\mathbf{W}_1) = \operatorname{rank}(\mathbf{X}'\mathbf{V}^{-1}\mathbf{X}) = p.$$

Moreover

$$\begin{aligned}\operatorname{rank}(\mathbf{W}_1) &= \operatorname{rank}\begin{pmatrix}\mathbf{V} & \mathbf{I}\\ \mathbf{I} & \mathbf{V}^{-1}\end{pmatrix}\begin{pmatrix}\mathbf{A} & \mathbf{0}\\ \mathbf{0} & \mathbf{X}\end{pmatrix} = \operatorname{rank}\begin{pmatrix}\mathbf{VA} & \mathbf{X}\\ \mathbf{A} & \mathbf{V}^{-1}\mathbf{X}\end{pmatrix}\\ &= \operatorname{rank}\begin{pmatrix}\mathbf{I} & \mathbf{0}\\ -\mathbf{V}^{-1} & \mathbf{I}\end{pmatrix}\begin{pmatrix}\mathbf{VA} & \mathbf{X}\\ \mathbf{A} & \mathbf{V}^{-1}\mathbf{X}\end{pmatrix} = \operatorname{rank}\begin{pmatrix}\mathbf{VA} & \mathbf{X}\\ \mathbf{0} & \mathbf{0}\end{pmatrix}\\ &= \operatorname{rank}(\mathbf{VA} : \mathbf{X}) = \operatorname{rank}(\mathbf{MVA}) + \operatorname{rank}(\mathbf{X})\\ &= \operatorname{rank}(\mathbf{MVA}) + p, \end{aligned} \tag{6.3.23}$$

where $\mathbf{M} = \mathbf{I} - \mathbf{X}(\mathbf{X}'\mathbf{X})^{-1}\mathbf{X}' = \mathbf{I} - \mathbf{H}$. It follows at once from (6.3.23) that $\operatorname{rank}(\mathbf{W}_1) = p$ if and only $\mathbf{MVA} = \mathbf{0}$.

Moreover, it is easy to see that when $\mathbf{A}'\mathbf{X} = \mathbf{I}$ then $\operatorname{cov}(\mathbf{A}'\mathbf{y} - \tilde{\boldsymbol{\beta}}) = \operatorname{cov}(\mathbf{A}'\mathbf{y}) - \operatorname{cov}(\tilde{\boldsymbol{\beta}})$ and so $\operatorname{cov}(\mathbf{A}'\mathbf{y}) = \operatorname{cov}(\tilde{\boldsymbol{\beta}})$ if and only if [364]

$$\mathbf{A}'\mathbf{y} = \tilde{\boldsymbol{\beta}} \quad \text{with probability 1} \quad \Longleftrightarrow \quad \mathbf{MVA} = \mathbf{0}. \tag{6.3.24}$$

When $\mathbf{A}'\mathbf{y}$ is the OLS estimator $\hat{\boldsymbol{\beta}} = (\mathbf{X}'\mathbf{X})^{-1}\mathbf{X}'\mathbf{y}$ then (6.3.24) becomes the well-known condition for the equality of OLS and GLS, i.e.,

$$\hat{\boldsymbol{\beta}} = \tilde{\boldsymbol{\beta}} \quad \text{with probability 1} \quad \Longleftrightarrow \quad \mathbf{MVX} = \mathbf{0}, \tag{6.3.25}$$

or equivalently $\mathbf{HV} = \mathbf{VH}$. The result (6.3.25) is due to Rao [368] and Zyskind [473] and has been called the *Rao–Zyskind Theorem*, see, e.g., [26]. Many further equivalent conditions for the equality of OLS and GLS are given in the survey by Puntanen & Styan [361].

6.3.2 Inverting the X′X matrix

Let us consider the partitioned full rank linear model $\mathcal{M} = \{\mathbf{y}, \mathbf{X}\boldsymbol{\beta}, \sigma^2\mathbf{I}\}$, where the $n \times (k+1)$ model matrix $\mathbf{X}$ is partitioned as

$$\mathbf{X} = (\mathbf{e} : \mathbf{x}_1 : \ldots : \mathbf{x}_k) = (\mathbf{e} : \mathbf{X}_0) = (\mathbf{X}_1 : \mathbf{x}_k), \tag{6.3.26}$$

where $\mathbf{e} \in \mathbb{R}^n$ is vector of ones and $\mathbf{X}_0$ is an $n \times k$ matrix. Then

$$(\mathbf{X}'\mathbf{X})^{-1} = \begin{pmatrix}\mathbf{e}'\mathbf{e} & \mathbf{e}'\mathbf{X}_0\\ \mathbf{X}_0'\mathbf{e} & \mathbf{X}_0'\mathbf{X}_0\end{pmatrix}^{-1} = \begin{pmatrix}\cdot & \cdot\\ \cdot & (\mathbf{X}_0'\mathbf{C}\mathbf{X}_0)^{-1}\end{pmatrix} = \begin{pmatrix}\cdot & \cdot\\ \cdot & \mathbf{T}_1^{-1}\end{pmatrix}, \tag{6.3.27}$$

where $\mathbf{T}_1 = \mathbf{X}_0'\mathbf{C}\mathbf{X}_0 = \{t_{ij}\}$,

$$\mathbf{T}_1 = \mathbf{X}'\mathbf{X}/\mathbf{e}'\mathbf{e}, \quad \mathbf{C} = \mathbf{I}_n - \mathbf{J}, \quad \mathbf{J} = \tfrac{1}{n}\mathbf{e}\mathbf{e}' = \mathbf{P}_{\mathbf{e}}; \tag{6.3.28}$$

here $\mathbf{C}$ is the centering matrix. Denoting $\mathbf{T}_1^{-1} = \{t^{ij}\}$ and $(\mathbf{X}'\mathbf{X})^{-1} = \{\underline{t}^{ij}\}$ we obtain the last diagonal element of $\mathbf{T}_1^{-1}$, i.e., the last diagonal element of $(\mathbf{X}'\mathbf{X})^{-1}$:

$$t^{kk} = (\mathbf{x}_k'\mathbf{M}_1\mathbf{x}_k)^{-1} = \frac{1}{\mathbf{x}_k'(\mathbf{I}-\mathbf{P}_1)\mathbf{x}_k} = \frac{1}{\|(\mathbf{I}-\mathbf{P}_1)\mathbf{x}_k\|^2} = \underline{t}^{kk}, \tag{6.3.29}$$

which is well defined provided $\mathbf{x}_k \notin \mathcal{C}(\mathbf{X}_1)$. We may note in passing that in the spirit of the quotient property (6.0.26), we can write

$$\begin{aligned}\mathbf{x}_k'\mathbf{M}_1\mathbf{x}_k &= \mathbf{X}'\mathbf{X}/\mathbf{X}_1'\mathbf{X}_1 = \mathbf{X}_0'\mathbf{C}\mathbf{X}_0/\mathbf{X}_1'\mathbf{C}\mathbf{X}_1 \\ &= (\mathbf{X}'\mathbf{X}/\mathbf{e}'\mathbf{e})/(\mathbf{X}_1'\mathbf{X}_1/\mathbf{e}'\mathbf{e}),\end{aligned} \tag{6.3.30}$$

Let us denote $\mathbf{X}_* = (\mathbf{X}_0 : \mathbf{y})$, and consider the matrix

$$\mathbf{T} = \mathbf{X}_*'\mathbf{C}\mathbf{X}_* = \begin{pmatrix}\mathbf{X}_0'\mathbf{C}\mathbf{X}_0 & \mathbf{X}_0'\mathbf{C}\mathbf{y} \\ \mathbf{y}'\mathbf{C}\mathbf{X}_0 & \mathbf{y}'\mathbf{C}\mathbf{y}\end{pmatrix} = \begin{pmatrix}\mathbf{T}_1 & \mathbf{t}_2 \\ \mathbf{t}_2' & t_{yy}\end{pmatrix}. \tag{6.3.31}$$

Since $\operatorname{rank}(\mathbf{e} : \mathbf{X}_*) = 1 + \operatorname{rank}(\mathbf{C}\mathbf{X}_*) = 1 + \operatorname{rank}(\mathbf{T})$, we see that

$$\operatorname{rank}(\mathbf{T}) = k+1 \iff (\mathbf{e} : \mathbf{X}_0 : \mathbf{y}) \text{ has full column rank.} \tag{6.3.32}$$

If $\mathbf{T}$ is invertible, then the last diagonal element of $\mathbf{T}^{-1}$ is

$$t^{yy} = t_{yy\cdot\mathbf{x}}^{-1} = \frac{1}{t_{yy} - \mathbf{t}_2'\mathbf{T}_1^{-1}\mathbf{t}_2} = \frac{1}{\text{SSE}} = \frac{1}{\mathbf{T}/\mathbf{T}_1}. \tag{6.3.33}$$

Here SSE refers to the residual error sum of squares $\text{SSE} = \mathbf{y}'(\mathbf{I}_n - \mathbf{P}_\mathbf{X})\mathbf{y} = \mathbf{y}'\mathbf{M}\mathbf{y}$. To confirm that SSE indeed can be expressed as in (6.3.33), we note that

$$\begin{aligned}\text{SSE} &= \mathbf{y}'(\mathbf{I}_n - \mathbf{P}_\mathbf{X})\mathbf{y} = \mathbf{y}'(\mathbf{I}_n - \mathbf{J} - \mathbf{P}_{\mathbf{C}\mathbf{X}_0})\mathbf{y} = \mathbf{y}'(\mathbf{C} - \mathbf{P}_{\mathbf{C}\mathbf{X}_0})\mathbf{y} \\ &= \mathbf{y}'\mathbf{C}\mathbf{y} - \mathbf{y}'\mathbf{C}\mathbf{X}_0(\mathbf{X}_0'\mathbf{C}\mathbf{X}_0)^{-1}\mathbf{X}_0'\mathbf{C}\mathbf{y} = t_{yy\cdot\mathbf{x}}.\end{aligned} \tag{6.3.34}$$

Let $\mathbf{R}$ be the (partitioned) sample correlation matrix of the variables $x_1, x_2, \ldots, x_k, y$. Then

$$\mathbf{R} = \begin{pmatrix}\mathbf{R}_1 & \mathbf{r}_2 \\ \mathbf{r}_2' & 1\end{pmatrix} = (\operatorname{diag}\mathbf{T})^{-1/2}\mathbf{T}(\operatorname{diag}\mathbf{T})^{-1/2}; \tag{6.3.35}$$

here we assume that all variables have nonzero variances. Now the last diagonal element of $\mathbf{R}^{-1}$ is

$$r^{yy} = \frac{t_{yy}}{\text{SSE}} = \frac{\text{SST}}{\text{SSE}} = \frac{\mathbf{y}'\mathbf{C}\mathbf{y}}{\mathbf{y}'\mathbf{M}\mathbf{y}} = \frac{\sum_{i=1}^n (y_i - \bar{y})^2}{\sum_{i=1}^n (y_i - \hat{y}_i)^2} = \frac{1}{1-R^2}, \tag{6.3.36}$$

where $\{\hat{y}_i\} = \hat{\mathbf{y}} = \mathbf{H}\mathbf{y} = \mathbf{y} - \mathbf{M}\mathbf{y}$, SST refers to the total sum of squares, and R^2 is the sample multiple correlation squared (coefficient of determination) when y is explained by the variables $x_1, x_2, \dots, x_k$ (and a constant).

The last regression coefficient

$$\hat{\beta}_k = \frac{\mathbf{x}_k'\mathbf{M}_1\mathbf{y}}{\mathbf{x}_k'\mathbf{M}_1\mathbf{x}_k} \tag{6.3.37}$$

has variance

$$\operatorname{var}(\hat{\beta}_k) = \frac{\sigma^2}{\mathbf{x}_k'\mathbf{M}_1\mathbf{x}_k} = \frac{\sigma^2}{\|(\mathbf{I} - \mathbf{P}_1)\mathbf{x}_k\|^2}, \tag{6.3.38}$$

and so we may expect problems in estimation if the column vector $\mathbf{x}_k$ is almost in the column space $\mathcal{C}(\mathbf{X}_1)$ for then the denominator in (6.3.38) will be close to 0. As noted by Seber & Lee [412, Ex 3, p. 53] and by C. R. Rao [372, p. 236], it follows that

$$\operatorname{var}(\hat{\beta}_k) \geq \frac{\sigma^2}{\mathbf{x}_k'\mathbf{x}_k}, \tag{6.3.39}$$

with equality if and only if $\mathbf{X}_1'\mathbf{x}_k = \mathbf{0}$.

If we consider such a linear model where $\mathbf{x}_k$ is explained by all other x variables (plus a constant), that is, the model matrix is $\mathbf{X}_1$, then the residual sum of squares and the total sum of squares are

$$\operatorname{SSE}(k) = \mathbf{x}_k'\mathbf{M}_1\mathbf{x}_k, \quad \operatorname{SST}(k) = \mathbf{x}_k'\mathbf{C}\mathbf{x}_k = t_{kk}. \tag{6.3.40}$$

Moreover, the corresponding coefficient of determination is

$$R_k^2 = 1 - \frac{\mathbf{x}_k'\mathbf{M}_1\mathbf{x}_k}{t_{kk}}, \tag{6.3.41}$$

and $\mathbf{x}_k'\mathbf{M}_1\mathbf{x}_k = (1 - R_k^2)t_{kk}$. Hence the variance of $\hat{\beta}_k$ can be expressed as

$$\operatorname{var}(\hat{\beta}_k) = \frac{1}{1 - R_k^2} \cdot \frac{1}{t_{kk}}\,\sigma^2 := \frac{\operatorname{VIF}_k}{t_{kk}}\,\sigma^2, \tag{6.3.42}$$

where VIF_k is the *variance inflation factor*. As Belsley [43, p. 29] states, much interesting work on collinearity diagnostics has been done with VIF-like measures, although the concept masquerades under various names [427].

We may note that, corresponding to (6.3.36), the ith diagonal element of $\mathbf{R}_1^{-1}$ can be expressed as

$$r^{ii} = \frac{1}{1 - R^2_{i\cdot 1\dots i-1,i+1,\dots k}} = \operatorname{VIF}_i. \tag{6.3.43}$$

We see that the diagonal elements of the correlation matrix can be quite informative for data analysis [381].

Let's study the last two regressors under the full rank model $\{\mathbf{y},\, \mathbf{X}\boldsymbol{\beta},\, \sigma^2\mathbf{I}\}$ and set

$$\mathbf{X}_1 = (\mathbf{e} : \mathbf{x}_1 : \cdots : \mathbf{x}_{k-2}), \quad \mathbf{X} = (\mathbf{X}_1 : \mathbf{x}_{k-1} : \mathbf{x}_k) = (\mathbf{X}_1 : \mathbf{X}_2).$$

Now the (sample) partial correlation between $\mathbf{x}_{k-1}$ and $\mathbf{x}_k$ when variables in $\mathbf{X}_1$ are held constant, is defined as the (sample) correlation between the residual vectors

$$\mathbf{s}_{k-1\cdot 1} = (\mathbf{I} - \mathbf{P}_{\mathbf{X}_1})\mathbf{x}_{k-1} = \mathbf{M}_1\mathbf{x}_{k-1} \quad \text{and} \quad \mathbf{s}_{k\cdot 1} = (\mathbf{I} - \mathbf{P}_{\mathbf{X}_1})\mathbf{x}_k = \mathbf{M}_1\mathbf{x}_k.$$

Because these residuals are centered, their (sample) correlation is

$$\operatorname{corr}(\mathbf{s}_{k-1\cdot 1}, \mathbf{s}_{k\cdot 1}) = \frac{\mathbf{x}'_{k-1}\mathbf{M}_1\mathbf{x}_k}{\sqrt{\mathbf{x}'_{k-1}\mathbf{M}_1\mathbf{x}_{k-1}}\,\sqrt{\mathbf{x}'_k\mathbf{M}_1\mathbf{x}_k}}. \tag{6.3.44}$$

Since $\operatorname{cov}(\hat{\boldsymbol{\beta}}_2) = \sigma^2(\mathbf{X}'_2\mathbf{M}_1\mathbf{X}_2)^{-1}$, it is straightforward to conclude that

$$\operatorname{corr}(\hat{\beta}_{k-1}, \hat{\beta}_k) = -r_{k-1,k\cdot 1\ldots k-2}, \tag{6.3.45}$$

see, e.g., Belsley [43, p. 33], C. R. Rao [372, p. 270 (with a missing minus sign)], and Kanto & Puntanen [254].

6.3.3 The covariance matrix of the BLUE($\mathbf{X}\boldsymbol{\beta}$)

Using (6.3.11) it is easy to introduce the following general representations of the covariance matrix of the best linear unbiased estimator (BLUE) of $\mathbf{X}\boldsymbol{\beta}$ under the model $\mathcal{M} = \{\mathbf{y},\, \mathbf{X}\boldsymbol{\beta},\, \mathbf{V}\}$:

$$\operatorname{cov}(\mathbf{X}\tilde{\boldsymbol{\beta}}) = \mathbf{HVH} - \mathbf{HVM}(\mathbf{MVM})^{-}\mathbf{MVH} \tag{6.3.46a}$$

$$= \mathbf{V} - \mathbf{VM}(\mathbf{MVM})^{-}\mathbf{MV} \tag{6.3.46b}$$

$$= \mathbf{X}(\mathbf{X}'\mathbf{W}^{-}\mathbf{X})^{-}\mathbf{X}' - \mathbf{XUX}', \tag{6.3.46c}$$

see also [27]. If $\mathbf{X}$ has full column rank, then the covariance matrix of the BLUE($\boldsymbol{\beta}$) can be expressed as

$$\begin{aligned}\operatorname{cov}(\tilde{\boldsymbol{\beta}}) &= (\mathbf{X}'\mathbf{X})^{-1}[\mathbf{X}'\mathbf{VX} - \mathbf{X}'\mathbf{VZ}(\mathbf{Z}'\mathbf{VZ})^{-}\mathbf{Z}'\mathbf{VX}](\mathbf{X}'\mathbf{X})^{-1} \\ &= \operatorname{cov}(\hat{\boldsymbol{\beta}}) - (\mathbf{X}'\mathbf{X})^{-1}\mathbf{X}'\mathbf{VZ}(\mathbf{Z}'\mathbf{VZ})^{-}\mathbf{Z}'\mathbf{VX}(\mathbf{X}'\mathbf{X})^{-1}, \end{aligned} \tag{6.3.47}$$

where $\mathcal{C}(\mathbf{Z}) = \mathcal{C}(\mathbf{M})$ and so

$$\begin{aligned}\operatorname{cov}(\hat{\boldsymbol{\beta}}) - \operatorname{cov}(\tilde{\boldsymbol{\beta}}) &= (\mathbf{X}'\mathbf{X})^{-1}\mathbf{X}'\mathbf{VZ}(\mathbf{Z}'\mathbf{VZ})^{-}\mathbf{Z}'\mathbf{VX}(\mathbf{X}'\mathbf{X})^{-1} \\ &= (\mathbf{X}'\mathbf{X})^{-1}\mathbf{X}'\mathbf{VM}(\mathbf{MVM})^{-}\mathbf{MVX}(\mathbf{X}'\mathbf{X})^{-1}. \end{aligned} \tag{6.3.48}$$

Because the joint covariance matrix of the OLS fitted values $\mathbf{Hy}$ and the residuals $\mathbf{My}$ is

$$\operatorname{cov}\begin{pmatrix}\mathbf{Hy}\\ \mathbf{My}\end{pmatrix} = \begin{pmatrix}\mathbf{HVH} & \mathbf{HVM}\\ \mathbf{MVH} & \mathbf{MVM}\end{pmatrix} = (\mathbf{H}:\mathbf{M})'\mathbf{V}(\mathbf{H}:\mathbf{M}) := \mathbf{\Sigma}_{\mathrm{HM}}, \tag{6.3.49}$$

and the covariance matrix

$$\operatorname{cov}\begin{pmatrix}\mathbf{y}\\ \mathbf{My}\end{pmatrix} = \begin{pmatrix}\mathbf{V} & \mathbf{VM}\\ \mathbf{MV} & \mathbf{MVM}\end{pmatrix} := \mathbf{\Sigma}_{\mathrm{IM}}, \tag{6.3.50}$$

we observe immediately that the BLUE's covariance matrix can be interpreted as a Schur complement:

$$\operatorname{cov}(\mathrm{BLUE}(\mathbf{X}\boldsymbol{\beta})) = \mathbf{\Sigma}_{\mathrm{HM}}/\mathbf{MVM} = \mathbf{\Sigma}_{\mathrm{IM}}/\mathbf{MVM}. \tag{6.3.51}$$

To see that the BLUE's covariance matrix is a specific Schur complement we consider the general linear model $\{\mathbf{y}, \mathbf{X}\boldsymbol{\beta}, \mathbf{V}\}$. Then $\mathbf{Gy}$ is unbiased for $\mathbf{X}\boldsymbol{\beta}$ whenever $\mathbf{G}$ satisfies the equation $\mathbf{GX} = \mathbf{X}$. The general solution $\mathbf{G}$ to the unbiasedness condition can be expressed as $\mathbf{G} = \mathbf{H} - \mathbf{FM}$, where the matrix $\mathbf{F}$ is free to vary [377, Th. 2.3.2]. In other words, all unbiased linear estimators of $\mathbf{X}\boldsymbol{\beta}$ can be generated through $\mathbf{Hy} - \mathbf{FMy}$ by varying $\mathbf{F}$. If we now apply the minimization result (6.2.26) to the covariance matrix

$$\operatorname{cov}\begin{pmatrix}\mathbf{Hy}\\ \mathbf{My}\end{pmatrix} = \begin{pmatrix}\mathbf{HVH} & \mathbf{HVM}\\ \mathbf{MVH} & \mathbf{MVM}\end{pmatrix} = \mathbf{\Sigma}_{\mathrm{HM}}, \tag{6.3.52}$$

we obtain

$$\operatorname{cov}[\mathbf{Hy} - \mathbf{HVM}(\mathbf{MVM})^{-}\mathbf{My}] \leq_{\mathrm{L}} \operatorname{cov}(\mathbf{Hy} - \mathbf{FMy}) \text{ for all } \mathbf{F}, \tag{6.3.53}$$

and so the minimal covariance matrix is, according to (6.2.27), the Schur complement $\mathbf{\Sigma}_{\mathrm{HM}}/\mathbf{MVM}$. Note that (6.3.53) indeed means that

$$\mathbf{Hy} - \mathbf{HVM}(\mathbf{MVM})^{-}\mathbf{My} = \mathrm{BLUE}(\mathbf{X}\boldsymbol{\beta}). \tag{6.3.54}$$

We note that since the rank is additive on the Schur complement

$$\operatorname{rank}[\operatorname{cov}(\mathbf{X}\tilde{\boldsymbol{\beta}})] = \operatorname{rank}(\mathbf{V}) - \operatorname{rank}(\mathbf{VM}) = \dim \mathcal{C}(\mathbf{X}) \cap \mathcal{C}(\mathbf{V}), \tag{6.3.55}$$

and $\mathcal{C}[\operatorname{cov}(\mathbf{X}\tilde{\boldsymbol{\beta}})] = \mathcal{C}(\mathbf{X}) \cap \mathcal{C}(\mathbf{V})$.

6.3.4 The matrix $\dot{\mathbf{M}} = \mathbf{M}(\mathbf{MVM})^{-}\mathbf{M}$

We now look briefly at some properties of the matrix

$$\dot{\mathbf{M}} = \mathbf{M}(\mathbf{MVM})^{-}\mathbf{M}, \tag{6.3.56}$$

which appears in several formulas above. We observe that $\dot{\mathbf{M}}$ is unique if and only if $\mathcal{C}(\mathbf{M}) \subset \mathcal{C}(\mathbf{MV})$, which is equivalent to $\mathbb{R}^n = \mathcal{C}(\mathbf{X} : \mathbf{V})$. Even though $\dot{\mathbf{M}}$ is not necessarily unique, the matrix product

$$\mathbf{P_V}\dot{\mathbf{M}}\mathbf{P_V} = \mathbf{P_V}\mathbf{M}(\mathbf{MVM})^{-}\mathbf{M}\mathbf{P_V} := \bar{\mathbf{M}} \tag{6.3.57}$$

is, however, clearly invariant for any choice of $(\mathbf{MVM})^{-}$. Moreover, let $\mathbf{W} = \mathbf{V} + \mathbf{XAA'X'}$ be a nonnegative definite matrix such that $\mathcal{C}(\mathbf{W}) = \mathcal{C}(\mathbf{X} : \mathbf{V})$, and let $\dot{\mathbf{M}}_{\mathrm{W}} = \mathbf{M}(\mathbf{MWM})^{-}\mathbf{M} = \dot{\mathbf{M}}$. Let

$$\bar{\mathbf{M}}_{\mathrm{W}} = \mathbf{P_W}\dot{\mathbf{M}}_{\mathrm{W}}\mathbf{P_W} = \mathbf{P_W}\dot{\mathbf{M}}\mathbf{P_W} = \mathbf{P_W}\mathbf{M}(\mathbf{MVM})^{-}\mathbf{M}\mathbf{P_W}.$$

The matrices $\dot{\mathbf{M}}$, $\bar{\mathbf{M}}$, and $\bar{\mathbf{M}}_{\mathrm{W}}$ are very useful in many considerations related to linear model $\mathcal{M} = \{\mathbf{y}, \mathbf{X}\boldsymbol{\beta}, \sigma^2\mathbf{V}\}$. For example, the following properties hold:

$$\bar{\mathbf{M}}_{\mathrm{W}} = \mathbf{P_W}\dot{\mathbf{M}}\mathbf{P_W} = \mathbf{W}^{+} - \mathbf{W}^{+}\mathbf{X}(\mathbf{X'W^{+}X})^{-}\mathbf{X'W^{+}}, \tag{6.3.58}$$

$$\dot{\mathbf{M}}_{\mathrm{W}} = \mathbf{M}\bar{\mathbf{M}}_{\mathrm{W}}\mathbf{M} = \mathbf{MW^{+}M} - \mathbf{MW^{+}X}(\mathbf{X'W^{+}X})^{-}\mathbf{X'W^{+}M}. \tag{6.3.59}$$

Formula (6.3.58) can be proved by putting $\mathbf{F} = (\mathbf{W}^{+})^{1/2}\mathbf{X}, \mathbf{L} = \mathbf{W}^{1/2}\mathbf{M}$ and noting that $\mathbf{F'L} = \mathbf{0}$ implies $\mathbf{P_F} + \mathbf{P_L} = \mathbf{P_W}$; inserting explicit expressions of $\mathbf{P_F}$ and $\mathbf{P_L}$ yields (6.3.58). Decomposition (6.3.58) can also be obtained using Corollary 3 in [27]. Note that clearly we have also

$$\bar{\mathbf{M}}_{\mathrm{W}} = \mathbf{P_W}\mathbf{Z}(\mathbf{Z'VZ})^{-}\mathbf{Z'}\mathbf{P_W}, \tag{6.3.60}$$

where $\mathcal{C}(\mathbf{Z}) = \mathcal{C}(\mathbf{M})$. We now see immediately that $\bar{\mathbf{M}}_{\mathrm{W}}$ is a specific Schur complement. If we write

$$\mathbf{S} = (\mathbf{X} : \mathbf{M})'\mathbf{W}^{+}(\mathbf{X} : \mathbf{M}) = \begin{pmatrix} \mathbf{X'W^{+}X} & \mathbf{X'W^{+}M} \\ \mathbf{MW^{+}X} & \mathbf{MW^{+}M} \end{pmatrix}, \tag{6.3.61}$$

it follows at once that

$$\bar{\mathbf{M}}_{\mathrm{W}} = \mathbf{P_W}\dot{\mathbf{M}}\mathbf{P_W} = \mathbf{S}/\mathbf{X'W^{+}X}. \tag{6.3.62}$$

From (6.3.62), we can conclude that the equality

$$\mathbf{B}'\bar{\mathbf{M}}_{\mathrm{W}}\mathbf{C} = \mathbf{B}'\dot{\mathbf{M}}\mathbf{C} = \mathbf{B}'(\mathbf{S}/\mathbf{X'W^{+}X})\mathbf{C} \tag{6.3.63}$$

holds for all matrices $\mathbf{B}$ and $\mathbf{C}$ such that $\mathcal{C}(\mathbf{B}) \subset \mathcal{C}(\mathbf{W})$, $\mathcal{C}(\mathbf{C}) \subset \mathcal{C}(\mathbf{W})$.

Premultiplying (6.3.58) by $\mathbf{W}$ yields

$$\begin{aligned} \mathbf{X}(\mathbf{X'W^{+}X})^{-}\mathbf{X'W^{+}} &= \mathbf{P_W} - \mathbf{WM}(\mathbf{MVM})^{-}\mathbf{MP_W} \\ &= \mathbf{P_W} - \mathbf{VM}(\mathbf{MVM})^{-}\mathbf{MP_W}. \end{aligned} \tag{6.3.64}$$

Hence for every $\mathbf{y} \in \mathcal{C}(\mathbf{X} : \mathbf{V}) = \mathcal{C}(\mathbf{W})$, we have

$$\begin{aligned}\mathbf{X}(\mathbf{X}'\mathbf{W}^+\mathbf{X})^-\mathbf{X}'\mathbf{W}^+\mathbf{y} &= \mathbf{y} - \mathbf{VM}(\mathbf{MVM})^-\mathbf{My} = \mathbf{y} - \mathbf{V}\dot{\mathbf{M}}\mathbf{y} \\ &= \mathbf{y} - \mathbf{V}(\mathbf{S}/\mathbf{X}'\mathbf{W}^+\mathbf{X})\mathbf{y} = \mathrm{BLUE}(\mathbf{X}\boldsymbol{\beta}), \end{aligned} \tag{6.3.65}$$

with probability 1.

Note also that postmultiplying (6.3.64) with $\mathbf{W}$ yields

$$\mathbf{X}(\mathbf{X}'\mathbf{W}^+\mathbf{X})^-\mathbf{X}' = \mathbf{W} - \mathbf{VM}(\mathbf{MVM})^-\mathbf{MV}, \tag{6.3.66}$$

and thereby

$$\mathbf{X}(\mathbf{X}'\mathbf{W}^+\mathbf{X})^-\mathbf{X}' - \mathbf{XAA}'\mathbf{X} = \mathbf{V} - \mathbf{VM}(\mathbf{MVM})^-\mathbf{MV}, \tag{6.3.67}$$

showing the equality of two representations for the covariance matrix of $\mathrm{BLUE}(\mathbf{X}\boldsymbol{\beta})$.

The BLUE's residual can be conveniently expressed as

$$\tilde{\mathbf{s}} = \mathbf{y} - \mathrm{BLUE}(\mathbf{X}\boldsymbol{\beta}) = \mathbf{V}\dot{\mathbf{M}}\mathbf{y}, \tag{6.3.68}$$

and the weighted sum of squares of errors can be written as

$$\mathrm{SSE}(V) = \tilde{\mathbf{s}}'\mathbf{W}^-\tilde{\mathbf{s}} = \tilde{\mathbf{s}}'\mathbf{V}^-\tilde{\mathbf{s}} = \mathbf{y}'\dot{\mathbf{M}}\mathbf{V}\mathbf{W}^-\mathbf{V}\dot{\mathbf{M}}\mathbf{y} = \mathbf{y}'\dot{\mathbf{M}}\mathbf{y}. \tag{6.3.69}$$

On the other hand, from (6.3.58) it follows that for every $\mathbf{y} \in \mathcal{C}(\mathbf{X} : \mathbf{V})$,

$$\mathbf{y}'\mathbf{P}_\mathbf{W}\dot{\mathbf{M}}\mathbf{P}_\mathbf{W}\mathbf{y} = \mathbf{y}'\dot{\mathbf{M}}\mathbf{y} = \mathbf{y}'[\mathbf{W}^+ - \mathbf{W}^+\mathbf{X}(\mathbf{X}'\mathbf{W}^+\mathbf{X})^-\mathbf{X}'\mathbf{W}^+]\mathbf{y}, \tag{6.3.70}$$

i.e., the weighted residual sum of squares can now be expressed as

$$\begin{aligned}\mathrm{SSE}(V) &= \mathbf{y}'\dot{\mathbf{M}}\mathbf{y} = \mathbf{y}'[\mathbf{W}^+ - \mathbf{W}^+\mathbf{X}(\mathbf{X}'\mathbf{W}^+\mathbf{X})^-\mathbf{X}'\mathbf{W}^+]\mathbf{y} \\ &= \mathbf{y}'(\mathbf{S}/\mathbf{X}'\mathbf{W}^+\mathbf{X})\mathbf{y}. \end{aligned} \tag{6.3.71}$$

We may note that the equality

$$\mathbf{y}'\mathbf{M}\mathbf{y} = \mathbf{y}'\dot{\mathbf{M}}\mathbf{y} \qquad \forall\, \mathbf{y} \in \mathcal{C}(\mathbf{X} : \mathbf{V}) \tag{6.3.72}$$

was studied by Groß [191]. He showed that (6.3.72) holds if and only if $\mathbf{VM}$ is idempotent.

Denoting $\mathbf{F} = (\mathbf{V}^+)^{1/2}\mathbf{X}, \mathbf{L} = \mathbf{V}^{1/2}\mathbf{M}$, we observe that if the condition $\mathbf{F}'\mathbf{L} = \mathbf{X}'\mathbf{P}_\mathbf{V}\mathbf{M} = \mathbf{0}$ holds, then $\bar{\mathbf{M}}$ has the expressions

$$\bar{\mathbf{M}} = \mathbf{P}_\mathbf{V}\dot{\mathbf{M}}\mathbf{P}_\mathbf{V} = \mathbf{V}^+ - \mathbf{V}^+\mathbf{X}(\mathbf{X}'\mathbf{V}^+\mathbf{X})^-\mathbf{X}'\mathbf{V}^+, \tag{6.3.73}$$

$$\bar{\mathbf{M}} = \mathbf{M}\bar{\mathbf{M}}\mathbf{M} = \mathbf{M}\mathbf{V}^+\mathbf{M} - \mathbf{M}\mathbf{V}^+\mathbf{X}(\mathbf{X}'\mathbf{V}^+\mathbf{X})^-\mathbf{X}'\mathbf{V}^+\mathbf{M}. \tag{6.3.74}$$

We denote the number of unit canonical correlations between $\mathbf{Hy}$ and $\mathbf{My}$ [359, Lemma 3.4.1] by

$$u = \operatorname{rank}(\mathbf{X}'\mathbf{P}_\mathbf{V}\mathbf{M}) = \operatorname{rank}(\mathbf{H}\mathbf{P}_\mathbf{V}\mathbf{M}) = \dim \mathcal{C}(\mathbf{VH}) \cap \mathcal{C}(\mathbf{VM}). \quad (6.3.75)$$

When $\mathbf{X}'\mathbf{X} = \mathbf{I}_p$ (and $u = 0$), it can be shown that

$$(\mathbf{X}'\mathbf{V}^+\mathbf{X})^+ = \mathbf{X}'\mathbf{V}\mathbf{X} - \mathbf{X}'\mathbf{V}\mathbf{Z}(\mathbf{Z}'\mathbf{V}\mathbf{Z})^-\mathbf{Z}'\mathbf{V}\mathbf{X}, \quad (6.3.76)$$

and hence, in this particular situation,

$$\mathbf{X}(\mathbf{X}'\mathbf{V}^+\mathbf{X})^+\mathbf{X}' = \mathbf{\Sigma}_{\mathrm{HM}}/\mathbf{MVM} = \operatorname{cov}(\mathrm{BLUE}(\mathbf{X}\boldsymbol{\beta})). \quad (6.3.77)$$

When $\mathbf{V}$ is positive definite, we obtain

$$\begin{aligned}\dot{\mathbf{M}} &= \mathbf{M}(\mathbf{MVM})^-\mathbf{M} = \mathbf{M}(\mathbf{MVM})^+\mathbf{M} = (\mathbf{MVM})^+ \\ &= \mathbf{V}^{-1} - \mathbf{V}^{-1}\mathbf{X}(\mathbf{X}'\mathbf{V}^{-1}\mathbf{X})^-\mathbf{X}'\mathbf{V}^{-1} = \mathbf{V}^{-1}(\mathbf{I} - \mathbf{P}_{\mathbf{X};\mathbf{V}^{-1}}). \quad (6.3.78)\end{aligned}$$

Denoting

$$\mathbf{S}_{\mathbf{V}^{-1}} = (\mathbf{X} : \mathbf{M})'\mathbf{V}^{-1}(\mathbf{X} : \mathbf{M}), \quad \mathbf{R}_{\mathbf{V}^{-1}} = (\mathbf{X} : \mathbf{I})'\mathbf{V}^{-1}(\mathbf{X} : \mathbf{I}), \quad (6.3.79)$$

we see that $\dot{\mathbf{M}}$ is two Schur complements

$$\dot{\mathbf{M}} = \mathbf{S}_{\mathbf{V}^{-1}}/\mathbf{X}'\mathbf{V}^{-1}\mathbf{X} = \mathbf{R}_{\mathbf{V}^{-1}}/\mathbf{X}'\mathbf{V}^{-1}\mathbf{X}. \quad (6.3.80)$$

We can of course replace $\mathbf{X}$ with $\mathbf{H}$ in (6.3.79) and (6.3.80).

6.3.5 The covariance matrix as a shorted matrix

The BLUE's covariance matrix can be interpreted as a *shorted matrix*, as shown by Mitra & Puntanen [318]. Let $\mathbf{V}$ be a given $n \times n$ nonnegative definite matrix and $\mathbf{X}$ an $n \times p$ matrix. Consider the following set of nonnegative definite matrices:

$$\mathcal{U} = \{\mathbf{U} : \mathbf{0} \leq \mathbf{U} \leq \mathbf{V},\ \mathcal{C}(\mathbf{U}) \subset \mathcal{C}(\mathbf{X})\}. \quad (6.3.81)$$

The maximal element $\mathbf{U}$ in $\mathcal{U}$ is the shorted matrix $S(\mathbf{V} \mid \mathbf{X})$ of $\mathbf{V}$ with respect to $\mathbf{X}$, introduced by Kreĭn [269]; see also W. N. Anderson [10].

Mitra & Puri [320] obtained explicit expressions for the shorted operator of a nonnegative definite symmetric operator in terms of a minimum-seminorm generalized inverse and a semi-least-squares inverse of a complex matrix. This paper [320] may be considered, therefore, as the starting point for statistical applications of the shorted matrix and the shorted operator. For more on shorted matrices and their applications in statistics, see [319].

As shown by W. N. Anderson [10] and by W. N. Anderson & Trapp [13], the set $\mathcal{U}$ in (6.3.81) has a maximal element and it, the shorted matrix, is unique. The shorted matrix of $\mathbf{V}$ with respect to $\mathbf{X}$ is, therefore, the unique nonnegative definite matrix which is "as close to $\mathbf{V}$ as possible" in the Löwner partial ordering, but whose column space is in that of $\mathbf{X}$. We note that, in general, the concept of closeness between matrices is not uniquely defined, but closeness in the Löwner sense is quite natural, especially from the point of view of statistical applications.

Consider now the general linear model $\mathcal{M} = \{\mathbf{y}, \mathbf{X}\boldsymbol{\beta}, \mathbf{V}\}$. Then Mitra & Puntanen [318] proved that

$$\operatorname{cov}(\text{BLUE}(\mathbf{X}\boldsymbol{\beta})) = \mathcal{S}(\mathbf{V} \mid \mathbf{X}), \tag{6.3.82}$$

which shows the close connection between the shorted matrix and the Schur complement. Mitra, Puntanen & Styan [319, Th. 3.2] also showed that the following five statements are equivalent when considering the linear model $\{\mathbf{y}, \mathbf{X}\boldsymbol{\beta}, \mathbf{V}\}$ with $\mathbf{G}$ a generalized inverse of $\mathbf{X}$:

(i) $\mathbf{XGVG'X'} \leq_{\mathsf{L}} \mathbf{V}$,

(ii) $\mathbf{G'}$ is a minimum-$\mathbf{V}$-seminorm generalized inverse of $\mathbf{X'}$,

(iii) $\mathbf{XGy}$ is the BLUE of $\mathbf{X}\boldsymbol{\beta}$,

(iv) $\mathbf{XGVG'X'} \leq_{\mathsf{rs}} \mathbf{V}$,

(v) $\mathbf{XGVG'X'} = \mathcal{S}(\mathbf{V} \mid \mathbf{X})$,

In (iv) the symbol $\leq_{\mathsf{rs}}$ denotes the rank-subtractivity partial ordering; see, e.g., [205]; see also [117, Lemma 1.2], [118, Lemma 1.2].

6.3.6 The Watson efficiency of the OLSE

The ordinary least squares estimator (OLSE) and the best linear unbiased estimator (BLUE) of $\boldsymbol{\beta}$ under the full rank model $\mathcal{M} = \{\mathbf{y}, \mathbf{X}\boldsymbol{\beta}, \mathbf{V}\}$ have the covariance matrices

$$\operatorname{cov}(\hat{\boldsymbol{\beta}}) = (\mathbf{X'X})^{-1}\mathbf{X'VX}(\mathbf{X'X})^{-1}, \quad \operatorname{cov}(\tilde{\boldsymbol{\beta}}) = (\mathbf{X'V}^{-1}\mathbf{X})^{-1}, \tag{6.3.83}$$

and hence we have the Löwner ordering

$$(\mathbf{X'X})^{-1}\mathbf{X'VX}(\mathbf{X'X})^{-1} - (\mathbf{X'V}^{-1}\mathbf{X})^{-1} \geq_{\mathsf{L}} \mathbf{0}. \tag{6.3.84}$$

It is natural to ask how bad the OLSE could be with respect to the BLUE. There is no unique way to measure this. One frequently used measure [454, Section 3.3], [455, p. 330] is the *Watson efficiency*

$$\phi = \frac{\det \operatorname{cov}(\tilde{\boldsymbol{\beta}})}{\det \operatorname{cov}(\hat{\boldsymbol{\beta}})} = \frac{\left(\det(\mathbf{X}'\mathbf{X})\right)^2}{\det(\mathbf{X}'\mathbf{V}\mathbf{X}) \cdot \det(\mathbf{X}'\mathbf{V}^{-1}\mathbf{X})}. \tag{6.3.85}$$

Clearly $0 < \phi \leq 1$, with $\phi = 1$ if and only if OLSE = BLUE.

Another measure of efficiency, introduced by Bloomfield & Watson [61], is based on the Euclidean size of the commutator $\mathbf{HV} - \mathbf{VH}$

$$\psi = \tfrac{1}{2}\|\mathbf{HV} - \mathbf{VH}\|^2 = \|\mathbf{HVM}\|^2 = \operatorname{tr}(\mathbf{HVMV}). \tag{6.3.86}$$

Clearly $\psi = 0$ whenever OLSE = BLUE. Rao [375] suggested the difference of the traces (or equivalently the trace of the difference) of the covariance matrices of the OLSE and BLUE of $\mathbf{X}\boldsymbol{\beta}$:

$$\operatorname{tr}\operatorname{cov}(\mathbf{X}\hat{\boldsymbol{\beta}}) - \operatorname{tr}\operatorname{cov}(\mathbf{X}\tilde{\boldsymbol{\beta}}) = \operatorname{tr}\mathbf{HVH} - \operatorname{tr}\mathbf{X}(\mathbf{X}'\mathbf{V}^{-1}\mathbf{X})^{-1}\mathbf{X}', \tag{6.3.87}$$

while Krämer [268] suggested the ratio of the traces

$$\frac{\operatorname{tr}\operatorname{cov}(\mathbf{X}\hat{\boldsymbol{\beta}})}{\operatorname{tr}\operatorname{cov}(\mathbf{X}\tilde{\boldsymbol{\beta}})} = \frac{\operatorname{tr}\mathbf{HVH}}{\operatorname{tr}\mathbf{X}(\mathbf{X}'\mathbf{V}^{-1}\mathbf{X})^{-1}\mathbf{X}'}. \tag{6.3.88}$$

When $\mathbf{X}$ has full column rank, then, according to (6.3.47), the covariance matrix of the BLUE of $\boldsymbol{\beta}$ can be expressed as

$$\begin{aligned}\operatorname{cov}(\tilde{\boldsymbol{\beta}}) &= (\mathbf{X}'\mathbf{X})^{-1}[\mathbf{X}'\mathbf{V}\mathbf{X} - \mathbf{X}'\mathbf{V}\mathbf{Z}(\mathbf{Z}'\mathbf{V}\mathbf{Z})^{-}\mathbf{Z}'\mathbf{V}\mathbf{X}](\mathbf{X}'\mathbf{X})^{-1} \\ &= \operatorname{cov}(\hat{\boldsymbol{\beta}}) - (\mathbf{X}'\mathbf{X})^{-1}\mathbf{X}'\mathbf{V}\mathbf{Z}(\mathbf{Z}'\mathbf{V}\mathbf{Z})^{-}\mathbf{Z}'\mathbf{V}\mathbf{X}(\mathbf{X}'\mathbf{X})^{-1},\end{aligned} \tag{6.3.89}$$

where $\mathbf{Z}$ is a matrix such that $\mathcal{C}(\mathbf{Z}) = \mathcal{C}(\mathbf{M})$. Substituting (6.3.89) into (6.3.85) yields

$$\begin{aligned}\phi = \frac{\det \operatorname{cov}(\tilde{\boldsymbol{\beta}})}{\det \operatorname{cov}(\hat{\boldsymbol{\beta}})} &= \frac{\det\left(\mathbf{X}'\mathbf{V}\mathbf{X} - \mathbf{X}'\mathbf{V}\mathbf{Z}(\mathbf{Z}'\mathbf{V}\mathbf{Z})^{-}\mathbf{Z}'\mathbf{V}\mathbf{X}\right)}{\det \mathbf{X}'\mathbf{V}\mathbf{X}} \\ &= \frac{\det(\boldsymbol{\Sigma}_{\mathbf{XZ}}/\mathbf{Z}'\mathbf{V}\mathbf{Z})}{\det \mathbf{X}'\mathbf{V}\mathbf{X}},\end{aligned} \tag{6.3.90}$$

where

$$\boldsymbol{\Sigma}_{\mathbf{XZ}} = \operatorname{cov}\begin{pmatrix}\mathbf{X}'\mathbf{y} \\ \mathbf{Z}'\mathbf{y}\end{pmatrix} = \begin{pmatrix}\mathbf{X}'\mathbf{V}\mathbf{X} & \mathbf{X}'\mathbf{V}\mathbf{Z} \\ \mathbf{Z}'\mathbf{V}\mathbf{X} & \mathbf{Z}'\mathbf{V}\mathbf{Z}\end{pmatrix}. \tag{6.3.91}$$

It is interesting to note that in (6.3.90), the covariance matrix $\mathbf{V}$ need not be positive definite.

For the Watson efficiency ϕ to be defined using determinants as in (6.3.90), the matrix $\mathbf{X}'\mathbf{V}\mathbf{X}$ must be positive definite which happens if and only if $\mathcal{C}(\mathbf{X}) \cap \mathcal{C}(\mathbf{V})^{\perp} = \{\mathbf{0}\}$; since $\operatorname{rank}(\mathbf{X}'\mathbf{V}\mathbf{X}) = \operatorname{rank}(\mathbf{X}'\mathbf{V}) = \operatorname{rank}(\mathbf{X}) - \dim \mathcal{C}(\mathbf{X}) \cap \mathcal{C}(\mathbf{V})^{\perp}$. Moreover, since $\operatorname{rank}\operatorname{cov}(\tilde{\boldsymbol{\beta}}) = \dim \mathcal{C}(\mathbf{X}) \cap \mathcal{C}(\mathbf{V})$, we conclude that for the Watson efficiency ϕ to be defined using determinants as in (6.3.91), we must have

$$\mathcal{C}(\mathbf{X}) \subset \mathcal{C}(\mathbf{V}). \tag{6.3.92}$$

A linear model with $\mathbf{X}$ and $\mathbf{V}$ satisfying (6.3.92) is called a *weakly singular model* or a *Zyskind–Martin model.* We can rewrite (6.3.90) as

$$\begin{aligned}\phi &= \det(\mathbf{X}'\mathbf{V}\mathbf{X})^{-1} \cdot \det\big(\mathbf{X}'\mathbf{V}\mathbf{X} - \mathbf{X}'\mathbf{V}\mathbf{Z}(\mathbf{Z}'\mathbf{V}\mathbf{Z})^{-}\mathbf{Z}'\mathbf{V}\mathbf{X}\big) \\ &= \det\big(\mathbf{I} - (\mathbf{X}'\mathbf{V}\mathbf{X})^{-1}\mathbf{X}'\mathbf{V}\mathbf{Z}(\mathbf{Z}'\mathbf{V}\mathbf{Z})^{-}\mathbf{Z}'\mathbf{V}\mathbf{X}\big) \\ &:= \det(\mathbf{I} - \mathbf{L}). \end{aligned} \tag{6.3.93}$$

Since the eigenvalues of matrix $\mathbf{L}$ are the (squared) canonical correlations between random vectors $\mathbf{X}'\mathbf{y}$ and $\mathbf{Z}'\mathbf{y}$, $\operatorname{cc}_i^2(\mathbf{X}'\mathbf{y}, \mathbf{Z}'\mathbf{y})$, say, we have

$$\phi = \prod\big(1 - \operatorname{cc}_i^2(\mathbf{X}'\mathbf{y}, \mathbf{Z}'\mathbf{y})\big) = \prod\big(1 - \operatorname{cc}_i^2(\mathbf{H}\mathbf{y}, \mathbf{M}\mathbf{y})\big). \tag{6.3.94}$$

For related references, see [61, 37, 257, 359, 374, 414].

6.3.7 Adding a variable in multiple linear regression

Let us consider the partitioned linear model

$$\mathcal{M}_{12} = \{\mathbf{y}, \mathbf{X}\boldsymbol{\beta}, \sigma^2\mathbf{V}\} = \{\mathbf{y}, \mathbf{X}_1\boldsymbol{\beta}_1 + \mathbf{X}_2\boldsymbol{\beta}_2, \sigma^2\mathbf{V}\}, \tag{6.3.95}$$

where $\mathbf{X}_1$ is an $n \times p_1$ matrix and $\mathbf{X}_2$ is an $n \times p_2$; $p = p_1 + p_2$. We also denote

$$\mathcal{M}_1 = \{\mathbf{y}, \mathbf{X}_1\boldsymbol{\beta}_1, \sigma^2\mathbf{V}\}, \tag{6.3.96}$$

and $\mathcal{M}_2$ is defined correspondingly. The ordinary least squares estimator of $\boldsymbol{\beta}$ under $\mathcal{M}_{12}$ is ($\mathbf{X}$ having full column rank)

$$\operatorname{OLSE}(\boldsymbol{\beta}) = \hat{\boldsymbol{\beta}} = \begin{pmatrix}\hat{\boldsymbol{\beta}}_1 \\ \hat{\boldsymbol{\beta}}_2\end{pmatrix} = (\mathbf{X}'\mathbf{X})^{-1}\mathbf{X}'\mathbf{y} = \hat{\boldsymbol{\beta}}(\mathcal{M}_{12}), \tag{6.3.97}$$

where the notation $\hat{\boldsymbol{\beta}}(\mathcal{M}_{12})$ emphasizes that the estimator is calculated under the model $\mathcal{M}_{12}$. When dealing with regression models, it is often very helpful to have the explicit expression for subvector $\hat{\boldsymbol{\beta}}_2$ (or $\mathbf{X}_2\hat{\boldsymbol{\beta}}_2$) available. We recall that if $\mathbf{X}$ has no full column rank, then $\hat{\boldsymbol{\beta}}$ is merely a solution to normal equation.

Using the projector decomposition

$$\mathbf{H} = \mathbf{P}_{(\mathbf{X}_1 \,:\, \mathbf{X}_2)} = \mathbf{P}_{\mathbf{X}_1} + \mathbf{P}_{\mathbf{M}_1\mathbf{X}_2} = \mathbf{P}_1 + \mathbf{P}_{\mathbf{M}_1\mathbf{X}_2}, \tag{6.3.98a}$$

$$\mathbf{Hy} = \mathbf{X}_1\hat{\boldsymbol{\beta}}_1 + \mathbf{X}_2\hat{\boldsymbol{\beta}}_2 = \mathbf{P}_{\mathbf{X}_1}\mathbf{y} + \mathbf{P}_{\mathbf{M}_1\mathbf{X}_2}\mathbf{y}, \tag{6.3.98b}$$

and premultiplying (6.3.98b) by $\mathbf{M}_1$, we see that if $\operatorname{rank}(\mathbf{M}_1\mathbf{X}_2) = \operatorname{rank}(\mathbf{X}_2)$, then $\mathbf{X}_2\hat{\boldsymbol{\beta}}_2 = \mathbf{X}_2(\mathbf{X}_2'\mathbf{M}_1\mathbf{X}_2)^-\mathbf{X}_2'\mathbf{M}_1\mathbf{y}$, and if $\operatorname{rank}(\mathbf{M}_1\mathbf{X}_2) = p_2$, then

$$\hat{\boldsymbol{\beta}}_2 = (\mathbf{X}_2'\mathbf{M}_1\mathbf{X}_2)^{-1}\mathbf{X}_2'\mathbf{M}_1\mathbf{y} = \hat{\boldsymbol{\beta}}_2(\mathcal{M}_{12}). \tag{6.3.99}$$

We may note that, in view of [300, Corollary 6.2],

$$\operatorname{rank}(\mathbf{M}_1\mathbf{X}_2) = \operatorname{rank}(\mathbf{X}_2) - \dim \mathcal{C}(\mathbf{X}_2) \cap \mathcal{C}(\mathbf{X}_1), \tag{6.3.100}$$

and hence indeed $\operatorname{rank}(\mathbf{M}_1\mathbf{X}_2) = \operatorname{rank}(\mathbf{X}_2)$ if and only if $\mathcal{C}(\mathbf{X}_1) \cap \mathcal{C}(\mathbf{X}_2) = \{\mathbf{0}\}$. The disjointness condition is actually a condition for the estimability of $\mathbf{X}_2\boldsymbol{\beta}_2$, and $\operatorname{rank}(\mathbf{M}_1\mathbf{X}_2) = p_2$ means that $\boldsymbol{\beta}_2$ is estimable under $\mathcal{M}_{12}$. We note that if $\operatorname{rank}(\mathbf{M}_1\mathbf{X}_2) = \operatorname{rank}(\mathbf{X}_2)$, then we may write

$$\mathbf{Hy} = \mathbf{X}\hat{\boldsymbol{\beta}} = \mathbf{X}_1(\mathbf{X}_1'\mathbf{M}_2\mathbf{X}_1)^-\mathbf{X}_1'\mathbf{M}_2\mathbf{y} + \mathbf{X}_2(\mathbf{X}_2'\mathbf{M}_1\mathbf{X}_2)^-\mathbf{X}_2'\mathbf{M}_1\mathbf{y}, \tag{6.3.101}$$

and so, in view of (6.3.98b), we obtain that

$$\mathbf{X}_1\hat{\boldsymbol{\beta}}_1(\mathcal{M}_{12}) = \mathbf{X}_1\hat{\boldsymbol{\beta}}_1(\mathcal{M}_1) - \mathbf{P}_1\mathbf{X}_2\hat{\boldsymbol{\beta}}_2(\mathcal{M}_{12}).$$

Hence if disjointness $\mathcal{C}(\mathbf{X}_1) \cap \mathcal{C}(\mathbf{X}_2) = \{\mathbf{0}\}$ holds and $\mathbf{X}_1$ has full column rank, then

$$\hat{\boldsymbol{\beta}}_1(\mathcal{M}_{12}) = \hat{\boldsymbol{\beta}}_1(\mathcal{M}_1) - (\mathbf{X}_1'\mathbf{X}_1)^{-1}\mathbf{X}_1'\mathbf{X}_2\hat{\boldsymbol{\beta}}_2(\mathcal{M}_{12}). \tag{6.3.102}$$

Chu *et al.* [124, 125] recently considered the efficiency of the subvector $\hat{\boldsymbol{\beta}}_2$. Using the Schur determinant formula, they showed that the Watson efficiency of $\boldsymbol{\beta}$ under the partitioned weakly singular linear model $\mathcal{M}_{12} = \{\mathbf{y}, \mathbf{X}\boldsymbol{\beta}, \mathbf{V}\}$, where $\mathbf{X}$ has full column rank, can be expressed as a product

$$\phi(\hat{\boldsymbol{\beta}} \mid \mathcal{M}_{12}) = \phi(\hat{\boldsymbol{\beta}}_1 \mid \mathcal{M}_1) \cdot \phi(\hat{\boldsymbol{\beta}}_2 \mid \mathcal{M}_{12}) \cdot \alpha_1, \tag{6.3.103}$$

where $\phi(\cdot \mid \cdot)$ is an obvious notation and

$$\alpha_1 = \frac{\det(\mathbf{X}_2'\mathbf{M}_1\mathbf{V}\mathbf{M}_1\mathbf{X}_2)}{\det(\mathbf{X}_2'\mathbf{M}_1(\mathbf{M}_1\mathbf{V}^+\mathbf{M}_1)^-\mathbf{M}_1\mathbf{X}_2)}. \tag{6.3.104}$$

6.3.8 The Frisch–Waugh–Lovell Theorem

Let us consider the full rank partitioned linear model $\mathcal{M}_{12} = \{\mathbf{y}, \mathbf{X}\boldsymbol{\beta}, \sigma^2\mathbf{I}\}$:

$$\mathbf{y} = \mathbf{X}_1\boldsymbol{\beta}_1 + \mathbf{X}_2\boldsymbol{\beta}_2 + \boldsymbol{\varepsilon}. \tag{6.3.105}$$

Now we know that

$$\hat{\boldsymbol{\beta}}_2(\mathcal{M}_{12}) = (\mathbf{X}_2'\mathbf{M}_1\mathbf{X}_2)^{-1}\mathbf{X}_2'\mathbf{M}_1\mathbf{y}. \tag{6.3.106}$$

Premultiplying (6.3.105) by the orthogonal projector $\mathbf{M}_1$ yields the *reduced model*

$$\mathcal{M}_{12\cdot 1} = \{\mathbf{M}_1\mathbf{y}, \mathbf{M}_1\mathbf{X}_2\boldsymbol{\beta}_2, \sigma^2\mathbf{M}_1\}. \tag{6.3.107}$$

Taking a look at the models, we can immediately make an important conclusion: the OLS estimators of $\boldsymbol{\beta}_2$ under the models $\mathcal{M}_{12}$ and $\mathcal{M}_{12\cdot 1}$ coincide

$$\hat{\boldsymbol{\beta}}_2(\mathcal{M}_{12}) = \hat{\boldsymbol{\beta}}_2(\mathcal{M}_{12\cdot 1}) = (\mathbf{X}_2'\mathbf{M}_1\mathbf{X}_2)^{-1}\mathbf{X}_2'\mathbf{M}_1\mathbf{y}. \tag{6.3.108}$$

Davidson & MacKinnon [141, p. 19] and [142, p. 68] call (6.3.108) the *Frisch–Waugh–Lovell Theorem* "since those papers seem to have introduced, and then reintroduced, it to econometricians"; see Frisch & Waugh [176] and Lovell [292].

The covariance matrix in the model $\{\mathbf{M}_1\mathbf{y}, \mathbf{M}_1\mathbf{X}_2\boldsymbol{\beta}_2, \sigma^2\mathbf{M}_1\}$ is singular and hence there is a good reason to worry whether the OLSE($\boldsymbol{\beta}$) equals BLUE($\boldsymbol{\beta}$) under that model. The answer is positive, however, since the column space inclusion $\mathcal{C}(\mathbf{M}_1 \cdot \mathbf{M}_1\mathbf{X}_2) \subset \mathcal{C}(\mathbf{M}_1\mathbf{X}_2)$ holds, and hence the equality condition of Rao [368] and Zyskind [473] gives the result.

6.3.9 Partitioning the BLUE

We recall that the best linear unbiased estimator (BLUE) of $\boldsymbol{\beta}$ under the full rank model $\mathcal{M}_{12}$ is

$$\mathrm{BLUE}(\boldsymbol{\beta}) = \tilde{\boldsymbol{\beta}} = \begin{pmatrix}\tilde{\boldsymbol{\beta}}_1\\ \tilde{\boldsymbol{\beta}}_2\end{pmatrix} = (\mathbf{X}'\mathbf{V}^{-1}\mathbf{X})^{-1}\mathbf{X}'\mathbf{V}^{-1}\mathbf{y} = \tilde{\boldsymbol{\beta}}(\mathcal{M}_{12}), \tag{6.3.109}$$

while the covariance matrix is

$$\mathrm{cov}(\tilde{\boldsymbol{\beta}}) = \sigma^2(\mathbf{X}'\mathbf{V}^{-1}\mathbf{X})^{-1}. \tag{6.3.110}$$

Assume that the column space disjointness condition $\mathcal{C}(\mathbf{X}_1) \cap \mathcal{C}(\mathbf{X}_2) = \{\mathbf{0}\}$ holds and $\mathbf{X}_2$ has full column rank, i.e.,

$$\mathrm{rank}(\mathbf{M}_1\mathbf{X}_2) = p_2. \tag{6.3.111}$$

Then $\boldsymbol{\beta}$ is estimable and its BLUE is

$$\tilde{\boldsymbol{\beta}}_2(\mathcal{M}_{12}) = (\mathbf{X}_2'\dot{\mathbf{M}}_1\mathbf{X}_2)^{-1}\mathbf{X}_2'\dot{\mathbf{M}}_1\mathbf{y}, \tag{6.3.112}$$

where $\dot{\mathbf{M}}_1 = \mathbf{M}_1(\mathbf{M}_1\mathbf{V}\mathbf{M}_1)^-\mathbf{M}_1$; here (when $\mathbf{V}$ is positive definite), cf. (6.3.78))

$$\begin{aligned}\dot{\mathbf{M}}_1 &= \mathbf{M}_1(\mathbf{M}_1\mathbf{V}\mathbf{M}_1)^-\mathbf{M}_1 = \mathbf{M}_1(\mathbf{M}_1\mathbf{V}\mathbf{M}_1)^+\mathbf{M}_1 = (\mathbf{M}_1\mathbf{V}\mathbf{M}_1)^+ \\ &= \mathbf{V}^{-1} - \mathbf{V}^{-1}\mathbf{X}_1(\mathbf{X}_1'\mathbf{V}^{-1}\mathbf{X}_1)^-\mathbf{X}_1'\mathbf{V}^{-1}.\end{aligned} \tag{6.3.113}$$

Moreover, if $\mathbf{X}_1$ has full column rank, then

$$\tilde{\boldsymbol{\beta}}_1(\mathcal{M}_{12}) = \tilde{\boldsymbol{\beta}}_1(\mathcal{M}_1) - (\mathbf{X}_1'\mathbf{V}^{-1}\mathbf{X}_1)^{-1}\mathbf{X}_1'\mathbf{V}^{-1}\mathbf{X}_2\tilde{\boldsymbol{\beta}}_2(\mathcal{M}_{12}). \tag{6.3.114}$$

One way to introduce the above properties of the subvectors of $\tilde{\boldsymbol{\beta}}$ is to use the projector decomposition

$$\mathbf{P}_{(\mathbf{X}_1 : \mathbf{X}_2);\mathbf{V}^{-1}} = \mathbf{P}_{\mathbf{X}_1;\mathbf{V}^{-1}} + \mathbf{P}_{(\mathbf{I}-\mathbf{P}_{\mathbf{X}_1;\mathbf{V}^{-1}})\mathbf{X}_2;\mathbf{V}^{-1}}. \tag{6.3.115}$$

Just as $\dot{\mathbf{M}}$ in Section 6.3.4 was a Schur complement, so is also $\dot{\mathbf{M}}_1$ a specific Schur complement. Denoting

$$\mathbf{S}_{\mathbf{V}^{-1}}^{13} = (\mathbf{X}_1 : \mathbf{M}_1)'\mathbf{V}^{-1}(\mathbf{X}_1 : \mathbf{M}_1), \quad \mathbf{S}_{\mathbf{V}^{-1}}^{11} = \mathbf{X}'\mathbf{V}^{-1}\mathbf{X}, \tag{6.3.116}$$

we see at once that

$$\dot{\mathbf{M}}_1 = \mathbf{S}_{\mathbf{V}^{-1}}^{13}/\mathbf{X}_1\mathbf{V}^{-1}\mathbf{X}_1, \tag{6.3.117}$$

and, of course,

$$\mathbf{X}_2'\dot{\mathbf{M}}_1\mathbf{X}_2 = \mathbf{X}'\mathbf{V}^{-1}\mathbf{X}/\mathbf{X}_1'\mathbf{V}^{-1}\mathbf{X}_1 := \mathbf{S}_{\mathbf{V}^{-1}}^{11}/\mathbf{X}_1\mathbf{V}^{-1}\mathbf{X}_1. \tag{6.3.118}$$

Using the Schur complement notation we can express (6.3.117) and (6.3.118) as

$$\mathbf{X}'\mathbf{V}^{-1}\mathbf{X}/\mathbf{X}_1'\mathbf{V}^{-1}\mathbf{X}_1 = \mathbf{X}_2'[(\mathbf{X}_1 : \mathbf{M}_1)'\mathbf{V}^{-1}(\mathbf{X}_1 : \mathbf{M}_1)/\mathbf{X}_1'\mathbf{V}^{-1}\mathbf{X}_1]\mathbf{X}_2.$$

Let us next consider the expression for $\tilde{\boldsymbol{\beta}}_2(\mathcal{M}_{12})$ in the general situation allowing $\mathbf{V}$ to be singular. Let us denote

$$\mathbf{W} = \mathbf{V} + \mathbf{X}\mathbf{X}', \quad \mathbf{W}_i = \mathbf{V} + \mathbf{X}_i\mathbf{X}_i',\ i = 1, 2, \tag{6.3.119}$$

and hence $\mathcal{C}(\mathbf{W}) = \mathcal{C}(\mathbf{X} : \mathbf{V})$. Then, according to Rao [371], one general representation for the BLUE($\mathbf{X}\boldsymbol{\beta}$) is

$$\text{BLUE}(\mathbf{X}\boldsymbol{\beta}) = \mathbf{X}(\mathbf{X}'\mathbf{W}^-\mathbf{X})^-\mathbf{X}'\mathbf{W}^-\mathbf{y}. \tag{6.3.120}$$

It can be shown that if $\operatorname{rank}(\mathbf{M}_1\mathbf{X}_2) = p_2$, then

$$\tilde{\boldsymbol{\beta}}_2(\mathcal{M}_{12}) = (\mathbf{X}_2'\dot{\mathbf{M}}_{1W}\mathbf{X}_2)^{-1}\mathbf{X}_2'\dot{\mathbf{M}}_{1W}\mathbf{y}, \tag{6.3.121}$$

where $\dot{\mathbf{M}}_{1W} = \mathbf{M}_1(\mathbf{M}_1\mathbf{W}\mathbf{M}_1)^-\mathbf{M}_1 = \mathbf{M}_1(\mathbf{M}_1\mathbf{W}_2\mathbf{M}_1)^-\mathbf{M}_1$. If $\mathbf{X}$ has full column rank and $\mathbf{X}_2$ satisfies condition

$$\mathcal{C}(\mathbf{X}_2) \subset \mathcal{C}(\mathbf{X}_1 : \mathbf{V}) \;\text{ i.e., }\; \mathcal{C}(\mathbf{X}_1 : \mathbf{X}_2 : \mathbf{V}) = \mathcal{C}(\mathbf{X}_1 : \mathbf{V}), \tag{6.3.122}$$

then (6.3.121) simplifies to

$$\tilde{\boldsymbol{\beta}}_2(\mathcal{M}_{12}) = (\mathbf{X}_2'\dot{\mathbf{M}}_1\mathbf{X}_2)^{-1}\mathbf{X}_2'\dot{\mathbf{M}}_1\mathbf{y}, \tag{6.3.123}$$

and

$$\tilde{\boldsymbol{\beta}}_1(\mathcal{M}_{12}) = \tilde{\boldsymbol{\beta}}_1(\mathcal{M}_1) - (\mathbf{X}_1'\mathbf{W}_1^+\mathbf{X}_1)^{-1}\mathbf{X}_1'\mathbf{W}_1^+\mathbf{X}_2\tilde{\boldsymbol{\beta}}_2(\mathcal{M}_{12}). \tag{6.3.124}$$

The column space condition (6.3.122) is needed in order to avoid any contradiction between the consistencies of the full model $\mathcal{M}_{12}$ and the smaller model $\mathcal{M}_1$.

6.3.10 A generalized Frisch–Waugh–Lovell Theorem

Let us consider the partitioned linear model

$$\mathcal{M}_{12} = \{\mathbf{y},\, \mathbf{X}\boldsymbol{\beta},\, \sigma^2\mathbf{V}\} = \{\mathbf{y},\, \mathbf{X}_1\boldsymbol{\beta}_1 + \mathbf{X}_2\boldsymbol{\beta}_2,\, \sigma^2\mathbf{V}\}. \tag{6.3.125}$$

Premultiplying the model $\mathcal{M}_{12}$ by the orthogonal projector $\mathbf{M}_1$ yields the reduced model

$$\mathcal{M}_{12\cdot1} = \{\mathbf{M}_1\mathbf{y},\, \mathbf{M}_1\mathbf{X}_2\boldsymbol{\beta}_2,\, \sigma^2\mathbf{M}_1\mathbf{V}\mathbf{M}_1\}. \tag{6.3.126}$$

What about he BLUE of $\boldsymbol{\beta}_2$ in the reduced model $\mathcal{M}_{12\cdot1}$? In light of [371], one presentation of BLUE of $\boldsymbol{\beta}_2$ under $\mathcal{M}_{12\cdot1}$ is

$$\tilde{\boldsymbol{\beta}}_2(\mathcal{M}_{12\cdot1}) = (\mathbf{X}_2'\mathbf{M}_1\mathbf{W}_{12\cdot1}^-\mathbf{M}_1\mathbf{X}_2)^{-1}\mathbf{X}_2'\mathbf{M}_1\mathbf{W}_{12\cdot1}^-\mathbf{M}_1\mathbf{y}, \tag{6.3.127}$$

where $\mathbf{W}_{12\cdot1} = \mathbf{M}_1\mathbf{V}\mathbf{M}_1 + \mathbf{M}_1\mathbf{X}_2\mathbf{X}_2'\mathbf{M}_1 = \mathbf{M}_1\mathbf{W}_2\mathbf{M}_1 = \mathbf{M}_1\mathbf{W}\mathbf{M}_1$. Now, in view of (6.3.121), the formula (6.3.127) is just the same as presented for $\tilde{\boldsymbol{\beta}}_2$ under $\mathcal{M}_{12}$, and hence the equality

$$\tilde{\boldsymbol{\beta}}_2(\mathcal{M}_{12}) = \tilde{\boldsymbol{\beta}}_2(\mathcal{M}_{12\cdot1}) \tag{6.3.128}$$

is a *generalized Frisch–Waugh–Lovell Theorem*, which holds also for singular $\mathbf{V}$; see also [57, Th. 6.1], [119, Section 2.2], [164], [195, Th. 4], [194].

6.3.11 Deleting an observation in multiple linear regression

Let us consider three linear models:

$$\mathcal{M} = \{\mathbf{y},\, \mathbf{X}\boldsymbol{\beta},\, \sigma^2\mathbf{I}\}, \quad \mathcal{M}_{(i)} = \{\mathbf{y}_{(i)},\, \mathbf{X}_{(i)}\boldsymbol{\beta},\, \sigma^2\mathbf{I}_{n-1}\}, \tag{6.3.129a}$$

$$\mathcal{M}_Z = \{\mathbf{y},\, \mathbf{Z}\boldsymbol{\gamma},\, \sigma^2\mathbf{I}\} = \{\mathbf{y},\, \mathbf{X}\boldsymbol{\beta} + \mathbf{u}_i\delta,\, \sigma^2\mathbf{I}\}. \tag{6.3.129b}$$

By $\mathcal{M}_{(i)}$ we mean the version of $\mathcal{M}$ with the ith observation deleted; thus $\mathbf{y}_{(i)}$ has $n-1$ elements and $\mathbf{X}_{(i)}$ has $n-1$ rows. For notational simplicity we delete the last observation. Model $\mathcal{M}_Z$ is an extended version of $\mathcal{M}$:

$$\mathbf{Z} = (\mathbf{X} : \mathbf{u}_i), \quad \mathbf{u}_i = (0,\ldots,0,1)', \quad \boldsymbol{\gamma} = \begin{pmatrix} \boldsymbol{\beta} \\ \delta \end{pmatrix}. \tag{6.3.130}$$

We will use the following notation:

$$\hat{\boldsymbol{\beta}} = \hat{\boldsymbol{\beta}}(\mathcal{M}), \quad \hat{\boldsymbol{\beta}}_Z = \hat{\boldsymbol{\beta}}(\mathcal{M}_Z), \quad \hat{\delta} = \hat{\delta}(\mathcal{M}_Z), \quad \hat{\boldsymbol{\beta}}_{(i)} = \hat{\boldsymbol{\beta}}(\mathcal{M}_{(i)}). \tag{6.3.131}$$

Assuming that $\mathbf{Z}$ has full column rank we get

$$\hat{\boldsymbol{\beta}}_Z = [\mathbf{X}'(\mathbf{I} - \mathbf{u}_i\mathbf{u}_i')\mathbf{X}]^{-1}\mathbf{X}'(\mathbf{I} - \mathbf{u}_i\mathbf{u}_i')\mathbf{y}, \quad \hat{\delta} = \frac{\mathbf{u}_i'\mathbf{M}\mathbf{y}}{\mathbf{u}_i'\mathbf{M}\mathbf{u}_i} = \frac{r_i}{m_{ii}}, \tag{6.3.132}$$

in view of (6.3.99). Here $\mathbf{X}$ and $\mathbf{u}_i$ correspond to $\mathbf{X}_1$ and $\mathbf{X}_2$, and r_i refers to the ith element of the residual vector $\mathbf{r} = \mathbf{M}\mathbf{y}$. Furthermore,

$$\hat{\boldsymbol{\beta}}_Z = (\mathbf{X}_{(i)}'\mathbf{X}_{(i)})^{-1}\mathbf{X}_{(i)}'\mathbf{y}_{(i)} = \hat{\boldsymbol{\beta}}_{(i)}. \tag{6.3.133}$$

This result can be seen as a consequence of the Frisch–Waugh–Lovell Theorem: model $\mathcal{M}_Z$ corresponds to the full model $\mathcal{M}_{12}$, model $\mathcal{M}_{(i)}$ to the reduced model $\mathcal{M}_{12\cdot1}$ obtained by premultiplying $\mathcal{M}_Z$ by $\mathbf{I}-\mathbf{u}_i\mathbf{u}_i'$, and $\mathcal{M}$ is the small model $\mathcal{M}_1$. Hence $\hat{\boldsymbol{\beta}}_1(\mathcal{M}_{12}) = \hat{\boldsymbol{\beta}}_1(\mathcal{M}_1) - (\mathbf{X}_1'\mathbf{X}_1)^{-1}\mathbf{X}_1'\mathbf{X}_2\hat{\boldsymbol{\beta}}_2(\mathcal{M}_{12})$, and so $\hat{\boldsymbol{\beta}}_{(i)} = \hat{\boldsymbol{\beta}} - (\mathbf{X}'\mathbf{X})^{-1}\mathbf{X}'\mathbf{u}_i\hat{\delta}$, and [44, p. 13]

$$\text{DFBETA}_i = \hat{\boldsymbol{\beta}} - \hat{\boldsymbol{\beta}}_{(i)} = (\mathbf{X}'\mathbf{X})^{-1}\mathbf{X}'\mathbf{u}_i\,\hat{\delta} = \frac{r_i}{m_{ii}}(\mathbf{X}'\mathbf{X})^{-1}\mathbf{X}'\mathbf{u}_i. \tag{6.3.134}$$

Since $\mathbf{X}(\hat{\boldsymbol{\beta}} - \hat{\boldsymbol{\beta}}_{(i)}) = \mathbf{H}\mathbf{u}_i\hat{\delta}$, we obtain a representation for Cook's distance [127]:

$$D_i = \frac{1}{p\hat{\sigma}^2}(\hat{\boldsymbol{\beta}} - \hat{\boldsymbol{\beta}}_{(i)})'\mathbf{X}'\mathbf{X}(\hat{\boldsymbol{\beta}} - \hat{\boldsymbol{\beta}}_{(i)}) = \frac{h_{ii}\hat{\delta}^2}{p\hat{\sigma}^2}. \tag{6.3.135}$$

Furthermore, since

$$\mathbf{P}_\mathbf{Z} = \mathbf{u}_i\mathbf{u}_i' + \mathbf{P}_{(\mathbf{I}-\mathbf{u}_i\mathbf{u}_i')\mathbf{X}} = \begin{pmatrix} \mathbf{0} & \mathbf{0} \\ \mathbf{0}' & 1 \end{pmatrix} + \begin{pmatrix} \mathbf{P}_{\mathbf{X}_{(i)}} & \mathbf{0} \\ \mathbf{0}' & 0 \end{pmatrix}, \tag{6.3.136}$$

we observe that

$$\mathrm{SSE}_Z = \mathbf{y}'(\mathbf{I}_n - \mathbf{P}_\mathbf{Z})\mathbf{y} = \mathbf{y}'_{(i)}(\mathbf{I}_{n-1} - \mathbf{P}_{\mathbf{X}_{(i)}})\mathbf{y}_{(i)} = \mathrm{SSE}_{(i)}.$$

Now it is easy to conclude that the usual F-test statistic for testing the hypothesis $\delta = 0$ under the model $\mathcal{M}_Z$ becomes

$$t_i^2 = \frac{\mathbf{y}'\mathbf{P}_{\mathbf{Mu}_i}\mathbf{y}}{\frac{1}{n-p-1}\mathbf{y}'(\mathbf{M} - \mathbf{P}_{\mathbf{Mu}_i})\mathbf{y}} = \frac{r_i^2}{\frac{1}{n-p-1}\mathrm{SSE}_{(i)}m_{ii}}, \tag{6.3.137}$$

which is the externally Studentized residual squared.

We note that using the Schur determinant formula here gives

$$\det(\mathbf{Z}'\mathbf{Z}) = \det(\mathbf{X}'\mathbf{X}) \cdot \det(\mathbf{Z}'\mathbf{Z}/\mathbf{X}'\mathbf{X}) = \det(\mathbf{X}'\mathbf{X}) \cdot (1 - h_{ii}), \tag{6.3.138}$$

and hence

$$m_{ii} = \frac{\det(\mathbf{Z}'\mathbf{Z})}{\det(\mathbf{X}'\mathbf{X})} = \frac{\det(\mathbf{X}'_{(i)}\mathbf{X}_{(i)})}{\det(\mathbf{X}'\mathbf{X})}. \tag{6.3.139}$$

Note also that in view of the Woodbury inversion formula (6.0.11) or the Bartlett inversion formula (6.0.13), we have the identity

$$(\mathbf{X}'_{(i)}\mathbf{X}_{(i)})^{-1} = (\mathbf{X}'\mathbf{X})^{-1} + \frac{1}{1-h_{ii}}(\mathbf{X}'\mathbf{X})^{-1}\mathbf{x}_{(i)}\mathbf{x}'_{(i)}(\mathbf{X}'\mathbf{X})^{-1}, \tag{6.3.140}$$

where $\mathbf{x}'_{(i)}$ refers to the ith row of $\mathbf{X}$.

Let $\tilde{\boldsymbol\beta}$ and $\tilde{\boldsymbol\beta}_{(i)}$ denote the BLUEs of $\boldsymbol\beta$ under $\mathcal{M} = \{\mathbf{y}, \mathbf{X}\boldsymbol\beta, \sigma^2\mathbf{V}\}$ and under $\mathcal{M}_{(i)} = \{\mathbf{y}_{(i)}, \mathbf{X}_{(i)}\boldsymbol\beta, \sigma^2\mathbf{V}_{(i)}\}$, respectively. Using our generalized Frisch–Waugh–Lovell Theorem, it is straightforward to show that the generalized DFBETA$_i$ is

$$\mathrm{DFBETA}_i(V) = \tilde{\boldsymbol\beta} - \tilde{\boldsymbol\beta}_{(i)} = \frac{\dot r_i}{\dot m_{ii}}(\mathbf{X}'\mathbf{V}^{-1}\mathbf{X})^{-1}\mathbf{X}'\mathbf{V}^{-1}\mathbf{u}_i, \tag{6.3.141}$$

where

$$\begin{aligned}\dot m_{ii} &= \mathbf{u}_i'\dot{\mathbf{M}}\mathbf{u}_i = \mathbf{u}_i'\mathbf{M}(\mathbf{MVM})^-\mathbf{Mu}_i \\ &= \mathbf{u}_i'[\mathbf{V}^{-1} - \mathbf{V}^{-1}\mathbf{X}(\mathbf{X}'\mathbf{V}^{-1}\mathbf{X})^{-1}\mathbf{X}'\mathbf{V}^{-1}]\mathbf{u}_i,\end{aligned} \tag{6.3.142}$$

and $\dot{\mathbf{r}} = \dot{\mathbf{M}}\mathbf{y}$. The term $\dot m_{ii}$ corresponds now to m_{ii} in the formulas of the "original" Studentized residuals r_i and t_i. The generalized Cook's distance [398, p. 164], is

$$D_i(V) = \frac{1}{p\tilde\sigma^2}(\tilde{\boldsymbol\beta} - \tilde{\boldsymbol\beta}_{(i)})'\mathbf{X}'\mathbf{V}^{-1}\mathbf{X}(\tilde{\boldsymbol\beta} - \tilde{\boldsymbol\beta}_{(i)}), \tag{6.3.143}$$

where $\tilde\sigma^2 = \mathrm{SSE}(V)/\operatorname{rank}(\mathbf{VM})$, and $\mathrm{SSE}(V)$ refers to the weighted sum of squares of errors. For more about these so-called *deletion statistics* see [58, 123, 206, 209, 413].

6.3.12 Mixed linear models

A mixed linear model can be presented as

$$\mathbf{y} = \mathbf{X}\boldsymbol{\beta} + \mathbf{Z}\boldsymbol{\gamma} + \boldsymbol{\varepsilon}, \tag{6.3.144}$$

where $\mathbf{X}$ and $\mathbf{Z}$ are known matrices, $\boldsymbol{\beta}$ is a vector of unknown fixed effects, $\boldsymbol{\gamma}$ is an unobservable vector of random effects with $\mathrm{E}(\boldsymbol{\gamma}) = \mathbf{0}, \mathrm{cov}(\boldsymbol{\gamma}) = \mathbf{D}, \mathrm{cov}(\boldsymbol{\gamma}, \boldsymbol{\varepsilon}) = \mathbf{0}$ and $\mathrm{E}(\boldsymbol{\varepsilon}) = \mathbf{0}, \mathrm{cov}(\boldsymbol{\varepsilon}) = \mathbf{R}$. Writing

$$\boldsymbol{\xi} = \mathbf{Z}\boldsymbol{\gamma} + \boldsymbol{\varepsilon}, \quad \mathbf{V} = \mathrm{cov}(\boldsymbol{\xi}) = \mathbf{ZDZ}' + \mathbf{R}, \tag{6.3.145}$$

we can re-express (6.3.144) as

$$\mathbf{y} = \mathbf{X}\boldsymbol{\beta} + \boldsymbol{\xi}, \quad \mathrm{E}(\boldsymbol{\xi}) = \mathbf{0}, \quad \mathrm{cov}(\boldsymbol{\xi}) = \mathbf{V}. \tag{6.3.146}$$

Assuming that $\mathbf{V}$ is known and positive definite, and $\mathbf{X}$ has full column rank, we have

$$\mathrm{BLUE}(\boldsymbol{\beta}) = \tilde{\boldsymbol{\beta}} = (\mathbf{X}'\mathbf{V}^{-1}\mathbf{X})^{-1}\mathbf{X}'\mathbf{V}^{-1}\mathbf{y}. \tag{6.3.147}$$

As H. V. Henderson & Searle [219] point out, a difficulty with (6.3.147) in many applications, is that the matrix $\mathbf{V} = \mathbf{ZDZ}' + \mathbf{R}$ is often large and nondiagonal, so that inverting it is quite impractical. An alternative set of equations for solving for $\tilde{\boldsymbol{\beta}}$ is

$$\begin{pmatrix} \mathbf{X}'\mathbf{R}^{-1}\mathbf{X} & \mathbf{X}'\mathbf{R}^{-1}\mathbf{Z} \\ \mathbf{Z}'\mathbf{R}^{-1}\mathbf{X} & \mathbf{D}^{-1} + \mathbf{Z}'\mathbf{R}^{-1}\mathbf{Z} \end{pmatrix} \begin{pmatrix} \boldsymbol{\beta} \\ \boldsymbol{\gamma} \end{pmatrix} = \begin{pmatrix} \mathbf{X}'\mathbf{R}^{-1}\mathbf{y} \\ \mathbf{Z}'\mathbf{R}^{-1}\mathbf{y} \end{pmatrix}, \tag{6.3.148}$$

as suggested by C. R. Henderson [217]. If $\tilde{\boldsymbol{\beta}}$ and $\tilde{\boldsymbol{\gamma}}$ are solutions to (6.3.148), then $\mathbf{X}\tilde{\boldsymbol{\beta}}$ appears to be a BLUE of $\mathbf{X}\boldsymbol{\beta}$ and $\tilde{\boldsymbol{\gamma}}$ is a BLUP of $\boldsymbol{\gamma}$; see [218]. The BLUP of $\boldsymbol{\gamma}$ can be expressed as

$$\tilde{\boldsymbol{\gamma}} = \mathbf{DZ}'\mathbf{V}^{-1}(\mathbf{y} - \mathbf{X}\tilde{\boldsymbol{\beta}}). \tag{6.3.149}$$

The equations in (6.3.148) are known as *Henderson's mixed model equations.* The proof is based on the equality

$$(\mathbf{ZDZ}' + \mathbf{R})^{-1} = \mathbf{R}^{-1} - \mathbf{R}^{-1}\mathbf{Z}(\mathbf{Z}'\mathbf{R}^{-1}\mathbf{Z} + \mathbf{D}^{-1})^{-1}\mathbf{Z}'\mathbf{R}^{-1}, \tag{6.3.150}$$

which comes from the Duncan inversion formula (6.0.9). Note that (6.3.150) means that $\mathbf{XV}^{-1}\mathbf{X} = \mathbf{A}/\mathbf{A}_{22}$, where $\mathbf{A}$ refers to the left-most partitioned matrix in (6.3.148).

For further references to mixed model equations, see, e.g., [122, Section 12.3] and [383, 411].

6.4 Experimental design and analysis of variance

In experimental design the Schur complement plays a crucial role as the so-called C-matrix, introduced in 1947 by Bose [68] and first studied in detail in 1962 by Chakrabarti [111]; see also [112].

6.4.1 The C-matrix of a block design

Let us consider the two-way layout of analysis of variance with fixed effects and no interaction, see, e.g., Searle [410, Section 7.1], Latour & Styan [274]. Here we have observations which we label y_{ijk} arising from a random experiment involving two factors: the row effects, indexed by i, and the column effects, indexed by j. When the row effects identify *treatments* and the column effects *blocks* then the layout is often called a *block design*, see, e.g., Dey [145, ch. 2], John & Williams [242, Section 1], Raghavarao [366, ch. 4], Roy [387].

The random observation y_{ijk} comes from the kth occurrence (or replicate) of the ith row (treatment) and the jth column (block). We will assume that $k = 1, 2, \ldots, n_{ij}$, with $i = 1, 2, \ldots, r$ and $j = 1, 2, ..., c$. The n_{ij} are nonnegative integers and the associated $r \times c$ matrix

$$\mathbf{N} = \{n_{ij}\} \tag{6.4.1}$$

is known as the *incidence matrix*. We will write

$$n_{i\cdot} = \sum_{j=1}^{c} n_{ij};\ i = 1, 2, \ldots, r \quad \text{and} \quad n_{\cdot j} = \sum_{i=1}^{r} n_{ij};\ j = 1, 2, \ldots, c \tag{6.4.2}$$

and assume that

$$n_{i\cdot} \geq 1;\ i = 1, 2, \ldots, r \quad \text{and} \quad n_{\cdot j} \geq 1;\ j = 1, 2, \ldots, c \tag{6.4.3}$$

and so the diagonal matrices, respectively $r \times r$ and $c \times c$,

$$\mathbf{D}_r = \operatorname{diag}\{n_{i\cdot}\} \quad \text{and} \quad \mathbf{D}_c = \operatorname{diag}\{n_{\cdot j}\} \tag{6.4.4}$$

are both positive definite. The $r \times r$ Schur complement

$$\mathbf{S}_r = \mathbf{A}/\mathbf{D}_c = \mathbf{D}_r - \mathbf{N}\mathbf{D}_c^{-1}\mathbf{N}' \tag{6.4.5}$$

of $\mathbf{D}_c$ in the $(r + c) \times (r + c)$ matrix

$$\mathbf{A} = \begin{pmatrix} \mathbf{D}_r & \mathbf{N} \\ \mathbf{N}' & \mathbf{D}_c \end{pmatrix} \tag{6.4.6}$$

plays a key role in the analysis of the two-way layout and in the estimation of the r row (treatment) effects using ordinary least squares. When the two-way layout is a block design the Schur complement $\mathbf{S}_r$ is often known as the *C-matrix*.

As noted by Pukelsheim [357, p. 412], reference to a C-matrix is made implicitly by Bose [68, p. 12] and explicitly by Chakrabarti [112], where C-matrix appears in the title. Other early references include Chakrabarti [111], Raghavarao [366, p. 49], V. R. Rao [379], Thompson [440, 441]. The C-matrix $\mathbf{S}_r$ is called the *coefficient matrix of the design* by Roy [387, p. 182], the *contrast information matrix* by Pukelsheim [357, p. 94], and just the *information matrix* by John & Williams [242, p. 12].

We note that all the on-diagonal elements of $\mathbf{S}_r$ are nonnegative and all the off-diagonal elements are nonpositive: a matrix with elements satisfying this sign pattern is called *hyperdominant*, see, e.g., [253, 433], [435, p. 358].

Since $\mathbf{S}_r\mathbf{e} = \mathbf{D}_r\mathbf{e} - \mathbf{N}\mathbf{D}_c^{-1}\mathbf{N}'\mathbf{e} = \mathbf{0}$, where $\mathbf{e}$ is the $r \times 1$ column vector with each element equal to 1, we see that the row totals of the C-matrix $\mathbf{S}_r$ are all equal to 0 and so

$$\operatorname{rank}(\mathbf{S}_r) \leq r - 1. \tag{6.4.7}$$

Since $\mathbf{S}_r$ is symmetric, it follows that the column totals of $\mathbf{S}_r$ are also all equal to 0 and hence we say that the C-matrix $\mathbf{S}_r$ is *double-centered.* As shown by Sharpe & Styan [415], see also [366, Th. 4.2.4, p. 51], the first cofactors of a double-centered matrix $\mathbf{F}$, say, are all equal, say to f, and hence such a matrix may be called *equicofactor*; the nullity $\nu(\mathbf{F}) = 1$ if and only if $f > 0$. The first cofactors of the C-matrix $\mathbf{S}_r$ are, therefore, all equal, to g say, and so equality holds in (6.4.7) if and only if $g > 0$.

In electrical circuit theory, the *admittance matrix* is double-centered and like the C-matrix $\mathbf{S}_r$ is hyperdominant. The common value of the first cofactors of the admittance matrix is called the *network determinant*, see, e.g., Sharpe & Styan [415], Styan & Subak-Sharpe [433].

6.4.2 Connectedness

When equality holds in (6.4.7), we say that the layout is *connected.* Dey [145, Th. 2.1, p. 44] and Raghavarao [366, Th. 4.2.2, p. 50] present this result as a theorem, preferring as definition that the layout be connected whenever every *pair of treatments is connected* [145, Def. 2.2, p. 45], [366, Def. 4.2.3, p. 49].

Another definition of connectedness is based on statistical considerations. Let us assemble the random observations y_{ijk} into an $n \times 1$ column vector

$$\mathbf{y} = \begin{pmatrix} \mathbf{y}_{11} \\ \mathbf{y}_{21} \\ \cdots \\ \mathbf{y}_{rc} \end{pmatrix}, \quad \text{where} \quad \mathbf{y}_{ij} = \begin{pmatrix} y_{ij1} \\ y_{ij2} \\ \cdots \\ y_{ijn_{ij}} \end{pmatrix}, \quad i = 1, \ldots, r;\ j = 1, \ldots, c,$$

and so the vector $\mathbf{y}_{ij}$ is $n_{ij} \times 1$. When $n_{ij} = 0$ for a particular treatment i and a particular block j, then there is no vector $\mathbf{y}_{ij}$ in $\mathbf{y}$. We may express the expectation of the random vector $\mathbf{y}$ as

$$\mathrm{E}(\mathbf{y}) = \mathbf{X}_1\boldsymbol{\alpha} + \mathbf{X}_2\boldsymbol{\beta}, \tag{6.4.8}$$

where the $n \times r$ *design matrix for rows* $\mathbf{X}_1$ has full column rank r corresponding to the r row or treatment effects, the $n \times c$ *design matrix for columns* $\mathbf{X}_2$ has full column rank c corresponding to the c column or block effects, and $\mathbf{X}_1'\mathbf{X}_2 = \mathbf{N}$, the $r \times c$ incidence matrix introduced in (6.4.1).

A particular treatment i and a particular block j are said to be *associated* whenever the treatment i appears in the block j, i.e., when $n_{ij} \geq 1$. A pair of treatments is said to be connected whenever it is possible to pass from one to the other in the pair through a chain consisting alternatively of treatments and blocks such that any two members of a chain are associated.

The vectors $\boldsymbol{\alpha}$ and $\boldsymbol{\beta}$ are unknown and to be estimated using a single realization of $\mathbf{y}$. We will say that a linear function $\mathbf{h}'\boldsymbol{\alpha}$ is *estimable* whenever there exists a linear function $\mathbf{k}'\mathbf{y}$ such that $\mathrm{E}(\mathbf{k}'\mathbf{y}) = \mathbf{h}'\boldsymbol{\alpha}$. We define $\alpha_i - \alpha_{i'}$ to be an *elementary contrast*, where $i, i' = 1, \ldots, r\,(i \neq i')$. Here α_i is the ith element of $\boldsymbol{\alpha}$, $i = 1, \ldots, r$. Then the layout is connected whenever all the elementary contrasts are estimable, see, e.g., [145, Def. 2.1, p. 44], [366, Def. 4.2.2, p. 49].

Another property (or definition) of connectedness is that the $n \times (r + c)$ partitioned matrix

$$\mathbf{X} = (\mathbf{X}_1 : \mathbf{X}_2) \tag{6.4.9}$$

have rank $r + c - 1$. Here $\mathbf{X}_1$ and $\mathbf{X}_2$ are defined as in (6.4.8). We have

$$\mathbf{X}'\mathbf{X} = \begin{pmatrix} \mathbf{X}_1'\mathbf{X}_1 & \mathbf{X}_1'\mathbf{X}_2 \\ \mathbf{X}_2'\mathbf{X}_2 & \mathbf{X}_2'\mathbf{X}_2 \end{pmatrix} = \begin{pmatrix} \mathbf{D}_r & \mathbf{N} \\ \mathbf{N}' & \mathbf{D}_c \end{pmatrix} = \mathbf{A} \tag{6.4.10}$$

as in (6.4.6). Then

$$\mathrm{rank}(\mathbf{X}) = \mathrm{rank}(\mathbf{X}'\mathbf{X}) = \mathrm{rank}(\mathbf{X}'\mathbf{X}/\mathbf{D}_c) + \mathrm{rank}(\mathbf{D}_c) \tag{6.4.11}$$

since rank is additive on the Schur complement, see the "Guttman rank additivity formula" (0.9.2) in Chapter 0. Since $\text{rank}(\mathbf{D}_c) = c$, and since $\text{rank}(\mathbf{X}'\mathbf{X}/\mathbf{D}_c) = \text{rank}(\mathbf{S}_r) \leq r - 1$ from (6.4.7), we see that

$$\text{rank}(\mathbf{X}) = \text{rank}(\mathbf{S}_r) + c \leq r + c - 1. \tag{6.4.12}$$

Since the design is connected if and only if $\text{rank}(\mathbf{S}_r) = r - 1$, we see immediately that the design is connected if and only if $\text{rank}(\mathbf{X}) = r + c - 1$.

While we have seen that the C-matrix $\mathbf{S}_r$ is a Schur complement, we also note that $\mathbf{S}_r = \mathbf{X}_1'\mathbf{M}_2\mathbf{X}_1$, where the *residual matrix*

$$\mathbf{M}_2 = \mathbf{I}_n - \mathbf{X}_2(\mathbf{X}_2'\mathbf{X}_2)^{-1}\mathbf{X}_2' \tag{6.4.13}$$

is the Schur complement $\mathbf{B}/\mathbf{X}_2'\mathbf{X}_2$ of $\mathbf{X}_2'\mathbf{X}_2$ in

$$\mathbf{B} = \begin{pmatrix} \mathbf{I}_n & \mathbf{X}_2 \\ \mathbf{X}_2' & \mathbf{X}_2'\mathbf{X}_2 \end{pmatrix}. \tag{6.4.14}$$

Here $\mathbf{I}_n$ is the $n \times n$ identity matrix. The matrix $\mathbf{H}_2 = \mathbf{X}_2(\mathbf{X}_2'\mathbf{X}_2)^{-1}\mathbf{X}_2' = \mathbf{I}_n - \mathbf{M}_2$ is called the *hat matrix*. The residual matrix $\mathbf{M}_2$ is the orthogonal projector on the orthocomplement of the column space (range) of $\mathbf{X}_2$.

We may use the Guttman rank additivity formula to evaluate the rank of the Schur complement $\mathbf{B}/\mathbf{X}_2'\mathbf{X}_2 = \mathbf{M}_2$. We note first that $\text{rank}(\mathbf{B}) = \text{rank}(\mathbf{X}_2'\mathbf{X}_2) + \text{rank}(\mathbf{B}/\mathbf{X}_2'\mathbf{X}_2) = \text{rank}(\mathbf{I}_n) + \text{rank}(\mathbf{B}/\mathbf{I}_n) = c + \text{rank}(\mathbf{M}_2) = n$ since the other Schur complement $\mathbf{B}/\mathbf{I}_n = \mathbf{0}$. Hence $\text{rank}(\mathbf{M}_2) = n - c$.

6.4.3 Balance

A connected design is said to be *balanced* whenever all the nonzero eigenvalues of the C-matrix $\mathbf{S}_r$ are equal. As already observed the design is connected whenever $\text{rank}(\mathbf{S}_r) = r - 1$ and so $\mathbf{S}_r$ has $r - 1$ nonzero eigenvalues. It is easy to show that these $r - 1$ eigenvalues are all equal if and only if $\mathbf{S}_r$ is a nonzero multiple of the $r \times r$ *centering matrix*

$$\mathbf{C}_r = \mathbf{I}_r - \frac{1}{r}\mathbf{e}\mathbf{e}', \tag{6.4.15}$$

where $\mathbf{I}_r$ is the $r \times r$ identity matrix and each element of the $r \times 1$ column vector $\mathbf{e}$ is equal to 1. Clearly the centering matrix $\mathbf{C}_r$ is idempotent with rank $r - 1$ and so there are $r - 1$ nonzero eigenvalues all equal to 1.

When $\mathbf{S}_r = a\mathbf{C}_r$ then $\mathbf{S}_r$ has $r-1$ nonzero eigenvalues all equal to a. To go the other way, suppose that $\mathbf{S}_r$ has $r - 1$ nonzero eigenvalues all equal to b, say. Then $\mathbf{S}_r^2 = b\mathbf{S}_r$ and so

$$(\mathbf{S}_r - b\mathbf{C}_r)(\mathbf{S}_r + \mathbf{ee}') = \mathbf{S}_r^2 - b\mathbf{C}_r\mathbf{S}_r + \mathbf{S}_r\mathbf{ee}' - b\mathbf{C}_r\mathbf{ee}'$$

$$= \mathbf{S}_r^2 - b\mathbf{C}_r\mathbf{S}_r = \mathbf{S}_r^2 - b\mathbf{S}_r = \mathbf{0} \tag{6.4.16}$$

since $\mathbf{S}_r\mathbf{e} = \mathbf{C}_r\mathbf{e} = \mathbf{0}$. Moreover $\mathbf{S}_r\mathbf{e} = \mathbf{0}$ implies that $\operatorname{rank}(\mathbf{S}_r + \mathbf{ee}') = \operatorname{rank}(\mathbf{S}_r) + \operatorname{rank}(\mathbf{ee}') = (r-1) + 1 = r$ and so $\mathbf{S}_r + \mathbf{ee}'$ is nonsingular and we may postmultiply (6.4.16) by $(\mathbf{S}_r + \mathbf{ee}')^{-1}$ to yield $\mathbf{S}_r = b\mathbf{C}_r$ as desired.

It is interesting to note that the centering matrix $\mathbf{C}_r$ is the Schur complement $\mathbf{E}/r$ of the scalar r in the matrix

$$\mathbf{E} = \begin{pmatrix} \mathbf{I}_r & \mathbf{e} \\ \mathbf{e}' & r \end{pmatrix}. \tag{6.4.17}$$

When the covariance matrix of the random vector $\mathbf{y}$ as defined above in (6.4.8) is $\operatorname{cov}(\mathbf{y}) = \sigma^2\mathbf{I}_n$, for some unknown positive scalar σ^2, then it was shown in 1958 by V. R. Rao [379] that the design is balanced if and only if all the elementary treatment contrasts are estimated by ordinary least squares with the same variance [145, Th. 2.2, p. 52], [366, Th. 4.3.1, p. 5].

When a connected design is not balanced then Chakrabarti [112] suggested the *measure of imbalance*

$$\psi_r = \frac{\operatorname{tr}\mathbf{S}_r^2}{(\operatorname{tr}\mathbf{S}_r)^2} = \frac{\sum_{i=1}^{r-1}\lambda_i^2}{\left(\sum_{i=1}^{r-1}\lambda_i\right)^2}, \tag{6.4.18}$$

where λ_i is the ith largest eigenvalue of $\mathbf{S}_r$. It is easy to see that

$$\frac{1}{r-1} \le \psi_r \le 1. \tag{6.4.19}$$

Equality holds on the left of (6.4.19) if and only if the design is balanced. Equality holds on the right if and only if $r = 2$ and then equality holds throughout (6.4.19). Thibaudeau & Styan [437] gave improved bounds for ψ_r in certain circumstances; see also Boothroyd [67, Section 3.5].

When the design is binary, i.e., when $n_{ij} = 0$ or 1, then the design is balanced if and only if the C-matrix

$$\mathbf{S}_r = \frac{n-c}{r-1}\mathbf{C}_r \tag{6.4.20}$$

and so all the nonzero eigenvalues of $\mathbf{S}_r$ are then equal to $(n-c)/(r-1)$. The proof is easy. We know that $\mathbf{S}_r = a\mathbf{C}_r$ for some scalar a. Taking traces yields

$$\begin{aligned}\operatorname{tr}\mathbf{S}_r = a(r-1) &= \operatorname{tr}(\mathbf{D}_r - \mathbf{N}\mathbf{D}_c^{-1}\mathbf{N}') = n - \operatorname{tr}\mathbf{N}\mathbf{D}_c^{-1}\mathbf{N}'.\\ &= n - \sum_{i=1}^{r}\sum_{j=1}^{c} n_{ij}\frac{1}{n_{\cdot j}}n'_{ji} = n - \sum_{j=1}^{c}\frac{1}{n_{\cdot j}}\sum_{i=1}^{r} n_{ij}^2.\\ &= n - \sum_{j=1}^{c}\frac{1}{n_{\cdot j}}\sum_{i=1}^{r} n_{ij} = n - c\end{aligned}$$

since $n_{ij}^2 = n_{ij}$, the design being binary. Hence $\operatorname{tr}\mathbf{S}_r = n - c = a(r-1)$ which establishes (6.4.20).

6.4.4 The average efficiency factor and canonical efficiency factors

With $\mathbf{X}_1, \mathbf{X}_2$ and $\mathbf{y}$ defined as in

$$\mathrm{E}(\mathbf{y}) = \mathbf{X}_1\boldsymbol{\alpha} + \mathbf{X}_2\boldsymbol{\beta}, \tag{6.4.21}$$

see (6.4.8) above, it follows that the vectors

$$\mathbf{X}_1'\mathbf{y} = \Big\{\sum_{j,k} y_{ijk}\Big\} = \mathbf{y}_{\mathrm{rt}} \quad\text{and}\quad \mathbf{X}_2'\mathbf{y} = \Big\{\sum_{i,k} y_{ijk}\Big\} = \mathbf{y}_{\mathrm{ct}}, \tag{6.4.22}$$

say, contain, respectively, the row (treatment) and column (block) totals of the observations. When $\operatorname{cov}(\mathbf{y}) = \sigma^2\mathbf{I}$, we see that the joint covariance matrix of the vectors of row and column totals of the observations

$$\operatorname{cov}\begin{pmatrix}\mathbf{y}_{\mathrm{rt}}\\ \mathbf{y}_{\mathrm{ct}}\end{pmatrix} = \operatorname{cov}\begin{pmatrix}\mathbf{X}_1'\mathbf{y}\\ \mathbf{X}_2'\mathbf{y}\end{pmatrix} = \sigma^2\begin{pmatrix}\mathbf{X}_1'\mathbf{X}_1 & \mathbf{X}_1'\mathbf{X}_2\\ \mathbf{X}_2'\mathbf{X}_1 & \mathbf{X}_2'\mathbf{X}_2\end{pmatrix} = \sigma^2\begin{pmatrix}\mathbf{D}_r & \mathbf{N}\\ \mathbf{N}' & \mathbf{D}_c\end{pmatrix},$$

and so the canonical correlations ρ_h, say, between $\mathbf{y}_{\mathrm{rt}}$ and $\mathbf{y}_{\mathrm{ct}}$ are the positive square roots of the eigenvalues of $\mathbf{D}_r^{-1}\mathbf{N}\mathbf{D}_c^{-1}\mathbf{N}'$, see, e.g., T. W. Anderson [9, ch. 12]. These canonical correlations ρ_h are equivalently the singular values of the $r \times c$ matrix

$$\mathbf{D}_r^{-1/2}\mathbf{N}\mathbf{D}_c^{-1/2} = \Big\{\frac{n_{ij}}{\sqrt{n_{i\cdot}n_{\cdot j}}}\Big\},$$

see Latour & Styan [274, p. 227]. The quantities $1 - \rho_h^2$ are called the *canonical efficiency factors* of the design, see James & Wilkinson [238], John & Williams [242, § 2.3].

Let m denote the number of nonzero canonical correlations ρ_h or equivalently the number of canonical efficiency factors $1-\rho_h^2$ not equal to 1. Then $m = \text{rank}(\mathbf{N}) \leq \min(r, c)$.

Let u denote the number of canonical correlations ρ_h equal to 1 or equivalently the number of canonical efficiency factors $1-\rho_h^2$ equal to 0, and let t denote the number of positive canonical correlations ρ_h not equal to 1, or equivalently the number of positive canonical efficiency factors $1-\rho_h^2$ equal to 1. Then

$$1 \leq \text{rank}(\mathbf{N}) = m = u + t \leq \min(r, c). \tag{6.4.23}$$

It follows that

$$\begin{aligned} u &= \nu(\mathbf{I} - \mathbf{D}_r^{-1}\mathbf{N}\mathbf{D}_c^{-1}\mathbf{N}') \\ &= \nu\big(\mathbf{D}_r^{-1}(\mathbf{D}_r - \mathbf{N}\mathbf{D}_c^{-1}\mathbf{N}')\big) = \nu(\mathbf{S}_r) \geq 1 \end{aligned} \tag{6.4.24}$$

from (6.4.7). It follows at once from (6.4.24) that the layout is connected if and only if $u = 1$.

The vectors

$$\mathbf{z}_r = \mathbf{M}_2\mathbf{y}_{\text{rt}} \quad \text{and} \quad \mathbf{z}_c = \mathbf{M}_1\mathbf{y}_{\text{ct}} \tag{6.4.25}$$

are called, respectively, vectors of *adjusted* row (treatment) and column (block) totals. Latour & Styan [274] have shown that the canonical correlations between $\mathbf{z}_r$ and $\mathbf{z}_c$ are precisely the t canonical correlations between $\mathbf{y}_{\text{rt}}$ and $\mathbf{y}_{\text{ct}}$ that are not equal to 1. Hence

$$t = \text{rank}(\mathbf{X}_1'\mathbf{M}_2\mathbf{M}_1\mathbf{X}_2) \tag{6.4.26}$$

and since $m = u + t$, see (6.4.23) above, it follows that here

$$\text{rank}(\mathbf{X}_1'\mathbf{X}_2) = r + c - \text{rank}(\mathbf{X}_1 : \mathbf{X}_2) + \text{rank}(\mathbf{X}_1'\mathbf{M}_2\mathbf{M}_1\mathbf{X}_2).$$

Baksalary & Styan [28, Lemma 1] prove the more general rank equality

$$\text{rank}(\mathbf{A}^*\mathbf{B}) = \text{rank}(\mathbf{A}) + \text{rank}(\mathbf{B}) - \text{rank}(\mathbf{A} : \mathbf{B}) + \text{rank}(\mathbf{A}^*\mathbf{M}_\mathbf{B}\mathbf{M}_\mathbf{A}\mathbf{B}), \tag{6.4.27}$$

where the complex matrices $\mathbf{A}$ and $\mathbf{B}$ have the same number of rows, and the residual matrices $\mathbf{M}_\mathbf{A}$ and $\mathbf{M}_\mathbf{B}$ are the orthogonal projectors on the orthocomplements of the column spaces of, respectively, $\mathbf{A}$ and $\mathbf{B}$.

Chakrabarti [111, p. 19] calls the design *orthogonal* whenever the vectors $\mathbf{z}_r$ and $\mathbf{z}_c$ are uncorrelated, i.e., whenever $t = 0$. It follows that a design is connected and orthogonal if and only if $m = 1$, i.e., if and only if the incidence matrix $\mathbf{N}$ has rank equal to 1, and then the $r \times c$ incidence matrix

$$\mathbf{N} = \{n_{ij}\} = \left\{\frac{n_{i\cdot}n_{\cdot j}}{n}\right\}.$$

6.4.5 Proper, equireplicate and BIB designs

When the numbers $n_{\cdot j}$ of treatments per block $j = 1, \ldots, c$ in a binary design are all equal, say to d, then the design is said to be *proper* and when the numbers $n_{i\cdot}$ of blocks per treatment $i = 1, \ldots, r$ are all equal, say to s, then the design is said to *equireplicate*. Equivalently, the design is proper whenever $\mathbf{D}_c = d\mathbf{I}_c$ and equireplicate whenever $\mathbf{D}_r = s\mathbf{I}_r$.

V. R. Rao [379] showed that a binary design which is both balanced and proper must be equireplicate, but that a binary design that is balanced and equireplicate need not be proper.

When the design is equireplicate, the C-matrix

$$\mathbf{S}_r = s\mathbf{I}_r - \mathbf{N}\mathbf{D}_c^{-1}\mathbf{N}' \tag{6.4.28}$$

and its eigenvalues are $s(1-\rho_h^2)$, i.e., s times the canonical efficiency factors.

A binary design is said to be incomplete whenever $n < rc$, i.e., at least one $n_{ij} = 0$. An incomplete binary design that is proper and equireplicate, and for which each pair of treatments occurs in the same positive number λ, say, of blocks is said to be a *balanced incomplete block* (BIB) design. Here the adjective "balanced" may be interpreted as balance in that each pair of treatments occurs in the same positive number of blocks. In 1936 Yates [464] introduced BIB designs into statistics although the existence of such designs and their properties had been studied much earlier, e.g., almost a hundred years earlier in [260, 261] by the Reverend Thomas Penyngton Kirkman (1806–1895).

It is straightforward to show that for a BIB design the C-matrix

$$\mathbf{S}_r = \frac{r\lambda}{d}\mathbf{C}_r, \tag{6.4.29}$$

and so a BIB design is also balanced in the sense defined above in §6.1.2. In (6.4.29) λ is the common number of blocks in which each pair of treatments occurs and $d = n_{\cdot j}$, the common number of treatments per block $j = 1, \ldots, c$. The matrix $\mathbf{C}_r$ is the centering matrix defined in (6.4.15).

6.5 Broyden's matrix problem and mark-scaling algorithm

In the "Problems and Solutions" section of *SIAM Review*, C. G. Broyden posed [96] and solved [98] the following problem:

> Let $\mathbf{Z} = \{z_{ij}\}$ be an $r \times c$ matrix with no null columns and with $n_i \geq 1$ nonzero elements in its ith row, $i = 1, \ldots, r$. Let the $r \times r$ matrix $\mathbf{D}_r = \text{diag}\{n_i\}$, and let the $c \times c$ matrix $\mathbf{D}_z$ be the diagonal matrix whose diagonal elements are equal to the corresponding diagonal elements of $\mathbf{Z}'\mathbf{Z}$. Determine the conditions under which the matrix $\mathbf{D}_z - \mathbf{Z}'\mathbf{D}_r^{-1}\mathbf{Z}$ is (a) positive definite, (b) positive semidefinite.

As Broyden noted in [96], this problem arose in connection with an algorithm for scaling examination marks, which is described in [97] as an "algorithm designed to compensate for the varying difficulty of examination papers when options are permitted. The algorithm is non-iterative, and does not require a common paper to be taken by all students."

Let the $r \times c$ binary incidence matrix $\mathbf{N} = \{n_{ij}\}$ be defined by

$$\left.\begin{array}{ccccccc} n_{ij} & = & 1 & \Leftrightarrow & z_{ij} & \neq & 0 \\ n_{ij} & = & 0 & \Leftrightarrow & z_{ij} & = & 0 \end{array}\right\} \quad i = 1, \ldots, r;\ j = 1, \ldots, c. \tag{6.5.1}$$

Then $n_i = \Sigma_{j=1}^{c} n_{ij}$, $i = 1, \ldots, r$.

In his solution, Broyden [98] established the nonnegative definiteness of the *Broyden matrix*

$$\mathbf{B} = \mathbf{D}_z - \mathbf{Z}'\mathbf{D}_r^{-1}\mathbf{Z} \tag{6.5.2}$$

from the nonnegativity of the quadratic form

$$\mathbf{u}'\mathbf{B}\mathbf{u} = \mathbf{e}'\mathbf{D}_u(\mathbf{D}_z - \mathbf{Z}'\mathbf{D}_r^{-1}\mathbf{Z})\mathbf{D}_u\mathbf{e} = \sum_{i=1}^{r} \mathbf{e}_i'\mathbf{Z}\mathbf{D}_u\mathbf{E}_i\mathbf{D}_u\mathbf{Z}'\mathbf{e}_i, \tag{6.5.3}$$

where the $c \times c$ matrix $\mathbf{E}_i = \mathbf{I} - (1/n_i)\mathbf{N}'\mathbf{e}_i\mathbf{e}_i'\mathbf{N}$ is symmetric idempotent with rank equal to $c - 1$ (for all $i = 1, \ldots, r$). In (6.5.3) $\mathbf{D}_u = \text{diag}(\mathbf{u})$ is the $c \times c$ diagonal matrix formed from the $c \times 1$ vector $\mathbf{u}$, while $\mathbf{e}$ is the $c \times 1$ vector with each element equal to 1; the vector $\mathbf{e}_i$ is the $c \times 1$ vector with its ith element equal to 1 and the rest zero.

In addition, Broyden [98] showed that $\mathbf{B}$ is positive semidefinite when there exist scalars $a_1, \ldots, a_r, u_1, \ldots, u_c$, all nonzero, so that

$$z_{ij}u_j = a_i n_{ij} \text{ for all } i = 1, \ldots, r \text{ and for all } j = 1, \ldots, c. \quad (6.5.4)$$

Broyden [98] also observed that $\mathbf{B}$ is positive definite unless there exist scalars $a_1, \ldots, a_r, u_1, \ldots, u_c$, with at least one of the u_j nonzero, so that (6.5.4) holds. (At least one of the a_i must then also be nonzero when (6.5.4) holds or $\mathbf{Z}$ would have to have a null column.) These conditions do not, however, completely characterize the singularity of the Broyden matrix $\mathbf{B}$. Moreover, Broyden [96, 97, 98] did not consider the rank (or nullity) of $\mathbf{B}$.

Following Styan [432, Section 3], we solve Broyden's matrix problem by constructing an analysis-of-covariance linear statistical model in which the Broyden matrix $\mathbf{B}$ arises naturally as a Schur complement. This will enable us to completely characterize the rank of $\mathbf{B}$ from the structure of the matrix $\mathbf{Z}$ and its associated binary incidence matrix $\mathbf{N}$. When $\mathbf{Z} \equiv \mathbf{N}$ our analysis-of-covariance model reduces to the usual two-way layout as in Section 6.4.2 above.

6.5.1 An analysis-of-covariance model associated with the Broyden matrix

Consider the linear statistical model defined by

$$\mathrm{E}(y_{ij}) = \alpha_i n_{ij} + z_{ij}\gamma_j \quad (i = 1, \ldots, r;\, j = 1, \ldots, c), \quad (6.5.5)$$

where the n_{ij} are, as above, $(0,1)$-indicators of the z_{ij}, and so the n_{ij} and z_{ij} are zero only simultaneously and then the corresponding observation y_{ij} has zero mean (we could just as well have replaced y_{ij} in (6.5.5) with $n_{ij}y_{ij}$ and then the (i,j)th cell of the $r \times c$ layout would be missing whenever $n_{ij} = z_{ij} = 0$; such y_{ij} play no role in what follows).

The observations y_{ij} in (6.5.5) may be arranged in a two-way layout with r rows and c columns. The α_i may be taken to represent row effects, but the column effects in the usual two-way layout, §6.4.2 above, are here replaced by regression coefficients γ_i on each of c covariates on each of which we have (at most) r observations. This is the analysis-of-covariance model considered, for example, by Scheffé [399, p. 200]; in many analysis-of-covariance models, however, the γ_i are all taken to be equal.

We may rewrite (6.5.5) as

$$\mathrm{E}(\mathbf{y}_j) = \mathbf{Q}_j\boldsymbol{\alpha} + \gamma_j\mathbf{z}_j \quad (j = 1, \ldots, c), \quad (6.5.6)$$

where the $r \times 1$ vectors $\boldsymbol{\alpha} = \{\alpha_i\}$, $\mathbf{y}_j = \{y_{ij}\}$ and $\mathbf{z}_j = \{z_{ij}\}$. The $r \times r$ diagonal matrix $\mathbf{Q}_j = \mathrm{diag}\{n_{ij}\} = \mathrm{diag}(n_{1j}, \ldots, n_{rj})$ is symmetric idempotent with rank equal to $\mathrm{tr}(\mathbf{Q}_j) = \sum_{i=1}^r n_{ij}$ $(j = 1, \ldots, c)$. Moreover

$\mathbf{Q}_j\mathbf{z}_j = \mathbf{z}_j$ ($j = 1, \ldots, c$) and $\sum_{j=1}^{c} \mathbf{Q}_j = \operatorname{diag}\{n_i\} = \mathbf{D}_r$. And so we may combine (6.5.5) and (6.5.6) to write

$$\mathrm{E}(\mathbf{y}) = \mathbf{X}_1\boldsymbol{\alpha} + \mathbf{X}_2\boldsymbol{\gamma} = \mathbf{X}\boldsymbol{\beta}, \tag{6.5.7}$$

where $\mathbf{X} = (\mathbf{X}_1 : \mathbf{X}_2)$, $\boldsymbol{\beta} = \begin{pmatrix} \boldsymbol{\alpha} \\ \boldsymbol{\gamma} \end{pmatrix}$ and

$$\mathbf{y} = \begin{pmatrix} \mathbf{y}_1 \\ \vdots \\ \mathbf{y}_c \end{pmatrix}, \quad \mathbf{X}_1 = \begin{pmatrix} \mathbf{Q}_1 \\ \vdots \\ \mathbf{Q}_c \end{pmatrix}, \quad \text{and} \quad \mathbf{X}_2 = \begin{pmatrix} \mathbf{z}_1 & & & \\ & \mathbf{z}_2 & & \\ & & \ddots & \\ & & & \mathbf{z}_c \end{pmatrix}. \tag{6.5.8}$$

Then

$$\mathbf{X}'\mathbf{X} = \begin{pmatrix} \mathbf{D}_r & \mathbf{Z} \\ \mathbf{Z}' & \mathbf{D}_z \end{pmatrix}, \tag{6.5.9}$$

and so the Broyden matrix

$$\mathbf{B} = \mathbf{X}'\mathbf{X}/\mathbf{D}_r = \mathbf{D}_z - \mathbf{Z}'\mathbf{D}_r^{-1}\mathbf{Z} \tag{6.5.10}$$

is the Schur complement of $\mathbf{D}_r$ in $\mathbf{X}'\mathbf{X}$. To see that (6.5.9) follows from (6.5.7), we note that $\mathbf{X}_1'\mathbf{X}_1 = \sum_{j=1}^{c} \mathbf{Q}_j'\mathbf{Q}_j = \sum_{j=1}^{c} \mathbf{Q}_j = \mathbf{D}_r$ using $\mathbf{Q}_j' = \mathbf{Q}_j = \mathbf{Q}_j^2$. Moreover $\mathbf{X}_1'\mathbf{X}_2 = (\mathbf{Q}_1'\mathbf{z}_1 : \cdots : \mathbf{Q}_c'\mathbf{z}_c) = \mathbf{Z}$, since $\mathbf{Q}_j'\mathbf{z}_j = \mathbf{Q}_j\mathbf{z}_j = \mathbf{z}_j$, while $\mathbf{X}_2'\mathbf{X}_2 = \operatorname{diag}\{\mathbf{z}_j'\mathbf{z}_j\} = \operatorname{diag}(\mathbf{Z}'\mathbf{Z}) = \mathbf{D}_z$.

6.5.2 Nonnegative definiteness

Let u denote the nullity of the $(r+c) \times (r+c)$ matrix $\mathbf{X}'\mathbf{X}$ defined by (6.5.9). Then using the Haynsworth inertia additivity formula we see that the inertia

$$\begin{aligned} \operatorname{In}\mathbf{B} &= \operatorname{In}(\mathbf{X}'\mathbf{X}/\mathbf{D}_r) = \operatorname{In}(\mathbf{X}'\mathbf{X}) - \operatorname{In}\mathbf{D}_r \\ &= \{r + c - u, 0, u\} - \{r, 0, 0\} = \{c - u, 0, u\} \end{aligned} \tag{6.5.11}$$

and so the Broyden matrix $\mathbf{B}$ is nonnegative definite with nullity

$$u = \nu(\mathbf{B}) = \nu(\mathbf{X}'\mathbf{X}) = \nu(\mathbf{X}), \tag{6.5.12}$$

the number of unit canonical correlations between the $r \times 1$ vector $\mathbf{y}_{\mathrm{rt}} = \mathbf{X}_1'\mathbf{y} = \{\sum_{j=1}^{c} y_{ij}\}$ of row totals and the $c \times 1$ vector $\mathbf{y}_{\mathrm{ct}}^{(z)} = \mathbf{X}_2'\mathbf{y} = \{\sum_{i=1}^{r} z_{ij}y_{ij}\}$ of *weighted* column totals of the y_{ij} weighted by the z_{ij}.

Let us now consider the other Schur complement

$$\mathbf{S} = \mathbf{X}'\mathbf{X}/\mathbf{D}_z = \mathbf{D}_r - \mathbf{Z}\mathbf{D}_z^{-1}\mathbf{Z}', \tag{6.5.13}$$

which is, of course, also nonnegative definite. Moreover

$$\nu(\mathbf{S}) = \nu(\mathbf{X}'\mathbf{X}/\mathbf{D}_z) = \nu(\mathbf{X}'\mathbf{X}/\mathbf{D}_r) = u = \nu(\mathbf{B}). \tag{6.5.14}$$

6.5.3 All cells filled

When all the cells are filled, i.e., when $z_{ij} \neq 0 \Leftrightarrow n_{ij} = 1$ for all $i = 1, \ldots, r$ and all $j = 1, \ldots, c$ then

$$\mathbf{D}_r = c\mathbf{I}_r \tag{6.5.15}$$

and the model is equireplicate and the Schur complement

$$\mathbf{S} = \mathbf{X}'\mathbf{X}/\mathbf{D}_z = \mathbf{D}_r - \mathbf{Z}\mathbf{D}_z^{-1}\mathbf{Z}' = c\mathbf{I}_r - \mathbf{Z}\mathbf{D}_z^{-1}\mathbf{Z}' \tag{6.5.16}$$

and $u = \nu(\mathbf{S})$ is the number of unit eigenvalues of $(1/c)\mathbf{Z}\mathbf{D}_z^{-1}\mathbf{Z}'$. Since $\mathbf{Z}\mathbf{D}_z^{-1}\mathbf{Z} \geq_{\mathsf{L}} \mathbf{0}$, it follows at once that

$$\left.\begin{array}{cccccccc} u \geq 1 & \Leftrightarrow & u & = & 1 & \Leftrightarrow & \operatorname{rank}(\mathbf{Z}) & = 1 \\ & \text{and} & u & = & 0 & \Leftrightarrow & \operatorname{rank}(\mathbf{Z}) & \geq 2 \end{array}\right\} \tag{6.5.17}$$

and so, when all the cells are filled, the Broyden matrix $\mathbf{B}$ is

$$\text{positive definite} \Leftrightarrow \operatorname{rank}(\mathbf{B}) = c \Leftrightarrow \operatorname{rank}(\mathbf{Z}) \geq 2$$

$$\text{positive semidefinite singular} \Leftrightarrow \operatorname{rank}(\mathbf{B}) = c - 1 \Leftrightarrow \operatorname{rank}(\mathbf{Z}) = 1.$$

6.5.4 At least one cell empty

When at least one of the cells is empty, i.e., when $z_{ij} = 0 \Leftrightarrow n_{ij} = 0$ for at least one $i = 1, \ldots, r$ and at least one $j = 1, \ldots, c$, then the characterization above of the positive (semi)definiteness of the Broyden matrix $\mathbf{B}$ is much more complicated that when all the cells are filled (6.5.17). We may write

$$\mathbf{S} = \mathbf{Z}'\mathbf{Z}/\mathbf{D}_z = \mathbf{D}_r - \mathbf{Z}\mathbf{D}_z^{-1}\mathbf{Z}' = \sum_{j=1}^{c}\Big(\mathbf{Q}_j - \frac{1}{\mathbf{z}_j'\mathbf{z}_j}\mathbf{z}_j\mathbf{z}_j'\Big) = \sum_{j=1}^{c}\mathbf{G}_j,$$

say. Here the matrix $\mathbf{G}_j$ is symmetric and idempotent with $\operatorname{rank}(\mathbf{G}_j) = \operatorname{rank}(\mathbf{Q}_j) - 1 = n_j - 1$ $(j = 1, \ldots, c)$.

When the $r \times r$ Schur complement $\mathbf{S}$ is positive definite,

$$r = \operatorname{rank}(\mathbf{S}) = \operatorname{rank}(\sum_{j=1}^{c} \mathbf{G}_j) \leq \sum_{j=1}^{c} \operatorname{rank}(\mathbf{G}_j)$$
$$= \sum_{j=1}^{c} (n_{\cdot j} - 1) = n_{\cdot\cdot} - c, \qquad (6.5.18)$$

where $n_{\cdot\cdot}$ is the number of filled cells. The inequality (6.5.18) then shows that a necessary condition for $\mathbf{B}$ to be positive definite is that there be at least $r + c$ filled cells. We recall that $\nu(\mathbf{S}) = \nu(\mathbf{B})$, see (6.5.14).

We will divide our presentation of necessary and sufficient conditions for the nullity of the Broyden matrix $\mathbf{B}$ to be a particular number u into the two situations when the layout is either (a) connected or (b) not connected.

We will first suppose that the layout is connected; this may be characterized by the nullity

$$\nu(\mathbf{S}_r) = \nu(\mathbf{D}_r - \mathbf{N}\mathbf{D}_c^{-1}\mathbf{N}') = 1, \qquad (6.5.19)$$

where $\mathbf{D}_c = \operatorname{diag}\{\sum_{i=1}^{r} n_{ij}\}$.

THEOREM 6.6.1. *When the layout is connected, the Broyden matrix* $\mathbf{B}$ *is positive semidefinite and singular if and only if it is possible to write the matrix*

$$\mathbf{Z} = \{z_{ij}\} = \{a_i b_j n_{ij}\} = \mathbf{D}_a \mathbf{N} \mathbf{D}_b \qquad (6.5.20)$$

for some nonsingular diagonal matrices $\mathbf{D}_a$ *and* $\mathbf{D}_b$*, and then the nullity* $\nu(\mathbf{B}) = 1$.

We may interpret the condition (6.5.20) as follows: There exist non zero scalars a_i and b_j for each empty cell (i, j) with $z_{ij} = n_{ij} = 0$, so that the matrix $\mathbf{Z}^*$, say, formed from $\mathbf{Z}$ by replacing each empty cell (i, j) with $a_i b_j$ has rank equal to 1. Conversely there exists a matrix $\mathbf{Z}^*$, say, with rank equal to 1, so that $\mathbf{Z}$ can be formed from $\mathbf{Z}^*$ by changing some of the entries into 0. (We assume that neither $\mathbf{Z}$ nor $\mathbf{Z}^*$ have any null rows or null columns.) When all the cells are filled, the condition (6.5.20) reduces to $\operatorname{rank}(\mathbf{Z}) = 1$.

The formula (6.5.20) in Theorem 3.1 also yields the following sufficient condition for positive definiteness of Broyden's matrix $\mathbf{B}$: If there exists in the matrix $\mathbf{Z}$ a 2×2 nonsingular submatrix with all four elements nonzero then the matrix $\mathbf{B}$ is positive definite (since then no matrix $\mathbf{Z}^*$ of rank 1 can be constructed from $\mathbf{Z}$ which has rank at least equal to 2).

When (6.5.20) holds then $\operatorname{rank}(\mathbf{Z}) = \operatorname{rank}(\mathbf{N})$, but this equality of ranks is not sufficient to imply (6.5.20) when at least one of the cells is empty.

When all the cells are filled, however, rank($\mathbf{N}$) $= 1$ and so rank($\mathbf{Z}$) $=$ rank($\mathbf{N}$) does imply (6.5.20) but when at least one of the cells is empty, e.g., when $\mathbf{Z} = \left(\begin{smallmatrix} 1 & 2 & 0 \\ 3 & 4 & 5 \end{smallmatrix}\right)$ then rank($\mathbf{Z}$) $= 2 =$ rank($\mathbf{N}$), but (6.5.20) does not hold since the leading 2×2 submatrix in $\mathbf{Z}$ is nonsingular with all 4 elements nonzero.

When the layout associated with the matrix $\mathbf{Z}$ is not connected, then the rows and columns of $\mathbf{Z}$ may be arranged so that

$$\mathbf{Z} = \begin{pmatrix} \mathbf{Z}_1 & & \\ & \ddots & \\ & & \mathbf{Z}_q \end{pmatrix} \tag{6.5.21}$$

with $q \geq 2$, where the layout corresponding to each submatrix $\mathbf{Z}_h$ ($h = 1, \ldots, q$) is connected [410, p. 320]. The Broyden matrix $\mathbf{B}$ will then be positive definite if and only if none of the submatrices $\mathbf{Z}_h$ in (6.5.21) can be expressed in the form (6.5.20), and positive semidefinite with nullity u ($1 \leq u \leq l$) if and only if precisely u of the submatrices $\mathbf{Z}_h$ can be expressed in the form (6.5.20).

Chapter 7

Schur Complements and Applications in Numerical Analysis

7.0 Introduction

In numerical analysis, the solution of a system of linear equations is often involved, sometimes in a hidden way, in many methods and algorithms. In particular, linear systems appear under the form of a ratio of determinants in interpolation and extrapolation, in formal orthogonal polynomials, in continued fractions and Padé approximation, and in various methods and techniques related to the solution of systems of linear equations. Since they are related to such ratios by the Schur formula (1.1.4), Schur complements are playing a role, sometimes fundamental, in these topics. They also have important applications in recent techniques used in iterative methods for the solution of large sparse systems of linear equations. Obviously, we did not intend to enter too many details in this chapter, but only to show why and how Schur's complements are useful, and often unavoidable, in various topics of numerical analysis. For topics which are too technical or too specialized, we will only give references for the interested reader.

In order not to give too many references, we will not always cite the original source but mainly review papers and books. Of course, more recent references published after such reviews will be given.

An important tool for our purpose is *Sylvester's determinantal identity* whose proof can be obtained directly (see, for example, [76]). However, it also follows immediately from the quotient property as proved in [88]. Let us consider the partitioned matrices

$$A' = \begin{pmatrix} A & B \\ C & D \end{pmatrix}, \quad B' = \begin{pmatrix} B & E \\ D & F \end{pmatrix},$$

$$C' = \begin{pmatrix} C & D \\ G & H \end{pmatrix}, \quad D' = \begin{pmatrix} D & F \\ H & L \end{pmatrix}$$

and

$$M = \begin{pmatrix} A & B & E \\ C & D & F \\ G & H & L \end{pmatrix}.$$

Assuming that D'/D is nonsingular, the quotient property is

$$M/D' = (A'/D) - (B'/D)(D'/D)^{-1}(C'/D). \tag{7.0.1}$$

If A, E, G and L are numbers, the Schur complements in this identity are numbers (and thus identical to the corresponding determinant) and Schur's determinantal formula (1.1.4) immediately leads to Sylvester's identity

$$\det M \cdot \det D = \det A' \cdot \det D' - \det B' \cdot \det C'. \tag{7.0.2}$$

7.1 Formal orthogonality

Let c be a linear functional on the space of polynomials. It is entirely defined by the knowledge of its moments $c_i = c(\xi^i), i = 0, 1, \ldots$

Let $\{P_k\}$ be a family of polynomials such that, for all k, P_k has the degree k (for simplicity) and satisfies the orthogonality conditions

$$c(\xi^i P_k(\xi)) = 0, \quad i = 0, \ldots, k-1.$$

We say that $\{P_k\}$ is the family of *formal orthogonal polynomials* with respect to c (or to the sequence (c_i)). If the functional c is represented by a definite integral on the real axis with respect to a positive measure, then the usual orthogonal polynomials are recovered.

If we set $P_k(\xi) = a_0 + a_1\xi + \cdots + a_k\xi^k$, the orthogonality conditions write

$$c(\xi^i P_k(\xi)) = a_0 c_i + a_1 c_{i+1} + \cdots + a_k c_{i+k} = 0, \quad i = 0, \ldots, k-1.$$

This is a system of k equations in $k+1$ which defines P_k apart a multiplying normalization factor. This factor is determined by an additional condition. For example, P_k could be monic, or such that $P_k(0) = 1$, or satisfying $P_k(1) = 1$. In the case where P_k is monic, we have

$$a_0 c_i + a_1 c_{i+1} + \cdots + a_{k-1} c_{i+k-1} = -c_{i+k}, \quad i = 0, \ldots, k-1$$

and Cramer's rule gives (assuming, of course, that the determinant of the system is different from zero)

$$P_k(\xi) = \det \begin{pmatrix} c_0 & c_1 & \cdots & c_k \\ c_1 & c_2 & \cdots & c_{k+1} \\ \vdots & \vdots & & \vdots \\ c_{k-1} & c_k & \cdots & c_{2k-1} \\ 1 & \xi & \cdots & \xi^k \end{pmatrix} \Big/ \det \begin{pmatrix} c_0 & c_1 & \cdots & c_{k-1} \\ c_1 & c_2 & \cdots & c_k \\ \vdots & \vdots & & \vdots \\ c_{k-1} & c_k & \cdots & c_{2k-2} \end{pmatrix}.$$

So, we see that P_k is the Schur complement of the matrix of the denominator in the matrix of the numerator and that

$$P_k(\xi) = \xi^k - (1, \xi, \ldots, \xi^{k-1}) \begin{pmatrix} c_0 & \cdots & c_{k-1} \\ \vdots & & \vdots \\ c_{k-1} & \cdots & c_{2k-2} \end{pmatrix}^{-1} \begin{pmatrix} c_k \\ \vdots \\ c_{2k-1} \end{pmatrix}.$$

Formal orthogonal polynomials satisfy most of the algebraic properties of the usual orthogonal polynomials such as the three–term recurrence relationship, the Shohat–Favard theorem and its reciprocal, the Christoffel–Darboux identity and its variants, and some of the properties on the zeros. See [72].

Let us now define the linear functionals $c^{(n)}$ by $c^{(n)}(\xi^i) = c_{n+i}$ for $i = 0, 1, \ldots$. Obviously $c^{(n)}(\xi^{i+1}) = c^{(n+1)}(\xi^i)$. Let $P_k^{(n)}$ be the family of monic formal orthogonal polynomials with respect to $c^{(n)}$, that is such that $c^{(n)}(\xi^i P_k^{(n)}(\xi)) = 0$ for $i = 0, \ldots, k-1$. The preceding functional c and the polynomials P_k correspond to $n = 0$. These families are called *adjacent families.*

We denote by $N_k^{(n)}$ the numerator of $P_k^{(n)}$ and by $H_k^{(n)}$ its denominator

$$N_k^{(n)}(\xi) = \det \begin{pmatrix} c_n & \cdots & c_{n+k} \\ \vdots & & \vdots \\ c_{n+k-1} & \cdots & c_{n+2k-1} \\ 1 & \cdots & \xi^k \end{pmatrix},$$

$$H_k^{(n)} = \det \begin{pmatrix} c_n & \cdots & c_{n+k-1} \\ \vdots & & \vdots \\ c_{n+k-1} & \cdots & c_{n+2k-2} \end{pmatrix}.$$

There exist many recurrence relationships between polynomials belonging to two adjacent families. They can be obtained either by a direct inspection or by Sylvester's determinantal identity (7.0.2). We will not give all these identities but we will only show, on two of them, how they can be obtained. Applying Sylvester's identity to $N_k^{(n)}$ gives

$$H_{k-1}^{(n+2)} N_k^{(n)}(\xi) = \xi H_k^{(n)} N_{k-1}^{(n+2)}(\xi) - H_k^{(n+1)} N_{k-1}^{(n+1)}(\xi).$$

Let us put the last row $1, \ldots, \xi^k$ of $N_k^{(n)}$ as the first one and again apply Sylvester's identity. It comes out

$$(-1)^k H_{k-1}^{(n+1)} N_k^{(n)}(\xi) = (-1)^{k-1} H_k^{(n+1)} N_{k-1}^{(n)}(\xi) - (-1)^{k-1} \xi H_k^{(n)} N_{k-1}^{(n+1)}(\xi).$$

Applying alternately these two relations, allows to compute a descending staircase in the table of these polynomials.

Biorthogonal polynomials, as defined in [74], orthogonal polynomials on the circle, least-squares orthogonal polynomials [81], as well as other generalizations of orthogonal polynomials, can be similarly related to Schur complements. See [75]. Formal orthogonal polynomials and their generalizations are fundamental in Padé approximation.

7.2 Padé approximation

Let us consider a formal power series

$$f(z) = c_0 + c_1 z + c_2 z^2 + \cdots$$

We are looking for a rational function whose power series expansion agrees with that of f as far as possible. It means that its numerator $P(z) = b_0 + b_1 z + \cdots + b_p z^p$ and its denominator $Q(z) = a_0 + a_1 z + \cdots + a_q z^q$ must satisfy

$$P(z) - Q(z) f(z) = \mathcal{O}(z^{p+q+1}). \tag{7.2.3}$$

Such a rational function $P(z)/Q(z)$ is called a *Padé approximant* and it is denoted by $[p/q]_f(z)$.

From (7.2.3), it is easy to see, by identification of the coefficients and with $c_i = 0$ for $i < 0$, that

$$\begin{array}{llcll}
\text{degree } 0 & \Longrightarrow & a_0 & = & b_0 c_0 \\
\text{degree } 1 & \Longrightarrow & a_1 & = & b_0 c_1 + b_1 c_0 \\
 & & & \vdots & \\
\text{degree } p & \Longrightarrow & a_p & = & b_0 c_p + \cdots + b_q c_{p-q} \\
\text{degree } p+1 & \Longrightarrow & 0 & = & b_0 c_{p+1} + \cdots + b_q c_{p-q+1} \\
 & & & \vdots & \\
\text{degree } p+q & \Longrightarrow & 0 & = & b_0 c_{p+q} + \cdots + b_q c_p a_0.
\end{array}$$

Setting $b_0 = 1$, the last q equations give the coefficients of the denominator (assuming that the system is nonsingular) and, then, those of the numerator can be directly computed.

Moreover, we have the following formula due to Jacobi

$$[p/q]_f(z) = \frac{\det \begin{pmatrix} z^q f_{p-q}(z) & z^{q-1} f_{p-q+1}(z) & \cdots & f_p(z) \\ c_{p-q+1} & c_{p-q+2} & \cdots & c_{p+1} \\ \vdots & \vdots & & \vdots \\ c_p & c_{p+1} & \cdots & c_{p+q} \end{pmatrix}}{\det \begin{pmatrix} z^q & z^{q-1} & \cdots & 1 \\ c_{p-q+1} & c_{p-q+2} & \cdots & c_{p+1} \\ \vdots & \vdots & & \vdots \\ c_p & c_{p+1} & \cdots & c_{p+q} \end{pmatrix}}$$

with

$$f_k(z) = \begin{cases} \displaystyle\sum_{i=0}^{k} c_i z^i & \text{if } k \geq 0 \\ 0 & \text{if } k < 0. \end{cases}$$

Dividing the numerator and the denominator by $H_q^{(p-q+2)}$ shows that $[p/q]_f(z)$ can be represented as a ratio of Schur complements. This formula can also be written as

$$\det W \cdot [p/q]_f(z) =$$

$$\det \begin{pmatrix} \sum_{i=0}^{p-q} c_i z^i & c_{p-q+1} & \cdots & c_p \\ -z^{p-q+1} c_{p-q+1} & c_{p-q+1} - zc_{p-q+2} & \cdots & c_p - zc_{p+1} \\ \vdots & \vdots & & \vdots \\ -z^{p-q+1} c_p & c_p - zc_{p+1} & \cdots & c_{p+q-1} - zc_{p+q} \end{pmatrix}$$

where

$$W = \begin{pmatrix} c_{p-q+1} - zc_{p-q+2} & \cdots & c_p - zc_{p+1} \\ \vdots & & \vdots \\ c_p - zc_{p+1} & \cdots & c_{p+q-1} - zc_{p+q} \end{pmatrix}.$$

Let $c = (c_{p-q+1}, \ldots, c_p)^T$. Taking the Schur complement of W in the numerator of $[p/q]_f(z)$ and applying Schur's formula (1.1.4) reveals

$$[p/q]_f(z) = \sum_{i=0}^{p-q} c_i z^i + z^{p-q+1} c^T W^{-1} c.$$

This formula is known as *Nuttall's compact formula.* A similar formula can be obtained for the vector Padé approximants of van Iseghem which allow to approximate simultaneously several series by rational functions with the same denominator or for the partial Padé approximants introduced by Brezinski. On these topics, see, for example, [93].

Padé approximants are closely related to formal orthogonal polynomials. Let us consider, for simplicity, the case $p = k-1$ and $q = k$ and let c be the linear functional on the space of polynomials defined, as above, by $c(\xi^i) = c_i$ for $i = 0, 1, \ldots$ Then, the denominator of $[k-1/k]_f(z)$ is $z^k P_k(z^{-1})$.

When the system of equations giving the coefficients of the denominator of $[p/q]_f(z)$ is singular, it means that there exists a common factor between the numerator and the denominator of the approximant which becomes identical to another approximant with lower degrees. Such a phenomenon is known as the block structure of the Padé table. Without entering into details, new recursive algorithms for computing Padé approximants have to be derived. As shown in [40], these new rules can be derived using Schur complements.

There exists several generalizations of Padé approximants (for vector and matrix series, series of functions, Laurent series, multivariate series, etc.). They are based on some kind of formal orthogonality and they can also be connected to ratio of determinants and, thus, to Schur complements. See [74, 93].

7.3 Continued fractions

A topic related to Padé approximation, but more general, concerns continued fractions. A *continued fraction* is an expression of the form

$$\begin{aligned} C &= b_0 + \cfrac{a_1}{b_1 + \cfrac{a_2}{b_2 + \cfrac{a_3}{b_3 + \ddots}}} \\ &= b_0 + \frac{a_1 |}{| b_1} + \frac{a_2 |}{| b_2} + \frac{a_3 |}{| b_3} + \cdots \end{aligned}$$

The *convergents* of this continued fraction are the rational fractions

$$C_n = b_0 + \frac{a_1 |}{| b_1} + \cdots + \frac{a_n |}{| b_n} = \frac{A_n}{B_n}, \quad n = 0, 1, \ldots .$$

It is well known that

$$A_n = \det \begin{pmatrix} b_0 & -1 & & & \\ a_1 & b_1 & -1 & & \\ & \ddots & \ddots & \ddots & \\ & & a_{n-1} & b_{n-1} & -1 \\ & & & a_n & b_n \end{pmatrix},$$

$$B_n = \det \begin{pmatrix} b_1 & -1 & & & \\ a_2 & b_2 & -1 & & \\ & \ddots & \ddots & \ddots & \\ & & a_{n-1} & b_{n-1} & -1 \\ & & & a_n & b_n \end{pmatrix}.$$

By Schur's determinantal formula (1.1.4), we have

$$\begin{aligned} A_n &= A_{n-1}[b_n - (0,\ldots,0,a_n)A_{n-1}^{-1}(0,\ldots,0,-1)^T] \\ &= A_{n-1}\left(b_n + a_n \frac{A_{n-2}}{A_{n-1}}\right) \end{aligned}$$

and a similar formula for B_n. So, the usual recurrence formulae for the successive convergents of a continued fraction have been recovered. Conversely, using these recurrence relations and Schur's determinantal formula allows to express the convergents as a ratio of determinants $C_n = A_n/B_n$.

Generalized continued fractions and their extensions to a noncommutative algebra make an extensive implicit use of Schur complements.

7.4 Extrapolation algorithms

A domain of numerical analysis where Schur complements play a central role is the domain of extrapolation algorithms for scalar, vector and matrix sequences. So, we will consider this topic in details.

When a sequence (S_n) converges slowly to a limit S, it can be transformed into a new sequence (T_n) by a *sequence transformation*. Of course, it is hoped that (T_n) will converge to S faster that (S_n), that is

$$\lim_{n\to\infty} \frac{T_n - S}{S_n - S} = 0.$$

Since it has been proved that a sequence transformation for accelerating the convergence of all sequences cannot exist, many sequence transformations have been defined and studied, each of them being only able to accelerate a particular class of sequences.

Building a sequence transformation could be based on the notion of *kernel*, that is the set of sequences which are transformed into a constant sequence $\forall n, T_n = S$ by the transformation. Although this property has never been proved, it has been noticed that if the sequence (S_n) to accelerate is not "too far away" from the kernel, its convergence is accelerated by the transformation.

The kernel of a transformation is usually defined by a relation, depending on unknown parameters, that the successive terms of the sequence have to satisfy for all n. This relation is written for different values of n in order to obtain as many equations as parameters. So, each successive term T_n of the transformed sequence is given as the solution of a system of equations and, thus, Schur complements are involved.

The E–algorithm is the most general sequence transformation known so far. Its kernel is defined by the relation

$$S_n = S + a_1 g_1(n) + \cdots + a_k g_k(n), \quad n = 0, 1, \ldots \tag{7.4.4}$$

where the $(g_i(n))$ are known sequences whose terms can depend on the sequence (S_n) itself and the a_i are unknown parameters. Writing this relation for the indexes $n, \ldots, n+k$ and solving it for the unknown S, we obtain, by Cramer's rule, a new sequence with terms denoted by $E_k^{(n)}$ instead of T_n

$$E_k^{(n)} = \frac{\det \begin{pmatrix} S_n & \cdots & S_{n+k} \\ g_1(n) & \cdots & g_1(n+k) \\ \vdots & & \vdots \\ g_k(n) & \cdots & g_k(n+k) \end{pmatrix}}{\det \begin{pmatrix} 1 & \cdots & 1 \\ g_1(n) & \cdots & g_1(n+k) \\ \vdots & & \vdots \\ g_k(n) & \cdots & g_k(n+k) \end{pmatrix}}. \tag{7.4.5}$$

The choice $g_i(n) = x_n^i$, where (x_n) is an auxiliary known sequence, leads to Richardson extrapolation (and Romberg method in a particular case), the choice $g_i(n) = S_{n+i} - S_{n+i-1}$ corresponds to Shanks transformation (that is the ε-algorithm). Levin's transformations and many other transformations can be put into this framework.

For this transformation to be effective, it is necessary to find a recursive algorithm for its implementation. Let us first define the auxiliary quantities $g_{k,i}^{(n)}$ by a ratio of determinants similar to (7.4.5) with the first row of the numerator replaced by $g_i(n), \ldots, g_i(n+k)$. Thus, $g_{k,i}^{(n)} = 0$ for $i < k$.

Applying Sylvester's determinantal identity (7.0.2) to the numerators and the denominators of $E_k^{(n)}$ and $g_{k,i}^{(n)}$, we obtain the recursive rules

$$\begin{aligned} E_k^{(n)} &= E_{k-1}^{(n)} - \frac{E_{k-1}^{(n+1)} - E_{k-1}^{(n)}}{g_{k-1,k}^{(n+1)} - g_{k-1,k}^{(n)}}\, g_{k-1,k}^{(n)}, \quad k, n = 0, 1, \ldots, \\ g_{k,i}^{(n)} &= g_{k-1,i}^{(n)} - \frac{g_{k-1,i}^{(n+1)} - g_{k-1,i}^{(n)}}{g_{k-1,k}^{(n+1)} - g_{k-1,k}^{(n)}}\, g_{k-1,k}^{(n)}, \quad k, n = 0, 1, \ldots; i > k, \end{aligned}$$

with, for $n = 0, 1, \ldots$, $E_0^{(n)} = S_n$ and $g_{0,i}^{(n)} = g_i(n)$, for $i \geq 1$.

The $E_k^{(n)}$ are usually displayed in a two dimensional table. The quantities with the same lower index are placed in a column, while those with the same upper index are in a descending diagonal. So, $E_k^{(n)}$ is computed from two quantities in the preceding column, one on the west and the other one on the north-west. Such a rule is called a *triangular recursion scheme* (see Section 7.9). A more economical scheme for the implementation of this transformation is given in [169]. It is an immediate consequence of the Sylvester's determinantal identity.

The quantities in the $(k+m)$th column of this table can be computed directly from those on the kth column. Indeed, we have

$$E_{k+m}^{(n)} = \frac{\det \begin{pmatrix} E_k^{(n)} & \Delta E_k^{(n)} & \cdots & \Delta E_k^{(n+m-1)} \\ g_{k,k+1}^{(n)} & \Delta g_{k,k+1}^{(n)} & \cdots & \Delta g_{k,k+1}^{(n+m-1)} \\ \vdots & \vdots & & \vdots \\ g_{k,k+m}^{(n)} & \Delta g_{k,k+m}^{(n)} & \cdots & \Delta g_{k,k+m}^{(n+m-1)} \end{pmatrix}}{\det \begin{pmatrix} \Delta g_{k,k+1}^{(n)} & \cdots & \Delta g_{k,k+1}^{(n+m-1)} \\ \vdots & & \vdots \\ \Delta g_{k,k+m}^{(n)} & \cdots & \Delta g_{k,k+m}^{(n+m-1)} \end{pmatrix}} \tag{7.4.6}$$

and a similar formula from $g_{k+m,i}^{(n)}$. From Schur formula, we get the recursive rule

$$E_{k+m}^{(n)} = E_k^{(n)} - [\Delta E_k^{(n)}, \ldots, \Delta E_k^{(n+m-1)}][D_k^{(n)}]^{-1} \begin{pmatrix} g_{k,k+1}^{(n)} \\ \vdots \\ g_{k,k+m}^{(n)} \end{pmatrix}, \tag{7.4.7}$$

with

$$D_k^{(n)} = \begin{pmatrix} \Delta g_{k,k+1}^{(n)} & \cdots & \Delta g_{k,k+1}^{(n+m-1)} \\ \vdots & & \vdots \\ \Delta g_{k,k+m}^{(n)} & \cdots & \Delta g_{k,k+m}^{(n+m-1)} \end{pmatrix}.$$

For $m = 1$, this relation reduces to the usual rule of the E–algorithm given above.

These rules can only be used if the denominator is different from zero. If not, a breakdown occurs in the algorithm. If the denominator is close to zero, a near-breakdown arises and a severe numerical instability can affect the algorithm. It is possible to jump over breakdowns or near-breakdowns by taking $m > 1$ in (7.4.7), thus leading to a more stable algorithm (see [73] for a numerical example). Breakdowns and near-breakdowns can be avoided by a similar technique in the computation of formal orthogonal polynomials [84] and in the implementation of other extrapolation methods [85]. Let us also mention that, taking $k = 0$ in (7.4.7) gives a Nuttall–type formula for the E–algorithm.

Other interpretations of the E-algorithm based on Schur complements are given in [88]. On extrapolation methods, see [83] and [458].

Let us now consider the vector case. Of course, for accelerating the convergence of a sequence of vectors, it is always possible to use a scalar procedure separately on each component. However, vector methods are preferable since, usually, the components of the vectors to be accelerated are not independent.

Let us first define the *vector Schur complement.* We first need to define a vector determinant. Let $y, x_1, x_2, \ldots$ be elements of a vector space on $\mathcal{K}$, $u \in \mathcal{K}^n$, and $A \in \mathcal{K}^{n \times n}$. We set $x = [x_1, \ldots, x_n]$. The vector determinant $\det M$ of the matrix

$$M = \begin{pmatrix} y & x \\ u & A \end{pmatrix}$$

denotes the linear combination of $y, x_1, \ldots, x_k$ obtained by expanding this determinant with respect to its first (vector) row by the classical rule for expanding a determinant. The *vector Schur identity* is

$$M/A = y - xA^{-1}u = \det \begin{pmatrix} y & x \\ u & A \end{pmatrix} \Big/ \det A.$$

From this relation, it can be proved that a vector Sylvester's identity for determinants having a first row consisting of elements of $\mathcal{K}^n$ holds.

Many vector sequence transformations are based on such generalizations of the Schur complement and determinantal identities. This is, for example, the case of the vector E–algorithm which is based on a kernel of the form (7.4.4) where S_n, S and the $g_i(n)$ are vectors and the a_i unknown scalars. If we set $a = (a_1, \ldots, a_k)^T$ and if $G_k(n)$ is the matrix with columns $g_1(n), \ldots, g_k(n)$, this kernel can be written as $S_n = S + G_k(n)a$. A vector formula similar to (7.4.5) holds and the rules of the vector E-algorithm are exactly the same as the scalar ones. The case of a sum of terms $G_k(n)a$ is studied in [89] by means of Schur complements. It is also possible to consider a kernel of the form $S_n = S + A_1 g_1(n) + \cdots + A_k g_k(n)$, where the A_i are unknown matrices. Schur complements are also involved in transformations for treating this case. Some of them used pseudo–Schur complements which are defined by replacing the inverse of a matrix by the pseudo–inverse [89]. Some of these transformations are also related to projection methods that will be discussed in Section 7.6.

Matrix sequence transformations can be treated quite similarly. We consider a kernel of the form $S_n = S + AD_n$, where $S_n, S, A \in \mathbb{R}^{p\times q}$ and $D_n \in \mathbb{R}^{q\times q}$. To find a transformation with this kernel, the matrix A has to be eliminated by solving the system $\Delta S_n = A\Delta D_n$. If ΔD_n is nonsingular, the transformation $T_n = S_n - \Delta S_n(\Delta D_n)^{-1}D_n$ will has the desired kernel.

Vector and matrix sequence transformations are based on the solution of a system of linear equations in a noncommutative algebra. For such systems, it is the notion of *designant*, introduced in [221], which replaces the notion of determinant. We consider a 2×2 system in a noncommutative algebra

$$\begin{aligned} x_1 a_{11} + x_2 a_{12} &= b_1 \\ x_1 a_{21} + x_2 a_{22} &= b_2. \end{aligned}$$

Assuming that a_{11} is nonsingular, we proceed as in Gaussian elimination. We multiply the first equation on the right by a_{11}^{-1}, then by a_{21}, and subtract it from the second equation. We obtain

$$x_2(a_{22} - a_{12}a_{11}^{-1}a_{21}) = b_2 - b_1 a_{11}^{-1}a_{21}.$$

The square matrix $a_{22} - a_{12}a_{11}^{-1}a_{21}$ is called a (right) designant and it is denoted by

$$\begin{vmatrix} a_{11} & a_{12} \\ a_{21} & a_{22} \end{vmatrix}_r$$

where the subscript r stands for right. We see that this designant is the Schur complement of a_{11} in the matrix of the system. If this designant is nonsingular, then

$$x_2 = (b_2 - b_1 a_{11}^{-1}a_{21})(a_{22} - a_{12}a_{11}^{-1}a_{21})^{-1} = \begin{vmatrix} a_{11} & b_1 \\ a_{21} & b_2 \end{vmatrix}_r \begin{vmatrix} a_{11} & a_{12} \\ a_{21} & a_{22} \end{vmatrix}_r^{-1}.$$

x_1 is given by a similar formula.

Designants of higher order play the role of determinants in the solution of systems of higher dimension in a noncommutative algebra. They are defined recursively. We consider the designant of order n

$$\Delta_n = \begin{vmatrix} a_{11} & \cdots & a_{1n} \\ \vdots & & \vdots \\ a_{n1} & \cdots & a_{nn} \end{vmatrix}_r$$

and let $A_{r,s}^{(n-2)}$ be the designants of order $n-1$ obtained by keeping the rows $1, \ldots, n-2, r$ and the columns $1, \ldots, n-2, s$ of Δ_n. By Sylvester's identity, we have

$$\Delta_n = \begin{vmatrix} A_{n-1,n-1}^{(n-2)} & A_{n-1,n}^{(n-2)} \\ A_{n,n-1}^{(n-2)} & A_{n,n}^{(n-2)} \end{vmatrix}_r = A_{n,n}^{(n-2)} - A_{n-1,n}^{(n-2)}[A_{n-1,n-1}^{(n-2)}]^{-1}A_{n,n-1}^{(n-2)}.$$

Left designants can be defined similarly. Designants, which are clearly related to Schur complements, are used in extrapolation methods for vector and matrix sequences and in Padé approximation of series with vector coefficients. In order not to enter into too technical details, we will only refer the interested readers to [77, 86, 89, 92, 188, 311, 389, 390, 391, 392, 393, 394, 395].

Many other extrapolation algorithms can be based on Schur complements since their recursive rule is quite similar to the rule of the E–algorithm (see, for example, [459, 458, 83]. Extrapolation methods will be again discussed in Section 7.9 in a more general setting.

Schur complements also appear in fixed point iterations since they are strongly related to sequence transformations. A *quasi–Newton method* for solving $f(x) = 0$, where $f : \mathbb{R}^n \longmapsto \mathbb{R}^n$ consists in the iterations

$$x_{k+1} = x_k - J_k^{-1} f(x_k), \quad k = 0, 1, \ldots$$

where J_k is an approximation of the Jacobian matrix of f at the point x_k. By the vector Schur formula, we have

$$x_{k+1} = \det \begin{pmatrix} x_k & I \\ f(x_k) & J_k \end{pmatrix} \Big/ \det J_k.$$

Depending on the choice of J_k, the various quasi–Newton methods are obtained. Among them are those of Barnes, Broyden, Henrici, Davidon–Fletcher–Powell, Broyden–Fletcher–Goldfarb–Shanno, Huang and Wolfe. Newton's method corresponds to $J_k = f'(x_k)$. On this topic, see [76, 80].

7.5 The bordering method

In some applications, one has to solve a sequence of systems of linear equations of increasing dimension obtained by adding new rows and columns to the previous matrix. As we saw above, this is the case in extrapolation methods such as the E–algorithm, in formal orthogonal polynomials, in Lanczos method, etc. The *block bordering method*, based on Theorem 1.2, allows to solve these systems recursively.

We consider the systems $A_k x_k = b_k$ and $A_{k+1} x_{k+1} = b_{k+1}$ with

$$A_{k+1} = \begin{pmatrix} A_k & u_k \\ v_k & a_k \end{pmatrix} \quad \text{and} \quad b_{k+1} = \begin{pmatrix} b_k \\ c_k \end{pmatrix}$$

where $A_k \in \mathbb{R}^{n_k \times n_k}$, $u_k \in \mathbb{R}^{n_k \times p_k}$, $v_k \in \mathbb{R}^{p_k \times n_k}$, $a_k \in \mathbb{R}^{p_k \times p_k}$, $b_k \in \mathbb{R}^{n_k}$, $c_k \in \mathbb{R}^{p_k}$. By the block bordering method, it holds

$$x_{k+1} = \begin{pmatrix} x_k \\ 0 \end{pmatrix} + \begin{pmatrix} -A_k^{-1} u_k \\ I_k \end{pmatrix} S_k^{-1}(c_k - v_k x_k)$$

where S_k is the Schur complement of A_k in A_{k+1}, $S_k = a_k - v_k A_k^{-1} u_k$. It is possible to avoid the computation and the storage of the matrix A_k^{-1} by solving recursively the system $A_k q_k = -u_k$ by the bordering method (a FORTRAN subroutine is given in [83]).

The *reverse bordering method* consists in computing x_k from x_{k+1}. We set

$$A_{k+1}^{-1} = \begin{pmatrix} A'_k & u'_k \\ v'_k & a'_k \end{pmatrix} \quad \text{and} \quad x_{k+1} = \begin{pmatrix} x'_k \\ y_k \end{pmatrix}.$$

It is easy to check that

$$\begin{aligned} A_k^{-1} &= A'_k - u'_k [a'_k]^{-1} v'_k \\ x_k &= x'_k - u'_k [a'_k]^{-1} y_k. \end{aligned}$$

When solving a sequence of linear systems by the bordering method, some intermediate systems can be singular or nearly singular. It is possible to jump over these (near) singularities by the block bordering method and to compute directly the solution of the first non–nearly singular system. Then, the reverse bordering method can be used to obtain the solution of the intermediate systems which were skipped. Such a procedure improves the numerical stability. It has been described in [82] where numerical examples concerning Padé approximants show its effectiveness.

7.6 Projections

The notion of projection has many important applications in numerical analysis. Let us first give some general results.

Let E_k and F_k be the subspaces of $\mathbb{R}^n$ spanned by $u_1, \ldots, u_k$ and $v_1, \ldots, v_k$, respectively. When it exists, the *oblique projection* $\mathcal{P}_k$ on E_k along $F_k^\perp$, that is the projection on E_k orthogonal to F_k, is represented by the matrix

$$P_k = U_k(V_k^T U_k)^{-1} V_k^T$$

where $U_k = [u_1, \ldots, u_k]$ and $V_k = [v_1, \ldots, v_k]$. $\mathcal{I} - \mathcal{P}_k$ is the oblique projection on $F_k^\perp$ along E_k. These projections are orthogonal if $u_i = v_i$, for $i = 1, \ldots, k$. Clearly, the matrices P_k and $I - P_k$ are Schur complements. Thus, if $y \in \mathbb{R}^n$,

$$P_k y = -\frac{\det\begin{pmatrix} (v_1, u_1) & \cdots & (v_1, u_k) & (v_1, y) \\ \vdots & & \vdots & \vdots \\ (v_k, u_1) & \cdots & (v_k, u_k) & (v_k, y) \\ u_1 & \cdots & u_k & 0 \end{pmatrix}}{\det\begin{pmatrix} (v_1, u_1) & \cdots & (v_1, u_k) \\ \vdots & & \vdots \\ (v_k, u_1) & \cdots & (v_k, u_k) \end{pmatrix}}.$$

On projections, see [76].

More generally, let E be a vector space on $\mathcal{K}$ and E^* its dual. We denote by $< \cdot, \cdot >$ the bilinear form of the duality between E and E^*. Let $x_0, x_1, \ldots \in E$ and $z_1, z_2, \ldots \in E^*$. We consider the scalar determinants

$$D_k^{(n)} = \det\begin{pmatrix} < z_1, x_{n+1} > & \cdots & < z_1, x_{n+k} > \\ \vdots & & \vdots \\ < z_k, x_{n+1} > & \cdots & < z_k, x_{n+k} > \end{pmatrix}$$

and the vector determinants

$$N_k^{(n)} = \det\begin{pmatrix} x_n & \cdots & x_{n+k} \\ < z_1, x_n > & \cdots & < z_1, x_{n+k} > \\ \vdots & & \vdots \\ < z_k, x_n > & \cdots & < z_k, x_{n+k} > \end{pmatrix}.$$

By Schur formula, it can be proved that the ratio $E_k = N_k^{(0)}/D_k^{(0)}$ can be recursively computed by the *Recursive Projection Algorithm* (RPA)

$$\begin{aligned} E_k &= E_{k-1} - \frac{< z_k, E_{k-1} >}{< z_k, g_{k-1,k} >} g_{k-1,k}, \quad k > 0 \\ g_{k,i} &= g_{k-1,i} - \frac{< z_k, g_{k-1,i} >}{< z_k, g_{k-1,k} >} g_{k-1,k}, \quad i > k > 0 \end{aligned}$$

with $E_0 = x_0$ and $g_{0,n} = x_n$ for $n \geq 1$. The $g_{k,i}$ are given by a ratio of determinant similar to the ratio for E_k.

The E_k can be computed by a more compact algorithm with only one rule. We consider the ratios $e_k^{(i)} = N_k^{(i)}/D_k^{(i)}$. Obviously $e_k^{(0)} = E_k$. It holds

$$e_k^{(i)} = e_{k-1}^{(i)} - \frac{< z_k, e_{k-1}^{(i)} >}{< z_k, e_{k-1}^{(i+1)} >} e_{k-1}^{(i+1)}$$

with $e_0^{(i)} = x_i$. This algorithm is called the *Compact Recursive Projection Algorithm* (CRPA). A formula similar to (7.4.7) allows to compute the $e_{k+m}^{(i)}$ directly from the $e_k^{(i)}$ without computing the intermediate vectors.

The RPA and the CRPA have applications in recursive interpolation and extrapolation for vector and matrix sequences, and in the solution of systems of linear equations. They also have connections with Fourier series, the Gram–Schmidt orthonormalization process and Rosen's method for nonlinear programming. See [74, 239, 307, 308, 309, 310, 311].

Let us now turn to the solution of the system of linear equations $Ax = b$. Many projection methods for their solution use, explicitly or implicitly, Schur complements.

Lanczos method for solving a system consists in constructing a sequence of vectors (x_k) defined by the two conditions

$$\begin{aligned} & x_k - x_0 \in \mathcal{K}_k(A, r_0) = \text{Span}(r_0, Ar_0, \ldots, A^{k-1}r_0), \\ & r_k = b - Ax_k \perp \mathcal{K}_k(A^T, y) = \text{Span}(y, A^T y, \ldots, (A^T)^{k-1}y), \end{aligned}$$

where x_0 is arbitrarily chosen, $r_0 = b - Ax_0$, and $y \neq 0$ is a given vector. A subspace of the form $\mathcal{K}_k$ is called a *Krylov subspace* and Lanczos method belongs to the class of *Krylov subspace methods*, a particular case of projection methods [76].

The vectors x_k are completely determined by the two preceding conditions. Indeed, the first condition writes

$$x_k - x_0 = -a_{k-1}r_0 - a_{k-2}Ar_0 - \cdots - a_0 A^{k-1} r_0.$$

Multiplying both sides by A, adding and subtracting b, leads to

$$r_k = r_0 + a_{k-1}Ar_0 + \cdots + a_0 A^k r_0 = P_k(A)r_0$$

with $P_k(\xi) = 1 + a_{k-1}\xi + \cdots + a_0\xi^k$. The second condition gives

$$(r_k, (A^T)^i y) = (P_k(A)r_0, (A^T)^i y) = (A^i P_k(A)r_0, y) = 0, \quad i = 0, \ldots, k-1,$$

that is

$$(A^i r_0 + a_{k-1}A^{i+1}r_0 + \cdots + a_0 A^{i+k} r_0, y) = 0, \quad i = 0, \ldots, k-1.$$

Let c be the linear functional defined by

$$c(\xi^i) = c_i = (A^i r_0, y).$$

Thus, since c is linear, $c(p) = (y, p(A)r_0)$ for any polynomial p, and the preceding conditions are equivalent to

$$c(\xi^i P_k(\xi)) = 0, \quad i = 0, \ldots, k-1,$$

that is

$$c_i + a_{k-1}c_{i+1} + \cdots + a_0 c_{i+k} = 0, \quad i = 0, \ldots, k-1.$$

So, if it exists, P_k is the formal orthogonal polynomial of degree at most k with respect to c, normalized by the condition $P_k(0) = 1$ and it follows

$$r_k = \det\begin{pmatrix} r_0 & Ar_0 & \cdots & A^k r_0 \\ c_0 & c_1 & \cdots & c_k \\ \vdots & \vdots & & \vdots \\ c_{k-1} & c_k & \cdots & c_{2k-1} \end{pmatrix} \Big/ \det\begin{pmatrix} c_1 & \cdots & c_k \\ \vdots & & \vdots \\ c_k & \cdots & c_{2k-1} \end{pmatrix},$$

where the determinant in the numerator is the linear combination of the vectors in its first row, obtained by the usual rules for expanding a determinant.

By Schur's formula (1.1.4), we see that r_k is also a Schur complement

$$r_k = r_0 - A(r_0, Ar_0, \ldots, A^{k-1}r_0)\begin{pmatrix} c_1 & \cdots & c_k \\ \vdots & & \vdots \\ c_k & \cdots & c_{2k-1} \end{pmatrix}^{-1}\begin{pmatrix} c_0 \\ \vdots \\ c_{k-1} \end{pmatrix}.$$

The vectors x_k can also be expressed as Schur complements.

From the practical point of view, the residuals r_k and the corresponding iterates x_k are not obtained from these determinantal formulae nor from the system satisfied by the coefficients of the polynomials P_k. Since these polynomials are formal orthogonal polynomials, they can be recursively computed (and then the $r_k = P_k(A)r_0$ and the x_k) by using the recurrence relations given in Section 7.2 for adjacent families of formal orthogonal polynomials. According to the relations used, several algorithms for the implementation of Lanczos method are obtained. Among them is the *conjugate gradient algorithm* when A is symmetric and positive definite, and the *biconjugate gradient algorithm* in the general case. See [312, 388, 445].

Since the various recursive algorithms for the implementation of Lanczos method are based on formal orthogonal polynomials, breakdowns and near-breakdowns can occur as explained above. They can be treated by jumping over these flaws and using more complicated recurrences which can be derived from the block bordering method described in Section 7.5. A similar look–ahead strategy can be used for other Lanczos–based methods such as the CGS and the BiCGStab. On these topics, see [90, 91] where additional references could be found.

Other projection methods fit into a similar framework.

Let us now discuss *block projection methods* for the solution of systems of equations with several right hand sides. We first need an extension of the notion of determinant, where the first row is formed by $n \times s$ rectangular matrices and the other rows by $s \times s$ square matrices. Such determinants are $n \times s$ matrices defined as follows. We consider the matrix

$$M = \begin{pmatrix} A & B_1 & \cdots & B_k \\ C_1 & D_{1,1} & \cdots & D_{1,k} \\ \vdots & \vdots & & \vdots \\ C_k & D_{k,1} & \cdots & D_{k,k} \end{pmatrix}$$

where $A, B_1, \ldots, B_k$ are $n \times s$ matrices, and the C_i and the $D_{i,j}$ are $s \times s$ matrices. Let D be the submatrix which consists of all the blocks $D_{i,j}$. It is a $ks \times ks$ matrix and its determinant is the usual one. The Schur complement M/D is an $n \times s$ matrix. Let us now give a meaning to the determinant of the rectangular matrix M, denoted by $\operatorname{Det} M$ (notice the capital D and Det $\neq$ det). It is the $n \times s$ matrix whose entries are

$$(\operatorname{Det} M)_{ij} = \det \begin{pmatrix} A_{ij} & (B_1)_i & \cdots & (B_k)_i \\ (C_1)^j & D_{1,1} & \cdots & D_{1,k} \\ \vdots & \vdots & & \vdots \\ (C_k)^j & D_{k,1} & \cdots & D_{k,k} \end{pmatrix}$$

where $(B_j)_i$ denotes the ith row of B_j, $(C_i)^j$ the jth column of C_i, and A_{ij} is the corresponding term of the matrix A. With this definition for the determinant of a rectangular matrix, the Schur determinantal formula justifies the extension formula

$$M/D = A - BD^{-1}C = \operatorname{Det} M/\det D.$$

Finally, let us consider the matrix

$$K = \begin{pmatrix} A & B & E \\ C & D & F \\ G & H & L \end{pmatrix}.$$

We set

$$B' = \begin{pmatrix} B & E \\ D & F \end{pmatrix}, \quad C' = \begin{pmatrix} C & D \\ G & H \end{pmatrix}, \quad D' = \begin{pmatrix} D & F \\ H & L \end{pmatrix}.$$

We assume that D and D' are square and nonsingular. The following quotient property holds [131]

$$\begin{aligned} K/D' &= ((K/D)/(D'/D)) \\ &= (M/D) - (B'/D)(D'/D)^{-1}(C'/D). \end{aligned} \tag{7.6.8}$$

If A, E, G and L are numbers, B and H row vectors, and C and F column vectors, then the Schur complements involved in (7.6.8) are numbers (obviously equal to ratios of determinants), and the quotient property reduces to Sylvester's determinantal identity. Thus (7.6.8) appears as a generalization of Sylvester's identity.

We consider a system of n linear equations with s right hand sides $AX = B$ where $A \in \mathbb{R}^{n \times n}$ and $B, X \in \mathbb{R}^{n \times s}$. A block projection method for solving this system consists of generating, from an arbitrary $n \times s$ matrix X_0, a sequence (X_k) of $n \times s$ matrices defined by the two conditions

$$X_k - X_0 \in \mathcal{K}_k, \tag{7.6.9}$$

$$R_k = B - AX_k \perp \mathcal{L}_k \tag{7.6.10}$$

where $\mathcal{K}_k$ and $\mathcal{L}_k$ are subspaces of dimension k of $n \times s$ matrices.

Let $v_0, \ldots, v_{k-1}$ ($w_0, \ldots, w_{k-1}$, resp.) be matrices of dimension $n \times s$ forming a basis of $\mathcal{K}_k$ ($\mathcal{L}_k$ resp.). Any matrix $v \in \mathcal{K}_k$ can be written as

$$v = \sum_{i=0}^{k-1} v_i \gamma_i$$

where the γ_i are $s \times s$ matrices. The definition of a matrix in $\mathcal{L}_k$ is similar. So, each column of a matrix in $\mathcal{K}_k$ ($\mathcal{L}_k$ resp.) is a linear combination of the columns of all the matrices $v_0, \ldots, v_{k-1}$ ($w_0, \ldots, w_{k-1}$, resp.). We consider the $n \times ks$ matrices

$$V_k = [v_0, \ldots, v_{k-1}], \quad W_k = [w_0, \ldots, w_{k-1}].$$

So, the conditions (7.6.9) and (7.6.10) give

$$R_k = (I - P_k)R_0 \tag{7.6.11}$$

with

$$P_k = AV_k(W_k^* AV_k)^{-1}W_k^*. \tag{7.6.12}$$

The matrix P_k represents the oblique projection on $A\mathcal{K}_k$ along $\mathcal{L}_k^\perp$, that is orthogonal to $\mathcal{L}_k$, and $I - P_k$ represents the projection on $\mathcal{L}_k^\perp$ along $A\mathcal{K}_k$.

Formula (7.6.11) shows that R_k is the Schur complement of $W_k^* AV_k$ in the matrix

$$M_k = \begin{pmatrix} R_0 & AV_k \\ W_k^* R_0 & W_k^* AV_k \end{pmatrix}.$$

Similarly, X_k is the Schur complement of $W_k^* AV_k$ in

$$\begin{pmatrix} X_0 & -V_k \\ W_k^* R_0 & W_k^* AV_k \end{pmatrix}.$$

From the definition of the determinant of a rectangular matrix given above, we have

$$X_k = \frac{\mathrm{Det}\begin{pmatrix} X_0 & -V_k \\ W_k^* R_0 & W_k^* AV_k \end{pmatrix}}{\det(W_k^* AV_k)}, \quad R_k = \frac{\mathrm{Det}\begin{pmatrix} R_0 & AV_k \\ W_k^* R_0 & W_k^* AV_k \end{pmatrix}}{\det(W_k^* AV_k)}.$$

Let us now express P_{k+1} in term of P_k. For simplicity, we set $A_k = W_k^* AV_k$. We have

$$A_{k+1} = \begin{pmatrix} W_k^* \\ w_k^* \end{pmatrix} A[V_k, v_k] = \begin{pmatrix} A_k & W_k^* Av_k \\ w_k^* AV_k & w_k^* Av_k \end{pmatrix}. \tag{7.6.13}$$

From the block bordering method, we have

$$R_{k+1} = R_k - A(I - A^{-1}P_k A)v_k S_k^{-1} w_k^* R_k$$

with $S_k = w_k^*(I - P_k)Av_k = w_k^* A(I - A^{-1}P_k A)v_k$. Since R_k can be expressed as the Schur complement $R_k = M_k/A_k$, it can be recursively computed by the matrix recursive projection algorithm (MRPA) [307], an extension of the RPA discussed above. We have

$$M_{k+1} = \begin{pmatrix} R_0 & AV_k & Av_k \\ W_k^* R_0 & W_k^* AV_k & W_k^* Av_k \\ w_k^* R_0 & w_k^* AV_k & w_k^* Av_k \end{pmatrix}.$$

We set

$$G_k = \begin{pmatrix} AV_k & Av_k \\ W_k^* AV_k & W_k^* Av_k \end{pmatrix}, \quad H_k = \begin{pmatrix} W_k^* R_0 & W_k^* AV_k \\ w_k^* R_0 & w_k^* AV_k \end{pmatrix}.$$

It follows from the expression (7.6.13) of A_{k+1} and from the quotient property (7.6.8) for the Schur complements

$$\begin{aligned} R_{k+1} &= M_{k+1}/A_{k+1} = ((M_{k+1}/A_k)/(A_{k+1}/A_k)) \\ &= (M_k/A_k) - (G_k/A_k)(A_{k+1}/A_k)^{-1}(H_k/A_k). \end{aligned}$$

If we set

$$G'_{k,i} = \begin{pmatrix} Av_i & AV_k \\ W_k^* Av_i & W_k^* AV_k \end{pmatrix}, \quad H'_k = \begin{pmatrix} w_k^* R_0 & w_k^* AV_k \\ W_k^* R_0 & W_k^* AV_k \end{pmatrix}$$

and $G_{k,i} = G'_{k,i}/A_k$, then $H'_k/A_k = w_k^* R_k$ and $A_{k+1}/A_k = w_k^* G_{k,k}$ and the preceding expression for R_{k+1} becomes

$$R_{k+1} = R_k - G_{k,k}(w_k^* G_{k,k})^{-1} w_k^* R_k.$$

Since $G_{k,i}$ has the same form as R_k after replacing R_0 by Av_i, a similar recurrence relationship holds for the auxiliary $n \times s$ matrices $G_{k+1,i}$, that is

$$G_{k+1,i} = G_{k,i} - G_{k,k}(w_k^* G_{k,k})^{-1} w_k^* G_{k,i}, \quad i = k+1, k+2, \ldots$$

with $G_{0,i} = Av_i$.

According to the choice of the subspaces $\mathcal{K}_k$ and $\mathcal{L}_k$ several particular block methods are recovered such as Lanczos, FOM and GMRES, see [87].

Block descent methods involving Schur complements can also be defined. They are described in [78] and are all related to the Schur complement. In such methods, the iterates are given by

$$X_{k+1} = X_k + Z_k \Lambda_k, \quad k = 0, 1, \ldots$$

where $Z_k \in \mathbb{R}^{n \times s}$ and where the $s \times s$ matrix Λ_k is chosen to minimize some expression.

In the block Richardson method, Λ_k is defined by the condition $R_{k+1} = B - AX_{k+1} = R_k - AZ_k\Lambda_k = 0$ in the least squares sense, which gives

$$\Lambda_k = [(AZ_k)^* AZ_k]^{-1}(AZ_k)^* R_k.$$

The matrix Λ_k also satisfies the orthogonality condition $(AZ_k)^* R_{k+1} = 0$. It follows

$$R_{k+1} = (I - P_k)R_k \quad \text{with} \quad P_k = AZ_k[(AZ_k)^* AZ_k]^{-1}(AZ_k)^*.$$

We have $P_k = P_k^2$ and $P_k = P_k^*$ which shows that P_k and $I - P_k$ are orthogonal projectors. Moreover

$$R_{k+1}^* R_{k+1} = R_k^*(I - P_k)R_k.$$

In the block Barzilai–Borwein method, Λ_k is taken so that $\Delta X_{k-1}\Lambda_k^{-1} + \Delta Z_{k-1} = 0$ in the least squares sense, that is

$$\Lambda_k = -(\Delta X_{k-1}^* \Delta Z_{k-1})^{-1} \Delta X_{k-1}^* \Delta X_{k-1}.$$

We have

$$\Delta X_{k-1}\Lambda_k^{-1} + \Delta Z_{k-1} = (I - P_k)\Delta Z_{k-1},$$

with $P_k = \Delta X_{k-1}(\Delta X_{k-1}^* \Delta X_{k-1})^{-1} \Delta X_{k-1}^*$. Thus P_k and $I - P_k$ represent orthogonal projections and we have

$$(\Delta X_{k-1}\Lambda_k^{-1} + \Delta Z_{k-1})^*(\Delta X_{k-1}\Lambda_k^{-1} + \Delta Z_{k-1}) = \Delta Z_{k-1}^*(I - P_k)\Delta Z_{k-1}.$$

Moreover $\Delta X_{k-1}^*(\Delta X_{k-1}\Lambda_k^{-1} + \Delta Z_{k-1}) = 0$. A second method of the same kind can be obtained by choosing Λ_k so that $\Delta X_{k-1} + \Delta Z_{k-1}\Lambda_k = 0$ in the least squares sense. We obtain

$$\Lambda_k = -(\Delta Z_{k-1}^* \Delta Z_{k-1})^{-1} \Delta Z_{k-1}^* \Delta X_{k-1}.$$

We have

$$\Delta X_{k-1} + \Delta Z_{k-1}\Lambda_k = (I - P_k)\Delta X_{k-1},$$

with $P_k = \Delta Z_{k-1}(\Delta Z_{k-1}^* \Delta Z_{k-1})^{-1} \Delta Z_{k-1}^*$. Thus P_k and $I - P_k$ are orthogonal projectors. Moreover

$$(\Delta X_{k-1} + \Delta Z_{k-1}\Lambda_k)^*(\Delta X_{k-1} + \Delta Z_{k-1}\Lambda_k) = \Delta X_{k-1}^*(I - P_k)\Delta X_{k-1},$$

and $\Delta Z_{k-1}^*(\Delta X_{k-1} + \Delta Z_{k-1}\Lambda_k) = 0$.

Other block descent methods are described in [78].

7.7 Preconditioners

Let us first discuss the block Gaussian elimination for solving a system of linear equations. We consider the system

$$\begin{pmatrix} A_{11} & A_{12} & \cdots & A_{1m} \\ \vdots & \vdots & & \vdots \\ A_{m1} & A_{m2} & \cdots & A_{mm} \end{pmatrix} \begin{pmatrix} x_1 \\ \vdots \\ x_m \end{pmatrix} = \begin{pmatrix} b_1 \\ \vdots \\ b_m \end{pmatrix}$$

where $A_{ij} \in \mathbb{R}^{m_i \times m_j}$, $x_i \in \mathbb{R}^{m_i}$ and $b_i \in \mathbb{R}^{m_i}$. The idea of block Gaussian elimination is first to eliminate x_1 from equations 2 to m. The first block equation is multiplied on the right by A_{11}^{-1}, then by A_{i1}, and it is subtracted from the i-th equation. The, x_2 is eliminated in the same way from equations 3 to m, and so on. We thus obtain a succession of systems $A^{(k)}x = b^{(k)}$, $k = 1, \ldots, m$, with $A^{(1)} = A$ and $b^{(1)} = b$, where

$$A^{(k)} = \begin{pmatrix} A_{11}^{(1)} & A_{12}^{(1)} & \cdots & A_{1k}^{(1)} & \cdots & A_{1m}^{(1)} \\ & A_{22}^{(2)} & \cdots & A_{2k}^{(2)} & \cdots & A_{2m}^{(2)} \\ & & \ddots & \vdots & & \vdots \\ & & & A_{kk}^{(k)} & \cdots & A_{km}^{(k)} \\ & & & A_{k+1,k}^{(k)} & \cdots & A_{k+1,m}^{(k)} \\ & & & \vdots & & \vdots \\ & & & A_{m,k}^{(k)} & \cdots & A_{m,m}^{(k)} \end{pmatrix}, \quad b^{(k)} = \begin{pmatrix} b_1^{(1)} \\ b_2^{(2)} \\ \vdots \\ b_k^{(k)} \\ b_{k+1}^{(k)} \\ \vdots \\ b_m^{(k)} \end{pmatrix}.$$

The blocks $A_{ij}^{(k)}$ are given by

$$\begin{aligned} A_{ij}^{(k+1)} &= A_{ij}^{(k)} - A_{ik}^{(k)}[A_{kk}^{(k)}]^{-1}A_{kj}^{(k)}, \quad i,j = k+1, \ldots, m \\ b_i^{(k+1)} &= b_i^{(k)} - A_{ik}^{(k)}[A_{kk}^{(k)}]^{-1}b_k^{(k)}, \quad i = k+1, \ldots, m, \end{aligned}$$

the other blocks remaining unchanged. The last system $A^{(m)}x = b^{(m)}$ is block upper triangular.

It is easy to see that Schur complements are involved in this elimination procedure. It corresponds to a block factorization of A as $A = LU$, where L is a block lower triangular matrix with identity matrices on its diagonal and U is a block upper triangular matrix. Let us mention that the block factorization and the Schur complement play a role in multifrontal methods for the solution of systems with sparse matrices on high–performance machines [150] and in fast algorithms for the treatment of dense structured matrices, such as Hankel and Toeplitz matrices which are involved in Padé and other approximation problems [59, 347].

The condition number of a matrix A is the number $\kappa(A) = \|A\| \cdot \|A^{-1}\| \geq 1$. If $\kappa(A)$ is large with respect to 1, a small perturbation in the coefficients of A or in the components of b can induce a large perturbation in the exact solution of the system $Ax = b$. In that case, the system is ill–conditioned. Moreover, the convergence rate of some iterative methods for its solution (such as the conjugate gradient algorithm) is slower for large values of $\kappa(A)$ and the propagation of rounding errors is more important. When a system is ill–conditioned, it can be replaced by another one with a smaller condition number. Since the condition number of the identity matrix is 1, one can consider the system $M^{-1}Ax = M^{-1}b$ where M is an approximation of A whose inverse is easy to compute (or, equivalently, a system with M as its matrix is easy to solve). This is called left preconditioning and the matrix M is called a *preconditioner.* We can also consider a right preconditioning of the form $AM'y = b$ with $x = M'y$, or a double one $M^{-1}AM'y = M^{-1}b$ with $x = M'y$. There is no universal preconditioner for all matrices and each case is a particular one, but preconditioning a system is, at least, as important as the numerical method used for the computation of its solution. In practice, one can construct approximations of A easy to invert or, directly, approximations of A^{-1}. Preconditioners can be obtained by incomplete block LU decomposition.

In many applications, mostly coming out from the discretization of partial differential equations (see the next Section), one has to deal with block tridiagonal matrices of the form

$$A = \begin{pmatrix} A_1 & B_1 & & & \\ C_2 & A_2 & B_2 & & \\ & \ddots & \ddots & \ddots & \\ & & C_{n-1} & A_{n-1} & B_{n-1} \\ & & & C_n & A_n \end{pmatrix},$$

where all blocks are $m \times m$. This matrix can be decomposed as

$$A = \begin{pmatrix} D_1 & & & \\ C_2 & D_2 & & \\ & \ddots & \ddots & \\ & & C_n & D_n \end{pmatrix} D^{-1} \begin{pmatrix} D_1 & B_1 & & \\ & \ddots & \ddots & \\ & & D_{n-1} & B_{n-1} \\ & & & D_n \end{pmatrix},$$

where D is the block diagonal matrix formed by the blocks $D_1, \ldots, D_m$ given by

$$\begin{aligned} D_1 &= A_1 \\ D_i &= A_i - C_i D_{i-1}^{-1} B_{i-1}, \quad i = 2, \ldots, m. \end{aligned}$$

Thus the matrices D_i can be interpreted as Schur complements. This decomposition can be written as $A = (L+D)D^{-1}(L+U)$ where L and U are the strictly lower and upper parts of A respectively. Once this factorization has been obtained, the solution of the system $Ax = b$ can be obtained by two successive block triangular solves.

For finite difference approximations of elliptic or parabolic partial differential equations, the matrices B_i and C_i are diagonal and the matrices A_i are tridiagonal. However, the matrices D_i can be dense. The idea for obtaining a preconditioner is to replace, in the preceding formulae, the matrices D_i by sparse approximations Δ_i, thus leading to a matrix M which approximates A and is easy to invert. The matrices Δ_i are computed by

$$\begin{aligned} \Delta_1 &= A_1 \\ \Delta_i &= A_i - C_i \; \textit{approx}\, (\Delta_{i-1}^{-1})\; B_{i-1}, \quad i = 2, \ldots, m, \end{aligned}$$

where *approx* (Δ_{i-1}^{-1}) is a sparse approximation of Δ_{i-1}^{-1}. There are many ways of defining these sparse approximations, one of the most efficient is to consider tridiagonal approximations of the inverses. Then we take $M = (L + \Delta)\Delta^{-1}(U + \Delta)$ where Δ is the block diagonal matrix with blocks $\Delta_1, \ldots, \Delta_m$. On this topic, see [312].

7.8 Domain decomposition methods

The idea of *domain decomposition methods* comes out from the solution of large systems of linear equations issued from the discretization of partial differential equations. They consist in decomposing the domain of integration into smaller subdomains, thus replacing the system by smaller ones, then to solve them separately (possibly in parallel), and finally to paste the results together to obtain the global solution. Since these methods are closely related to the geometry of the domain, it is not always possible to present them on a purely algebraic basis. There are two types of methods according whether the subdomains overlap or not.

We will only give the main ideas of domain decomposition methods in the case of two subdomains which do not overlap without entering into the details. For a general presentation of domain decomposition methods and for more details, see [312, 365, 388].

We denote by x_1 and x_2 the vectors of unknowns in each subdomain and by x_{12} the vector of unknowns on the interface between them. Ordering the unknowns in a certain quite technical way, the linear system $Ax = b$

can be written as

$$\begin{pmatrix} A_1 & 0 & E_1 \\ 0 & A_2 & E_2 \\ E_1^T & E_2^T & A_{12} \end{pmatrix} \begin{pmatrix} x_1 \\ x_2 \\ x_{12} \end{pmatrix} = \begin{pmatrix} b_1 \\ b_2 \\ b_{12} \end{pmatrix}. \tag{7.8.14}$$

The matrices A_1 and A_2 are block tridiagonal and they represent the coupling of the unknowns in the first and the second subdomain respectively. A_{12} is tridiagonal and it is related to the unknowns on the interface. E_1 and E_2 are sparse matrices representing the coupling of the unknowns of each subdomain with those of the interface.

The system (7.8.14) is solved by an iterative method which makes use of the solution in each subdomain. The subsystems with the matrices A_1 and A_2 can either be solved exactly or approximately, thus leading to two different procedures. We will discuss both of them.

If exact solvers for the subdomains are used, we obtain the reduced system $Sx_{12} = \widetilde{b}_{12}$ for the unknowns on the interface where $\widetilde{b}_{12} = b_{12} - E_1^T A_1^{-1} b_1 - E_2^T A_2^{-1} b_2$ and

$$S = A_{12} - E_1^T A_1^{-1} E_1 - E_2^T A_2^{-1} E_2.$$

The matrix S is the Schur complement of A_{12} in A. In practice, since the inverses of A_1 and A_2 are dense matrices, it is too costly to construct and to decompose S, and the reduced system with the matrix S on the interface is solved by an iterative method. In order to choose the most appropriate iterative method for this purpose, it is necessary to know the properties of this Schur complement. In many iterative methods, such as the conjugate gradient algorithm, it is not necessary to know explicitly the matrix S but only to be able to compute the product of S by a vector. However, a good preconditioner is usually needed and, thus, one has to find an approximation of the Schur complement S.

Let us now consider the case where the systems corresponding to the subdomains are not solved exactly. The system (7.8.14) is solved approximately by an iterative method. A preconditioner (that is an approximation of the matrix A) is needed. It has the form

$$\begin{pmatrix} M_1 & 0 & E_1 \\ 0 & M_2 & E_2 \\ E_1^T & E_2^T & M_{12} \end{pmatrix}$$

where M_1, M_2, and M_{12} approximate A_1, A_2, and A_{12} respectively. Since these matrices are block tridiagonal, the procedure described in the preceding Section is appropriate.

7.9 Triangular recursion schemes

In this Chapter, we saw several examples of a triangular recursion scheme. In numerical analysis, such schemes are not uncommon. They arise, for example, in interpolation (the Neville-Aitken scheme), divided differences, Romberg method, best approximation, Bernstein polynomials, and splines. We will now discuss them in their full generality, beginning by the theory and, then, giving some examples. Let us mention that a theory of extrapolation functionals leading to some of the results given below was propose by Schneider [400]. For more details, see [94, 450].

We first consider the sequence transformation

$$T_k^{(n)} = \sum_{i=n}^{n+k} a_k^{(n,i)} S_i, \quad k, n = 0, 1, \ldots \tag{7.9.15}$$

where the S_i are given numbers, vectors or matrices and the $a_k^{(n,i)}$ are numbers. On the other hand, we consider the *triangular recursion scheme*

$$T_k^{(n)} = \lambda_k^{(n)} T_{k-1}^{(n)} + \mu_k^{(n)} T_{k-1}^{(n+1)}, \quad k = 1, 2, \ldots ; n = 0, 1, \ldots \tag{7.9.16}$$

with $T_0^{(n)} = S_n$.

A triangular recursion scheme (7.9.16) can always be interpreted as a succession of sequence transformations of the form (7.9.15) where the coefficients $a_k^{(n,i)}$ are given by

$$\begin{array}{lcll} a_k^{(n,n)} & = & \lambda_k^{(n)} a_{k-1}^{(n,n)} & \\ a_k^{(n,i)} & = & \lambda_k^{(n)} a_{k-1}^{(n,i)} + \mu_k^{(n)} a_{k-1}^{(n+1,i)}, & i = n+1, \ldots, n+k-1 \\ a_k^{(n,n+k)} & = & \mu_k^{(n)} a_{k-1}^{(n+1,n+k)} & \end{array}$$

with $a_0^{(n,n)} = 1$. If, for all values of k and n, $\lambda_k^{(n)} + \mu_k^{(n)} = \gamma_k$ then

$$\sum_{i=n}^{n+k} a_k^{(n,i)} = \begin{cases} \prod_{j=1}^{k} \gamma_j, & k \geq 1 \\ 1, & k = 0 \end{cases}$$

Conversely, if the sum of the coefficients $a_k^{(n,i)}$ for $i = n$ to $n+k$ is independent of n, so is the sum $\lambda_k^{(n)} + \mu_k^{(n)}$.

Now, let $Z = \{z_0, z_1, \ldots\}$ where the z_i are distinct real points and let Σ be the set of functions σ defined on a certain subset of Z. The linear functional on Σ

$$T_k^{(n)}(\sigma) = \sum_{i=n}^{n+k} a_k^{(n,i)} \sigma(z_i)$$

is called the *reference functional* of the transformation (7.9.15) (or of the recursion (7.9.16)). Setting $S_i = \sigma(z_i)$, we obtain $\mathcal{T}_k^{(n)}(\sigma) = T_k^{(n)}$.

We will make use of the condensed notation

$$\det \begin{pmatrix} f_0 & f_1 & \cdots & f_k \\ z_0 & z_1 & \cdots & z_k \end{pmatrix} = \det \begin{pmatrix} f_0(z_0) & \cdots & f_0(z_k) \\ \vdots & & \vdots \\ f_k(z_0) & \cdots & f_k(z_k) \end{pmatrix}. \quad (7.9.17)$$

Assume that there exists a $(k+1)$ dimensional subspace $\Sigma_k = \mathrm{Span}(\sigma_0, \ldots, \sigma_k)$ of Σ such that

$$\det \begin{pmatrix} \sigma_0 & \sigma_1 & \cdots & \sigma_k \\ z_n & z_{n+1} & \cdots & z_{n+k} \end{pmatrix} \neq 0$$

and that

$$\mathcal{T}_k^{(n)}(\sigma_j) = \begin{cases} \omega_k^{(n)}, & j = 0 \\ 0, & j = 1, \ldots, k \end{cases} \quad (7.9.18)$$

where all numbers $\omega_k^{(n)}$ are different from zero, then every linear functional $\mathcal{L}$ of the form

$$\mathcal{L}(\sigma) = \sum_{i=n}^{n+k} b_k^{(n,i)} \sigma(z_i)$$

which satisfies (7.9.18) coincides with $\mathcal{T}_k^{(n)}$. Moreover the transformation (7.9.15) has the representation

$$\mathcal{T}_k^{(n)}(\sigma) = \frac{\det \begin{pmatrix} \sigma & \sigma_1 & \cdots & \sigma_k \\ z_n & z_{n+1} & \cdots & z_{n+k} \end{pmatrix}}{\det \begin{pmatrix} \sigma_0 & \sigma_1 & \cdots & \sigma_k \\ z_n & z_{n+1} & \cdots & z_{n+k} \end{pmatrix}} \omega_k^{(n)}. \quad (7.9.19)$$

Dividing the numerator and the denominator by

$$\det \begin{pmatrix} \sigma_1 & \cdots & \sigma_k \\ z_{n+1} & \cdots & z_{n+k} \end{pmatrix},$$

and assuming that these determinants are different from zero for all k and n, shows that $\mathcal{T}_k^{(n)}(\sigma)/\omega_k^{(n)}$ can be expressed as a ratio of Schur complements. Moreover, using the quotient property (7.0.1) or, equivalently, Sylvester identity (7.0.2), we obtain the recursive scheme, for $k = 1, 2, \ldots$ and $n = 0, 1, \ldots$,

$$\mathcal{T}_k^{(n)}(\sigma) = \frac{\mathcal{T}_{k-1}^{(n+1)}(\sigma_k)\mathcal{T}_{k-1}^{(n)}(\sigma) - \mathcal{T}_{k-1}^{(n)}(\sigma_k)\mathcal{T}_{k-1}^{(n+1)}(\sigma)}{\omega_{k-1}^{(n)}\mathcal{T}_{k-1}^{(n+1)}(\sigma_k) - \omega_{k-1}^{(n+1)}\mathcal{T}_{k-1}^{(n)}(\sigma_k)} \omega_k^{(n)}, \quad (7.9.20)$$

with $\mathcal{T}_0^{(n)} = \sigma(z_n)$. This scheme has the form (7.9.16) and (7.9.20) gives the expressions of $\lambda_k^{(n)}$ and $\mu_k^{(n)}$. Moreover, a necessary and sufficient condition that, for all n, $T_k^{(n)}/\omega_k^{(n)} = S$ is that, for all n, $S_n = S\sigma_0(z_n) + a_1\sigma_1(z_n) + \cdots + a_k\sigma_k(z_n)$. It can be shown that $\mathcal{T}_k^{(n)}(\sigma)$ can also be represented as a complex contour integral. The scheme (7.9.20) is a generalization of the E–algorithm which is recovered by setting $E_k^{(n)} = \mathcal{T}_k^{(n)}(\sigma)/\mathcal{T}_k^{(n)}(\sigma_0)$ and $g_{k,i}^{(n)} = \mathcal{T}_k^{(n)}(\sigma_i)/\mathcal{T}_k^{(n)}(\sigma_0)$. As in the E–algorithm $\mathcal{T}_{k+m}^{(n)}$ can be expressed in terms of $\mathcal{T}_k^{(n)}, \ldots, \mathcal{T}_k^{n+m)}$ by a formula similar to (7.4.6). A formula of the Nuttall's type as (7.4.7) also holds.

Of course, for a practical application, the main problem is to find the functions σ_i. They are determined by the conditions

$$\mathcal{T}_k^{(n)}(\sigma_0) = \omega_k^{(n)} \quad \text{and} \quad \mathcal{T}_k^{(n)}(\sigma_i) = 0, \quad i = 1, \ldots, k.$$

These conditions lead to a (usually) nonlinear difference equations of order k with nonconstant coefficients which is difficult to solve. However, since the functions σ_i are independent of k, only one new function has to be determined at each step of the algorithm.

Let us now discuss some applications of this theory. As seen above, formal orthogonal polynomials enter into this framework and thus Padé approximation, extrapolation methods, and Krylov subspace methods. In particular, many extrapolation algorithms for accelerating scalar and vector sequences have the form of (7.9.20); see [459, 458]. Some of these algorithms are related to continued fractions.

In what follows, some usual notations have been changed to be consistent with the preceding ones.

The first application concerns interpolation. Let $g_0, g_1, \ldots$ be a set of functions such that, for any set of distinct points Z and any k, the determinants (7.9.17) are different from zero. We consider the interpolation problem which consists in constructing $P_k^{(n)}(z) = a_0g_0(z) + \cdots + a_kg_k(z)$ satisfying $P_k^{(n)}(z_i) = f(z_i)$ for $i = n, \ldots, n+k$. Obviously, such a function $P_k^{(n)}$ can be written as a ratio of determinants and as a ratio of Schur complements. As in the case of the E–algorithm, applying the quotient property (7.0.1), or Sylvester identity (7.0.2), to the numerators and the denominators leads to the following recursive rule, due to Mühlbach [326]

$$
\begin{aligned}
P_k^{(n)}(z) &= P_{k-1}^{(n)}(z) - \frac{P_{k-1}^{(n+1)}(z) - P_{k-1}^{(n)}(z)}{g_{k-1,k}^{(n+1)}(z) - g_{k-1,k}^{(n)}(z)}\, g_{k-1,k}^{(n)}(z) \\
g_{k,i}^{(n)}(z) &= g_{k-1,i}^{(n)}(z) - \frac{g_{k-1,i}^{(n+1)}(z) - g_{k-1,i}^{(n)}(z)}{g_{k-1,k}^{(n+1)}(z) - g_{k-1,k}^{(n)}(z)}\, g_{k-1,i}^{(n)}(z), \quad i > k
\end{aligned}
$$

with

$$
P_0^{(n)}(z) = f(z_n)\frac{g_0(z)}{g_0(z_n)} \quad \text{and} \quad g_{0,i}^{(n)}(z) = g_i(z_n)\frac{g_0(z)}{g_0(x_n)} - g_i(z).
$$

Polynomial interpolation corresponds to the choice $g_i(z) = z^i$. In this case, the second rule provides $g_{k,i}^{(n)}(z) = z_n \cdots z_{n+k-1}$ and the first one reduces to the classical Neville–Aitken scheme. The general interpolation problem, as described in [143], also fits into this framework. It is connected to projection and biorthogonality [74].

Interpolation is related to divided differences. Let us only discuss such a connection in the polynomial case. Divided differences are recursively defined by the scheme

$$
D_k^{(n)} = \frac{D_{k-1}^{(n+1)} - D_{k-1}^{(n)}}{z_{n+k} - z_n}, \quad k = 1, 2, \ldots; n = 0, 1, \ldots
$$

with $D_0^{(n)} = f(z_n)$ for $n = 0, 1, \ldots$. This scheme has the form (7.9.16) and it corresponds to the reference functional defined by $\mathcal{T}_k^{(n)}(\sigma) = 1$ if $\sigma = z^k$ and 0 if σ is a polynomial of degree at most $k - 1$. So, the representation (7.9.19) holds with $\sigma_i = z^{i-1}$ for $i = 1, \ldots, k$ and $\sigma_0 = z^k$. The general interpolation problem can be treated similarly, see [74].

Our second application deals with B–splines. To simplify, we restrict ourselves to the equidistant case and take, without any loss of generality, $x_n = n$. B-splines are defined by the recursive scheme

$$
B_k^{(n)}(x) = (x-n)B_{k-1}^{(n)}(x) + (n+k-x)B_{k-1}^{(n+1)}(x), \quad k = 1, 2, \ldots; n = 0, 1, \ldots
$$

with

$$
B_0^{(n)}(x) = (x-n)_+^{-1} = \begin{cases} 0, & x \le n, \\ (x-n)^{-1}, & x > n. \end{cases}
$$

We have $\sigma_i(z) = (z - x)^{-i}$ and the corresponding reference functional satisfies

$$
\mathcal{T}_k^{(n)}(\sigma_i) = \begin{cases} k!, & i = 0, \\ 0, & i = 1, \ldots, k. \end{cases}
$$

Moreover, from the theory of triangular recursion schemes, we obtain

$$B_k^{(n)}(x) = \frac{\det\begin{pmatrix} \sigma & \sigma_1 & \cdots & \sigma_k \\ n & n+1 & \cdots & n+k \end{pmatrix}}{\det\begin{pmatrix} \sigma_0 & \sigma_1 & \cdots & \sigma_k \\ n & n+1 & \cdots & n+k \end{pmatrix}} k!$$

with $\sigma(z) = (z-x)_+^{-1}$.

Similar results hold for Chebyshevian B–splines [293].

Let us now consider the Bernstein polynomials defined by

$$P_k^{(n)}(x) = \frac{k!}{n!(k-n)!} x^n (1-x)^{k-n}$$

on $[0,1]$ and 0 elsewhere. We set $P_k^{(n)} \equiv 0$ for $n < 0$ or $n > k$. These polynomials satisfy the recurrence

$$P_k^{(n)}(x) = (1-x)P_{k-1}^{(n)}(x) + xP_{k-1}^{(n+1)}(x), \quad k = 1, 2, \ldots; n = 0, 1, \ldots$$

with $P_0^{(0)}(x) = 1$ and $P_0^{(n)}(x) = 0$ for $n > 0$. The coefficients $a_k^{(n,i)}$ of the reference functional are given in [94] where the case of Bernstein polynomials with noninteger exponents is also treated.

Other topics connected to splines enter into the framework of triangular recursion schemes. Let us mention the algorithms of de Casteljau and de Boor [163] and blossoming [303]. In particular, de Casteljau's algorithm can be interpreted as an extrapolation method [110]. See also [35].

Another application concerns the problem of best uniform approximation by functions from the Haar subspace $V_k = \mathrm{Span}(\sigma_0, \ldots, \sigma_{k-1}) \subset C[a,b]$. We set $M_{k,n} = \{z_n, \ldots, z_{n+k}\}$ where $a \leq z_0 < z_1 < \cdots \leq b$. We consider the linear functional $\mathcal{L}_k^{(n)}$ on $C[a,b]$ defined by

$$\mathcal{L}_k^{(n)}(\sigma) = \sum_{i=n}^{n+k} c_i \sigma(z_i),$$

where the coefficients c_i, which depend on n and k, are such that $c_n > 0$, $c_i \neq 0$ for $i = n, \ldots, n+k$, $\operatorname{sgn} c_i = (-1)^{i-n}$ and satisfy the conditions

$$\sum_{i=n}^{n+k} |c_i| = 1,$$

$$\sum_{i=n}^{n+k} c_i \sigma_j(z_i) = 0, \quad j = 0, \ldots, k-1.$$

The existence and uniqueness of such a functional is well known and it holds

$$|\mathcal{L}_k^{(n)}(\sigma)| = \inf_{f \in V_k} \max_{x \in M_{k,n}} |\sigma(x) - f(x)|.$$

By using Gaussian elimination for solving the system of linear equations

$$\sum_{i=0}^{k-1} a_i \sigma_i(z_j) + (-1)^j \lambda = \sigma(z_j), \quad j = 0, \ldots, k,$$

Meinardus and Taylor [305] proved that the functionals $\mathcal{L}_k^{(n)}$ can be recursively computed by the following scheme

$$\mathcal{L}_k^{(n)}(\sigma) = \frac{\mathcal{L}_{k-1}^{(n+1)}(\sigma_k)\mathcal{L}_{k-1}^{(n)}(\sigma) - \mathcal{L}_{k-1}^{(n)}(\sigma_k)\mathcal{L}_{k-1}^{(n+1)}(\sigma)}{\mathcal{L}_{k-1}^{(n+1)}(\sigma_k) - \mathcal{L}_{k-1}^{(n)}(\sigma_k)}$$

with $\mathcal{L}_0^{(n)}(\sigma) = \sigma(z_n)$, $n = 0, 1, \ldots$ This scheme is a particular case of (7.9.20). It is related to generalized divided differences and, thus, to Schur complements.

7.10 Linear control

Although it is not the main topic of this Chapter, we will say a few words about linear control theory and related subjects since Schur complements are often involved in its computational aspects. Consult [79].

A *system* is an interconnected set of devices which has to provide a desired function. It has *input* and *output* variables. If its behavior changes over time, we speak of a *dynamical system*. The input variables, called the *control*, influence the output variables which can be measured. Independently of the input and output variables, a system may have non–accessible internal variables, called *state variables*. To control a system consists in acting on the input variables so that the output variables possess a desired property. A system is governed by a system of ordinary or partial differential, functional, functional–differential or integral equations. In this Section, we will only be interested by the case where these equations are ordinary linear differential equations with constant coefficients

$$x'(t) = Ax(t) + Bu(t), \quad x(t_0) = x_0 \tag{7.10.21}$$

$$y(t) = Cx(t) \tag{7.10.22}$$

where $x \in \mathbb{R}^n, u \in \mathbb{R}^m, y \in \mathbb{R}^p$ and $A \in \mathbb{R}^{n \times n}, B \in \mathbb{R}^{n \times m}, C \in \mathbb{R}^{p \times n}$. Usually, in practice, $m < n$ and $p \leq n$. The integer n is called the *dimension* of the system and the variable t is the time.

The state–space approach consists of studying the differential equations (7.10.21–7.10.22) in the time domain. The control problem consists of acting on the input vector $u(t)$ so that the output vector $y(t)$ has a desirable time trajectory. Modifying the input $u(t)$ according to the output $y(t)$ which is observed or to the state vector $x(t)$ is called *feedback*.

The frequency domain approach consists of an equivalent representation of the state–space system. Assuming that $x(0) = 0$ (which does not restrict the generality since the differential system is autonomous) and taking the Laplace transform of (7.10.21) gives

$$s\widetilde{x}(s) = A\widetilde{x}(s) + B\widetilde{u}(s)$$

that is

$$\widetilde{x}(s) = (sI - A)^{-1}B\widetilde{u}(s).$$

Taking the Laplace transform of (7.10.22) and using the preceding expression of $\widetilde{x}$ yields

$$\begin{aligned} \widetilde{y}(s) = C\widetilde{x}(s) &= C(sI - A)^{-1}B\widetilde{u}(s) \\ &= G(s)\widetilde{u}(s) \end{aligned}$$

with

$$G(s) = C(sI - A)^{-1}B.$$

It holds

$$G(s) = C(sI - A)^{-1}B = \frac{1}{s}\sum_{i=0}^{\infty} C_i s^{-i}, \quad \text{with} \quad C_i = CA^iB \in \mathbb{R}^{p\times m}.$$

The matrix G, called the *transfer function matrix* of the system (7.10.21–7.10.22), has dimension $p \times m$ and it relates the Laplace transform of the output vector to that of the input vector in the frequency domain. Clearly, G can be interpreted as a Schur complement.

In linear control theory, many problems are related to the properties of the transfer function matrix and its approximation: realization, model reduction, stability analysis, poles and zeros, decoupling, and state estimation. Some of them involve the singular value decomposition (SVD) of a matrix which can also be interpreted as a Schur complement and Krylov subspace techniques as well.

Bibliography

[1] Abelin-Schur, Hilda. A story about my father. In Joseph, Melnikov & Rentschler [249, p. xli (2003)].

[2] Aitken, A. C. On least squares and linear combination of observations. *Proceedings of the Royal Society of Edinburgh*, 55 (1935), 42–48.

[3] Aitken, A. C. Studies in practical mathematics, I: The evaluation, with applications, of a certain triple product matrix. *Proceedings of the Royal Society of Edinburgh*, 57 (1937), 269–304.

[4] Aitken, A. C. *Determinants and Matrices.* University Mathematical Texts, Oliver & Boyd, Edinburgh, 1939. (2nd–9th editions, 1942–1956; 9th edition, reset & reprinted, 1967. Reprint edition: Greenwood Press, Westport, Connecticut, 1983.)

[5] Aitken, A. C.; Silverstone, H. On the estimation of statistical parameters. *Proceedings of the Royal Society of Edinburgh, Section A: Mathematical and Physical Sciences*, 61 (1942), 186–194.

[6] Albert, Arthur. Conditions for positive and nonnegative definiteness in terms of pseudoinverses. *SIAM Journal on Applied Mathematics*, 17 (1969), 434–440.

[7] Albert, Arthur. The Gauss–Markov theorem for regression models with possibly singular covariances. *SIAM Journal on Applied Mathematics*, 24 (1973), 182–187.

[8] Alpargu, Gülhan. *Schur Komplementleri İstatistik ve Nümerik Analizdeki Uygulamaları* (in Turkish: *Schur complements with Some Statistical and Numerical Analysis Applications*). M.Sc. thesis, Dept. of Mathematics, Çukurova University, Adana, Turkey, 1993.

[9] Anderson, T. W. *An Introduction to Multivariate Statistical Analysis*, 3rd edition. Wiley-Interscience, Hoboken, New Jersey, 2003. (Original version: Wiley, New York, 1958; 2nd edition: Wiley, New York, 1984.)

[10] Anderson, William N., Jr. Shorted operators. *SIAM Journal on Applied Mathematics*, 20 (1971), 520–525.

[11] Anderson, W. N., Jr.; Duffin, R. J. Series and parallel addition of matrices. *Journal of Mathematical Analysis and Applications*, 26 (1969), 576–594.

[12] Anderson, W. N., Jr.; Schreiber, M. The infimum of two projections. *Acta Universitatis Szegediensis: Acta Scientiarum Mathematicarum*, 33 (1972), 165–168.

[13] Anderson, W. N., Jr.; Trapp, G. E. Shorted operators, II. *SIAM Journal on Applied Mathematics*, 28 (1975), 60–71.

[14] Anderson, William N., Jr.; Trapp, G. E. Problem 76-8: A matrix inequality. *SIAM Review*, 18 (1976), 295.

[15] Ando, T. Lebesgue-type decomposition of positive operators, *Acta Universitatis Szegediensis: Acta Scientiarum Mathematicarum*, 38 (1976), 253–260.

[16] Ando, T. Concavity of certain maps on positive definite matrices and applications to Hadamard products. *Linear Algebra and its Applications*, 26 (1979), 203–241.

[17] Ando, T. Generalized Schur complements. *Linear Algebra and its Applications*, 27 (1979), 173–186.

[18] Ando, T. Totally positive matrices. *Linear Algebra and its Applications*, 90 (1987), 165–219.

[19] T. Ando, Operator-Theoretic Methods for Matrix Inequalities, Lecture Notes, March 1998.

[20] Ando, T. Problem of infimum in the positive cone. In Rassias [380, pp. 1–12 (1999)].

[21] Baker, H. F., ed. *The Collected Mathematical Papers of James Joseph Sylvester, Volume I (1837–1853); Volume II (1854–1873); Volume III (1870–1883); Volume IV (1882–1897)*. Cambridge University Press, 1904–1912. (Corrected reprint edition, Chelsea, New York, 1973.)

[22] Baksalary, Jerzy K. Solution 1 (to Problem 89-7 by Sobel [424, (1989)]). *The IMS Bulletin*, 19 (1990), 213–214.

[23] Baksalary, Jerzy K. An elementary development of the equation characterizing best linear unbiased estimators. *Linear Algebra and its Applications*, in press (available online 29 March 2003).

[24] Baksalary, Jerzy K.; Kala, Radoslaw Range invariance of certain matrix products. *Linear and Multilinear Algebra*, 14 (1983), 89–96.

[25] Baksalary, Jerzy K.; Puntanen, Simo. Weighted-least-squares estimation in the general Gauss–Markov model. In Dodge [146, pp. 355–368 (1989)].

[26] Baksalary, Jerzy K.; Puntanen, Simo. Characterizations of the best linear unbiased estimator in the general Gauss-Markov model with the use of matrix partial orderings. *Linear Algebra and its Applications*, 127 (1990), 363–370.

[27] Baksalary, Jerzy K.; Puntanen, Simo; Styan, George P. H. A property of the dispersion matrix of the best linear unbiased estimator in the general Gauss–Markov model. *Sankhyā, Series A*, 52 (1990), 279–296.

[28] Baksalary, Jerzy K.; Styan, George P. H. Around a formula for the rank for a matrix product with some statistical applications. In Rees [382, pp. 1–18 (1993)].

[29] Banachiewicz, T. Zur Berechnung der Determinanten, wie auch der Inversen, und zur darauf basierten Auflösung der Systeme linearer Gleichungen. *Acta Astronomica, Série C*, 3 (1937), 41–67.

[30] Bapat, R. B. A refinement of Oppenheim's inequality. In Uhlig & Grone [443, pp. 29–32 (1987)].

[31] Barnavi, Eli, ed. *A Historical Atlas of the Jewish People, from the Time of the Patriachs to the Present.* First American edition (translated from the French: *Juifs, une histoire universelle*), Eli Barnavi, General editor; Miriam Eliav-Feldon, English edition editor. Alfred A. Knopf, New York, 1992.

[32] Barnett, Stephen. *Matrices: Methods and Applications.* Oxford Applied Mathematics and Computing Science Series, Oxford University Press, 1990.

[33] Barnett, S.; Jury, E. I. Inners and Schur complement. *Linear Algebra and its Applications*, 22 (1978), 57–63.

[34] Barretta, Wayne; Pierce, Stephen. Null spaces of correlation matrices. *Linear Algebra and its Applications*, 368 (2003), 129–157.

[35] Barry, Phillip J.; Shankar, Subash. Observations on recursive evaluation algorithms in CAGD. *Annals of Numerical Mathematics*, 3 (1996), 1–12.

[36] Bartlett, M. S. An inverse matrix adjustment arising in discriminant analysis. *The Annals of Mathematical Statistics*, 22 (1951), 107–111.

[37] Bartmann, Flavio C.; Bloomfield, Peter. Inefficiency and correlation. *Biometrika*, 68 (1981), 67–71.

[38] Beare, Arlene. *A Guide to Jewish Genealogy in Latvia and Estonia.* The Jewish Genealogical Society of Great Britain, London, 2001.

[39] Beare, Arlene, ed. *History of Latvia and Courland* Web site accessed 1 March 2004: www.jewishgen.org/Latvia/SIG_History_of_Latvia_and_Courland.html

(This history is derived from a few sources including [38] but mainly edited from the presentation made by Ruvin Ferber at the 21st International Conference of Jewish Genealogy held in London in July 2001.)

[40] Beckermann, B.; Neuber, A.; Mühlbach, G. Shanks's transformation revisited. *Linear Algebra and its Applications*, 173 (1992), 191–219.

[41] Begehr, H. G. W.; Koch, H.; Kramer, J.; Schappacher, N.; Thiele, E.-J.; on behalf of the Berliner Mathematische Gesellschaft, eds. *Mathematics in Berlin.* Birkhäuser Verlag, Berlin, 1998.

[42] Bekker, Paul A. The positive semidefiniteness of partitioned matrices. *Linear Algebra and its Applications*, 111 (1988), 261–278.

[43] Belsley, David A. *Conditioning Diagnostics: Collinearity and Weak Data in Regression.* Wiley, New York, 1991.

[44] Belsley, David A.; Kuh, Edwin; Welsch, Roy E. *Regression Diagnostics: Identifying Influential Data and Sources of Collinearity.* Wiley, New York, 1980.

[45] Ben-Israel, Adi; Greville, Thomas N. E. *Generalized Inverses: Theory and Applications*, 2nd edition. CMS Books in Mathematics/Ouvrages de

Mathématiques de la SMC 15, Springer-Verlag, New York, 2003. (Original version: Wiley, New York, 1974; corrected reprint edition: Krieger, Huntington, New York, 1980.)

[46] Bentz, Edna M. *If I Can, You Can Decipher Germanic Records*. Edna M. Bentz [San Diego, California], 1982. (Second printing 1983.)

[47] Berenshteĭn, A. D.; Vaĭnshteĭn, A. D. A multiplicative analogue of the Bergstrom inequality for a matrix product in the sense of Hadamard (in Russian). *Uspekhi Matematicheskikh Nauk*, 42 (1987), no. 6(258), 181–182.

[48] Berman, Abraham; Shaked-Monderer, Naomi. *Completely Positive Matrices*. World Scientific, Singapore, 2003.

[49] Berman, Abraham; Plemmons, Robert J. *Nonnegative Matrices in the Mathematical Sciences*, Revised reprint of edition. Classics in Applied Mathematics 9, SIAM, Philadelphia, 1994. (Original version: Academic Press, New York, 1979.)

[50] Bertrand, Philip V.; Holder, Roger L. A quirk in multiple regression: the whole regression can be greater than the sum of its parts. *The Statistician*, 37 (1988), 371–374.

[51] Bhatia, Rajendra. *Matrix Analysis*. Springer-Verlag, New York, 1997.

[52] Bhatia, Rajendra; Davis, Chandler. More matrix forms of the arithmetic-geometric mean inequality. *SIAM Journal on Matrix Analysis and Applications*, 14 (1993), 132–136.

[53] Bhatia, Rajendra; Kittaneh, Fuad. On the singular values of a product of operators. *SIAM Journal on Matrix Analysis and Applications*, 11 (1990), 272–277.

[54] Bhatia, Rajendra; Kittaneh, Fuad. Norm inequalities for positive operators. *Letters in Mathematical Physics*, 43 (1998), 225–231.

[55] Bhatia, Rajendra; Rosenthal, Peter. How and why to solve the operator equation $AX - XB = Y$. *The Bulletin of the London Mathematical Society*, 29 (1997), 1–21.

[56] Bhimasankaram, P. On generalized inverses of partitioned matrices. *Sankhyā, Series A*, 33 (1971), 311–314.

[57] Bhimasankaram, Pochiraju; Sengupta, Debasis. The linear zero functions approach to linear models. *Sankhyā, Series B*, 58 (1996) 338–351.

[58] Bingham, Christopher. Some identities useful in the analysis of residuals from linear regression. Technical Report 300, School of Statistics, University of Minnesota, 1977.

[59] Bini, Dario; Pan, Victor Y. *Polynomial and Matrix Computations, Vol. 1: Fundamental Algorithms*. Birkhäuser, Boston, 1994.

[60] Blaushild, Immanuel. Libau. In Snyder [423, §1 (*c.* 1995)].

[61] Bloomfield, Peter; Watson, Geoffrey S. The inefficiency of least squares. *Biometrika*, 62 (1975), 121–128.

[62] Bodewig, E. Bericht über die verschiedenen Methoden zur Lösung eines Systems linearer Gleichungen mit reellen Koeffizienten, II. *Koninklijke Nederlandse Akademie van Wetenschappen: Proceedings of the Section of Sci-*

ences, 50 (1947), 1104–1116 & 51 (1948), 218–219 = *Indagationes Mathematicae ex actis quibus titulus*, 9 (1947), 518–530 & 10 (1948), 89–90.

[63] Bodewig, E. *Die Inversion geodätischer Matrizen.* Nederlandse Organisatie voor Toegepast-Natuurwetenschappelijk Onderzoek (TNO), Den Haag, 1954. (See also Forsythe [170] and Bodewig [65].)

[64] Bodewig, E. *Matrix Calculus.* North-Holland, Amsterdam, 1956. (Second revised and enlarged edition, 1959.)

[65] Bodewig, E. Die Inversion geodätischer Matrizen. *Bulletin Géodésique, Nouvelle Série*, 1956 (1956), 9–62. (See also Bodewig [63].)

[66] Boltz, H. *Entwickelungs-Verfahren zum Ausgleichen geodätischer Netze nach der Methode der kleinsten Quadrate.* Veröffentlichungen des Preussischen Geodätischen Institutes, Neue Folge 90, Druck und Verlag von P. Stankiewicz' Buchdruckerei, Berlin, 1923.

[67] Boothroyd, Derek Brian. *On Some Properties of Balanced Incomplete Block Designs.* M.Sc. thesis, Dept. of Mathematics and Statistics, McGill University, Montréal (Québec), Canada, 1988.

[68] Bose, R. C. The design of experiments. (Presidential address, Section on Statistics, Indian Science Congress.) In *Proceedings of the 34th Indian Science Congress*, pp. 1–25 (1947).

[69] Boullion, Thomas L.; Odell, Patrick L., eds. *Proceedings of the Symposium on Theory and Application of Generalized Inverses of Matrices: Held at the Department of Mathematics, Texas Technological College, Lubbock, Texas, March 1968.* Texas Technological College Mathematics Series 4, The Texas Tech Press, Lubbock, 1968.

[70] Brauer, Alfred. Gedenkrede auf Issai Schur. In Brauer & Rohrbach [71, *Band I*, pp. v–xiv (1973)]. (Translated into English in Ledermann & Neumann [277, pp. xlvi–lx (2003)].)

[71] Brauer, Alfred; Rohrbach, Hans, eds. *Issai Schur Gesammelte Abhandlungen: Band I; Band II; Band III.* Springer-Verlag, Berlin, 1973.

[72] Brezinski, Claude. *Padé-Type Approximation and General Orthogonal Polynomials.* International Series of Numerical Mathematics 50. Birkhäuser Verlag, Basel, 1980.

[73] Brezinski, Claude. Other manifestations of the Schur complement. *Linear Algebra and its Applications*, 111 (1988), 231–247.

[74] Brezinski, Claude. *Biorthogonality and its Applications to Numerical Analysis.* Marcel Dekker, New York, 1992.

[75] Brezinski, C. Formal orthogonality on an algebraic curve. *Annals of Numerical Mathematics*, 2 (1995), 21–33.

[76] Brezinski, Claude. *Projection Methods for Systems of Equations.* Studies in Computational Mathematics 7. North-Holland, Amsterdam, 1997.

[77] Brezinski, C. Vector sequence transformations: methodology and applications to linear systems. *Journal of Computational and Applied Mathematics*, 98 (1998), 149–175.

[78] Brezinski, C. Block descent methods and hybrid procedures for linear systems. *Numerical Algorithms*, 29 (2002), 21–32.

[79] Brezinski, Claude. *Computational Aspects of Linear Control.* Numerical Methods and Algorithms 1. Kluwer Academic Publishers, Dordrecht, 2002.

[80] Brezinski, C. A classification of quasi-Newton methods. *Numerical Algorithms*, 33 (2003), 123–135.

[81] Brezinski, Claude; Matos, Ana C. Least-squares orthogonal polynomials. *Journal of Computational and Applied Mathematics*, 46 (1993), 229–239.

[82] Brezinski, C.; Morandi Cecchi, M.; Redivo Zaglia, M. The reverse bordering method. *SIAM Journal on Matrix Analysis and Applications*, 15 (1994), 922–937.

[83] Brezinski, Claude; Redivo Zaglia, Michela. *Extrapolation Methods. Theory and Practice.* Studies in Computational Mathematics 2. North-Holland, Amsterdam, 1991.

[84] Brezinski, C.; Redivo Zaglia, M. Breakdowns in the computation of orthogonal polynomials. In Cuyt [137, pp. 49–59 (1994)].

[85] Brezinski, C.; Redivo-Zaglia, M. A look-ahead strategy for the implementation of some old and new extrapolation methods. *Numerical Algorithms*, 11 (1996), 35–55.

[86] Brezinski, C.; Redivo Zaglia, M. Vector and matrix sequence transformations based on biorthogonality. *Applied Numerical Mathematics*, 21 (1996), 353–373.

[87] Brezinski, C.; Redivo Zaglia, M. Block projection methods for linear systems. *Numerical Algorithms*, 29 (2002), 33–43.

[88] Brezinski, C.; Redivo Zaglia, M. A Schur complement approach to a general extrapolation algorithm. *Linear Algebra and its Applications*, 368 (2003), 279–301.

[89] Brezinski, C.; Redivo Zaglia, M. New vector sequence transformations. Submitted for publication, 2004.

[90] Brezinski, C.; Redivo Zaglia, M; Sadok, H. New look-ahead Lanczos-type algorithms for linear systems. *Numerische Mathematik*, 83 (1999), 53–85.

[91] Brezinski, C.; Redivo Zaglia, M; Sadok, H. A review of formal orthogonality in Lanczos-based methods. *Journal of Computational and Applied Mathematics*, 140 (2002), 81–98

[92] Brezinski, C.; Salam, A. Matrix and vector sequence transformations revisited. *Proceedings of the Edinburgh Mathematical Society. Series II*, 38 (1995), 495–510.

[93] Brezinski, Claude; Van Iseghem, Jeannette. Padé approximations. In Ciarlet & Lions [126, pp. 47–222 (1994)].

[94] Brezinski, Claude; Walz, Guido. Sequences of transformations and triangular recursion schemes, with applications in numerical analysis. *Journal of Computational and Applied Mathematics*, 34 (1991), 361–383.

[95] Brown, Lawrence D.; Ingram Olkin; Jerome Sacks; Henry P. Wynn, eds. *Jack Carl Kiefer Collected Papers III: Design of Experiments.* Springer-Verlag, New York, 1985.

[96] Broyden, C. G. Problem 82-6: A matrix problem. *SIAM Review*, 24 (1982), 223. (Solution by the proposer: Broyden [98, (1983)].)

[97] Broyden, C. G. A mark-scaling algorithm. *The Computer Journal*, 26 (1983), 109–112.

[98] Broyden, C. G. A matrix problem: solution by the proposer to Problem 82-6. *SIAM Review*, 25 (1983), 405. (Solution to Broyden [96, (1982)].)

[99] Brualdi, R. A.; Schneider, H. Determinantal identities: Gauss, Schur, Cauchy, Sylvester, Kronecker, Jacobi, Binet, Laplace, Muir, and Cayley. *Linear Algebra and its Applications*, 52/53 (1983), 765–791.

[100] Brüning, Jochen; Ferus, Dirk; Siegmund-Schultze; Reinhard. *Terror and Exile: Persecution and Expulsion of Mathematicians from Berlin between 1933 and 1945.* An Exhibition on the Occasion of the International Congress of Mathematicians, Technische Universität Berlin, August 19 to 27, 1998, Deutsche Mathematiker-Vereinigung, Berlin, 1998.

[101] Bujakiewicz-Korońska, Renata; Koroński, Jan. The Krakowian calculus and other mathematical ideas of Tadeusz Banachiewicz (in Polish). *Zeszyty Naukowe Politechniki Ślaskiej: Matematyka-Fizyka*, 76 (1996), 23–46.

[102] Burns, Fennell; Carlson, David; Haynsworth, Emilie; Markham, Thomas. Generalized inverse formulas using the Schur complement. *SIAM Journal on Applied Mathematics*, 26 (1974), 254–259.

[103] Cabillon, Julio Gonzalez. *Crelle's Journal* (was: Riemann's mathematical papers on the Web) Web site accessed 2 July 2004: sunsite.utk.edu/math_archives/.http/hypermail/historia/dec98/0098.html

[104] Carlson, David. Matrix decompositions involving the Schur complement. *SIAM Journal on Applied Mathematics*, 28 (1975), 577–587.

[105] Carlson, David. What are Schur complements, anyway? *Linear Algebra and its Applications*, 74 (1986), 257–275. (See also pp. 188–193 in "The 1983 Utah State University Department of Mathematics Conference, 9, 10, and 11 June 1983, Logan, Utah" by LeRoy Beasley & E. E. Underwood, *Linear Algebra and its Applications*, 59 (1984), 183–211.)

[106] Carlson, David; Haynsworth, Emilie; Markham, Thomas. A generalization of the Schur complement by means of the Moore–Penrose inverse. *SIAM Journal on Applied Mathematics*, 26 (1974), 169–175.

[107] Carlson, David; Markham, Thomas L. Schur complements of diagonally dominant matrices. *Czechoslovak Mathematical Journal*, 29/104 (1979), 246–251.

[108] Carlson, David; Markham; Thomas L.; Uhlig, Frank. Emilie Haynsworth, 1916–1985. *Linear Algebra and its Applications*, 75 (1986), 269–276.

[109] Carlson, David; Schneider, Hans. Inertia theorems for matrices: the semidefinite case. *Journal of Mathematical Analysis and Applications*, 6 (1963), 430–446.

[110] Carstensen, C.; Mühlbach, G.; Schmidt, G. De Casteljau's algorithm is an extrapolation method. *Computer Aided Geometric Design*, 12 (1995), 371–380.

[111] Chakrabarti, M. C. *Mathematics of Design and Analysis of Experiments*. Asia Publishing House, London, 1962.

[112] Chakrabarti, M. C. On the C-matrix in design of experiments. *Journal of the Indian Statistical Association*, 1 (1963), 8–23.

[113] Chandler, Bruce; Magnus, Wilhelm. *The History of Combinatorial Group Theory: A Case Study in the History of Ideas.* Studies in the History of Mathematics and Physical Sciences 9. Springer-Verlag, New York, 1982.

[114] Chatterjee, Samprit; Hadi, Ali S. Influential observations, high leverage points, and outliers in linear regression (with discussion). *Statistical Science*, 1 (1986), 379–416.

[115] Chatterjee, Samprit; Hadi, Ali S. *Sensitivity Analysis in Linear Regression.* Wiley, New York, 1988.

[116] Chipman, John S. The multi-sector multiplier. *Econometrica*, 18 (1950), 355–374.

[117] Chipman, John S. Specification problems in regression analysis. In Boullion & Odell [69, pp. 114–176 (1968)].

[118] Chipman, John S. Estimation and aggregation in econometrics: an application of the theory of generalized inverses. In Nashed [328, pp. 549–769 (1976)].

[119] Chipman, John S. The contributions of Ragnar Frisch to economics and econometrics. In Strøm [429, 58–108 (1998)].

[120] Choi, Man Duen. A Schwarz inequality for positive linear maps on C^*-algebras. *Illinois Journal of Mathematics*, 18 (1974), 565–574.

[121] Chollet, John. Some inequalities for principal submatrices. *The American Mathematical Monthly*, 104 (1997), 609–617.

[122] Christensen, Ronald. *Plane Answers to Complex Questions: The Theory of Linear Models*, 3rd edition. Springer-Verlag, New York, 2002. (Original version: 1987; 2nd edition 1996.)

[123] Christensen, Ronald; Pearson, Larry M.; Johnson, Wesley. Case-deletion diagnostics for mixed models. *Technometrics*, 34 (1992), 38–45.

[124] Chu, Ka Lok; Isotalo, Jarkko; Puntanen, Simo; Styan, George P. H. *On decomposing the Watson efficiency of ordinary least squares in a partitioned weakly singular linear model. Sankhyā*, 66 (2004), in press.

[125] Chu, Ka Lok; Isotalo, Jarkko; Puntanen, Simo; Styan, George P. H. Inequalities and equalities for the generalized efficiency function in orthogonally partitioned linear models. Invited paper for *Inequalities and Applications* edited by Themistocles M. Rassias, submitted for publication, June 2004.

[126] Ciarlet, P. G.; Lions, J.-L., eds. *Handbook of Numerical Analysis, Vol. III: Techniques of Scientific Computing.* Handbook of Numerical Analysis III. North-Holland, Amsterdam, 1994.

[127] Cook, R. Dennis. Detection of influential observation in linear regression. *Technometrics*, 19 (1977), 15–18 & 42 (2000), 65–68. (Additional correspondence, *Technometrics*, 19 (1977), 348–350.)

[128] Cottle, Richard W. Manifestations of the Schur complement. *Linear Algebra and its Applications*, 8 (1974), 189–211.

[129] Cottle, Richard W. On manifestations of the Schur complement. *Rendiconti del Seminario Matematico e Fisico di Milano*, 45 (1975), 31–40.

[130] Council of Biology Editors Style Manual Committee. *Scientific Style and Format: The CBE Manual for Authors, Editors, and Publishers*, 6th edition. Cambridge University Press, 1994. (Original version: *Style Manual for Biological Journals* prepared by the Committee on Form and Style of the Conference of Biological Editors, American Institute of Biological Sciences, 1960 (2nd edition: 1964); 3rd–5th editions: *CBE Style Manual: A Guide for Authors, Editors, and Publishers in the Biological Sciences*, 1972–1983.)

[131] Crabtree, Douglas E.; Haynsworth, Emilie V. An identity for the Schur complement of a matrix. *Proceedings of the American Mathematical Society*, 22 (1969), 364–366.

[132] Cramér, Harald. *Mathematical Methods of Statistics*. Almqvist & Wiksell, Uppsala, 1945. (Reprinted: Princeton Mathematical Series 9, Princeton University Press, 1946; Princeton Landmarks in Mathematics, Princeton University Press, 1999.)

[133] Csörgő, Miklós; Dawson, Donald A.; Rao, J. N. K.; Saleh, A. K. Md. E., eds., *Statistics and Related Topics: Proceedings of the International Symposium on Statistics and Related Topics: Ottawa, May 1980*. North-Holland, Amsterdam, 1981.

[134] Cuadras, C. M. Interpreting an inequality in multiple regression. *The American Statistician*, 47 (1993), 256–258.

[135] Currie, I.; Korabinski, A. Some comments on bivariate regression. *The Statistician*, 33 (1984), 283–293.

[136] Curtis, Charles W. *Pioneers of Representation Theory: Frobenius, Burnside, Schur, and Brauer*. History of Mathematics 15, American Mathematical Society & London Mathematical Society, 1999.

[137] Cuyt, Annie, ed. *Nonlinear Numerical Methods and Rational Approximation II: Proceedings of the International Conference held at the University of Antwerp, Wilrijk, September 5–11, 1993*. Mathematics and its Applications 296. Kluwer Academic Publishers Group, Dordrecht, 1994.

[138] Das Gupta, Somesh. Nonsingularity of the sample covariance matrix. *Sankhyā, Series A*, 33 (1971), 475–478.

[139] Das Gupta, S.; Ghosh, J. K.; Mitra, S. K.; Mukhopadhyay, A. C.; Rao, P. S. S. N. V. P.; Sarma, Y. R., eds. *Selected Papers of C. R. Rao: Volume One, Volume Two*. Wiley Eastern Limited, New Delhi, 1994.

[140] Das Gupta, S.; Ghosh, J. K.; Mitra, S. K.; Mukhopadhyay, A. C.; Prakasa Rao, B. L. S.; Rao, P. S. S. N. V. P.; Rao, S. B.; Sarma, Y. R., eds. *Selected Papers of C. R. Rao: Volume Three, Volume Four, Volume Five*. Indian Statistical Institute, Kolkata & New Age International (P) Limited, New Delhi, 1995 (vol. 3), 1999 (vol. 4), 2001 (vol. 5). Keywords: Studies 4) 28022425

[141] Davidson, Russell; MacKinnon, James G. *Estimation and Inference in Econometrics*. Oxford University Press, New York, 1993.

[142] Davidson, Russell; MacKinnon, James G. *Econometric Theory and Methods*. Oxford University Press, New York, 2004.

[143] Davis, Philip J. *Interpolation and Approximation.* Dover, New York, 1975. [Republication, with minor corrections, of the 1963 original, with a new preface and bibliography.]

[144] Davis, Philip J. *The Education of a Mathematician.* A K Peters, Natick, Mass., 2000.

[145] Dey, Aloke. *Theory of Block Designs.* Wiley Eastern, New Delhi & Halsted Press (Wiley), New York, 1986.

[146] Dodge, Yadolah, ed. *Statistical Data Analysis and Inference: Papers from the International Conference on Recent Developments in Statistical Data Analysis and Inference in Honor of C. Radhakrishna Rao held in Neuchâtel, August 21–24, 1989.* North-Holland, Amsterdam, 1989.

[147] Drury, S. W.; Liu, Shuangzhe; Lu, Chang-Yu; Puntanen, Simo; Styan, George P. H. Some comments on several matrix inequalities with applications to canonical correlations: historical background and recent developments. *Sankhyā, Series A*, 64 (2002), 453–507.

[148] Drygas, Hilmar. *The Coordinate-Free Approach to Gauss–Markov Estimation.* Lecture Notes in Operations Research and Mathematical Systems 40, Springer-Verlag, Berlin, 1970.

[149] Duda, Richard O.; Hart, Peter E.; Stork, David G. *Pattern Classification,* 2nd edition. Wiley, New York, 2001. (Original version: *Pattern Classification and Scene Analysis* by Richard O. Duda & Peter E. Hart, 1973.)

[150] Duff, Iain S. The impact of high-performance computing in the solution of linear systems: trends and problems. *Journal of Computational and Applied Mathematics*, 123 (2000), 515–530.

[151] Duncan, W. J. Some devices for the solution of large sets of simultaneous linear equations. (With an Appendix on the reciprocation of partitioned matrices.) *The London, Edinburgh, and Dublin Philosophical Magazine and Journal of Science, Seventh Series*, 35 (1944), 660–670.

[152] Dym, Harry; Katsnelson, Victor. Contributions of Issai Schur to analysis. In Joseph, Melnikov & Rentschler [249, pp. xci–clxxxiii (2003)].

[153] Elffers, Henk. Grenzen voor korrelatie coëfficienten in een trivariate verdeling (in Dutch: Bounds for correlation coefficients in a trivariate distribution). Report SW42, Mathematisch Centrum, Amsterdam, 1976.

[154] Elffers, Henk. On interpreting the product moment correlation coefficient. *Statistica Neerlandica*, 34 (1980), 3–11.

[155] Elffers, Henk. Opgave 87 (in Dutch: Problem 87). *Statistica Neerlandica*, 34 (1980), 167–168.

[156] Elffers, H. Solution 2 (to Problem 89-7 by Sobel [424, (1989)]). *The IMS Bulletin*, 19 (1990), 214.

[157] Elsner, L.; Ikramov, Kh. D.. Normal matrices: an update. *Linear Algebra and its Applications*, 285 (1998), 291–303.

[158] *Encyclopædia Britannica* article: Gymnasium. Web site accessed 29 December 2003: www.search.eb.com/eb/article?eu=39437

[159] Erdős, Paul. Issai Schur, 1875–1995: On the 120-th anniversary of the birth of Schur. *Geombinatorics*, 5 (1995), 4–5.

[160] Fan, Yizheng. Schur complements and its applications to symmetric nonnegative and Z-matrices. *Linear Algebra and its Applications*, 353 (2002), 289–307.

[161] Farebrother, R. W. A note on the Schur and Samuelson conditions. *The Manchester School of Economic and Social Studies*, 60 (1992), 79–81.

[162] Farebrother, R. W. *Fitting Linear Relationships: A History of the Calculus of Observations 1750–1900.* Springer-Verlag, New York, 1999.

[163] Farin, Gerald. *Curves and Surfaces for Computer Aided Geometric Design: A Practical Guide,* Second edition. With contributions by P. Bézier & W. Boehm. Academic Press, Boston, 1990.

[164] Fiebig, Denzil G.; Bartels, Robert; Krämer, Walter. The Frisch–Waugh theorem and generalized least squares. *Econometric Reviews*, 15 (1996), 431–443.

[165] Fiedler, Miroslav. *Special Matrices and their Applications in Numerical Mathematics.* Translated from the Czech by Petr Přikryl & Karel Segeth. Martinus Nijhoff, Dordrecht, 1986.

[166] Fiedler, Miroslav; Markham, Thomas L. A classification of matrices of class Z. *Linear Algebra and its Applications*, 173 (1992), 115–124.

[167] Fiedler, Miroslav; Markham, Thomas L. Some results on the Bergstrom and Minkowski inequalities. *Linear Algebra and its Applications*, 232 (1996), 199–211.

[168] FitzGerald, Carl H.; Horn, Roger A. On fractional Hadamard powers of positive definite matrices. *Journal of Mathematical Analysis and Applications*, 61 (1977), 633–642.

[169] Ford, William F.; Sidi, Avram. *SIAM Journal on Matrix Analysis and Applications*, 24 (1987), 1212–1232.

[170] Forsythe, George E. Review of Bodewig [63]. *Mathematical Reviews*, MR0066762 (16,629h), 1955.

[171] Frazer, Robert A.; Duncan, W. J.; Collar, A. R. *Elementary Matrices and Some Applications to Dynamics and Differential Equations.* Cambridge University Press, 1938. (Reprinted: 1947–1963. Reprint edition: AMS Press, New York, 1982.)

[172] Fréchet, Maurice. Sur l'extension de certaines évaluations statistiques au cas de petits échantillons. *Revue de l'Institut International de Statistique*, 11 (1943), 182–205.

[173] Fréchet, M.; Lévy, P.; Mandelbrojt, S.; Schwartz, L., eds. *Œuvres de Jacques Hadamard, Tome I, II, III, IV.* Éditions du Centre national de la recherche scientifique, Paris, 1968.

[174] Frei, Günter. Zur Geschichte des Crelleschen Journals. *Journal für die reine und angewandte Mathematik*, 500 (1998), 1–4.

[175] Freund, Rudolf J. When is $R^2 > r_{y1}^2 + r_{y2}^2$? (revisited). *The American Statistician*, 42 (1988), 89–90.

[176] Frisch, R.; Waugh, F. V. Partial time regressions as compared with individual trends. *Econometrica*, 1 (1933), 387–401.

[177] Fritzsche, Bernd; Kirstein, Bernd, eds. *Ausgewählte Arbeiten zu den Ursprüngen der Schur-Analysis: Gewidmet dem großen Mathematiker Issai Schur (1875–1941).* Edited and with a foreword and afterword by Bernd Fritzsche and Bernd Kirstein; with contributions by G. Herglotz, I. Schur, G. Pick, R. Nevanlinna, H. Weyl, W. Ledermann & W. K. Hayman. Teubner-Archiv zur Mathematik 16, B. G. Teubner Verlagsgesellschaft, Stuttgart, 1991.

[178] Gaffke, N.; Krafft, O. Optimum properties of Latin square designs and a matrix inequality. *Mathematische Operationsforschung und Statistik, Series Statistics*, 8 (1977), 345–350.

[179] Gaffke, N.; Krafft, O. Matrix inequalities in the Löwner ordering. In Korte [263, pp. 595–622 (1982)].

[180] Gantmacher, F. R. *The Theory of Matrices: Volume 1,* Reprint Edition. Translated from the Russian by K. A. Hirsch. AMS Chelsea Publishing, Providence, 1998. (Reprint of English version pub. Chelsea, New York, 1959.)

[181] Gigas, Erwin, ed. *Professor Hans Boltz und sein Werk.* Herausgegeben von Erwin Gigas, Direktor des Instituts für Erdmessung. Veröffentlichungen des Instituts für Erdmessung 3, Bamberger Verlagshaus Meisenbach & Co., Bamberg, 1949.

[182] Glass, Gene V.; Collins, James R. Geometric proof of the restriction on the possible values of r_{xy} when r_{xz} and r_{yz} are fixed. *Educational and Psychological Measurement*, 30 (1970), 37–39.

[183] Gohberg, I., ed. *I. Schur Methods in Operator Theory and Signal Processing.* Operator Theory: Advances and Applications OT18, Birkhäuser Verlag, Basel, 1986.

[184] Goldberger, Arthur S. Best linear unbiased prediction in the generalized linear regression model. *Journal of American Statistical Association*, 58 (1962), 369–375.

[185] Golub, Gene H.; Van Loan, Charles F. *Matrix Computations*, 3rd edition. Johns Hopkins Studies in the Mathematical Sciences, The Johns Hopkins University Press, Baltimore, 1996. (Original version: 1983, 2nd edition 1989.)

[186] Good, I. J. C53: Partial correlation and spherical trigonometry. *Journal of Statistical Computation and Simulation*, 9 (1979), 243–245.

[187] Good, I. J. Solutions 3 & 4 (to Problem 89-7 by Sobel [424, (1989)]). *The IMS Bulletin*, 19 (1990), 215.

[188] Graves-Morris, P.R.; Roberts, D.E.; Salam, A. The epsilon algorithm and related topics. *Journal of Computational and Applied Mathematics*, 122 (2000), 51–80.

[189] Grewal, Mohinder S.; Andrews, Angus P. *Kalman Filtering: Theory and Practice using* MATLAB, 2nd edition. Wiley, New York, 2001. (Original version: *Kalman Filtering: Theory and Practice*, Prentice-Hall, Englewood Cliffs, 1993.)

[190] Grewal, Mohinder S.; Weill, Lawrence R.; Andrews, Angus P. *Global Positioning Systems, Inertial Navigation, and Integration.* Wiley, New York, 2000.

[191] Groß, Jürgen. A note on equality of MINQUE and simple estimator in the general Gauss–Markov model. *Statistics & Probability Letters*, 35 (1997), 335–339.

[192] Groß, Jürgen. The Moore–Penrose inverse of a partitioned nonnegative definite matrix. *Linear Algebra and its Applications*, 321 (2000), 113–121.

[193] Groß, Jürgen. The general Gauß–Markov model with possibly singular covariance matrix. *Statistical Papers*, in press, 2004.

[194] Groß, Jürgen; Puntanen, Simo. Remark on pseudo-generalized least squares. *Econometric Reviews*, 19 (2000), 131–133.

[195] Groß, Jürgen; Puntanen, Simo. Estimation under a general partitioned linear model. *Linear Algebra and its Applications*, 321 (2000), 131–144.

[196] Groves, Theodore; Rothenberg, Thomas. A note on the expected value of an inverse matrix. *Biometrika*, 56 (1969), 690–691.

[197] Guttman, Louis. Enlargement methods for computing the inverse matrix. *The Annals of Mathematical Statistics*, 17 (1946), 336–343.

[198] Gyires, B., ed. *Analytic Function Methods in Probability Theory: Proceedings of the Colloquium on the Methods of Complex Analysis in the Theory of Probability and Statistics held at the Kossuth L. University, Debrecen, Hungary, August 29–September 2, 1977.* Colloquia Mathematica Societatis János Bolyai 21, János Bolyai, Budapest & North-Holland, Amsterdam, 1980.

[199] Hadamard, Jacques. Résolution d'une question relative aux déterminants. *Bulletin des sciences mathématiques*, 2 (1893), 240–248; reprinted in Fréchet *et al.* [173, *Tome I*, pp. 239–245 (1968)].

[200] Hager, William W. Updating the inverse of a matrix. *SIAM Review*, 31 (1989), 221–239.

[201] Hall, Peter G.; Hallin, Marc; Roussas, George G., eds. *Madan Lal Puri Selected Collected Works, Volume 3: Time Series, Fuzzy Analysis and Miscellaneous Topics.* VSP, Utrecht, 2003.

[202] Hamilton, David C. Sometimes $R^2 > r_{yx_1}^2 + r_{yx_2}^2$: correlated variables are not always redundant. *The American Statistician*, 41 (1987), 129–132. (See also Hamilton [203]).

[203] Hamilton, David C. Reply to comments on Hamilton [202]. *The American Statistician*, 42 (1988), 90–91. (Comments by Freund [175] and Mitra [321].)

[204] Hartwig, R. E. Block generalized inverses. *Archive for Rational Mechanics and Analysis*, 61 (1976), 197–251.

[205] Hartwig, Robert E.; Styan, George P. H. On some characterizations of the "star" partial ordering for matrices and rank subtractivity. *Linear Algebra and its Applications*, 82 (1986), 145–161.

[206] Haslett, Stephen. Updating linear models with dependent errors to include additional data and/or parameters. *Linear Algebra and its Applications*, 237/238 (1996), 329–349.

[207] Haubold, Rev Dr. A. *Nicolaikirche Leipzig: A Short Architectural History.* On the Web site www.nikolaikirche-leipzig.de accessed 19 January 2004.

[208] Haubrich, Ralf. Frobenius, Schur, and the Berlin algebraic tradition. In Begehr *et al.* [41, pp. 83–96 (1998)].

[209] Hayes, Kevin; Haslett, Stephen. Simplifying general least squares. *The American Statistician*, 53 (1999), 376–381.

[210] Haynsworth, E. V. On the Schur complement. *Basel Mathematical Notes*, #BMN 20, 17 pp., June 1968. (*Basle Mathematical Notes* on cover page.)

[211] Haynsworth, E. V. Determination of the inertia of a partitioned Hermitian matrix. *Linear Algebra and its Applications*, 1 (1968), 73–81.

[212] Haynsworth, Emilie V. Reduction of a matrix using properties of the Schur complement. *Linear Algebra and its Applications*, 3 (1970), 23–29.

[213] Haynsworth, Emilie V. Applications of an inequality for the Schur complement. *Proceedings of the American Mathematical Society*, 24 (1970), 512–516.

[214] Haynsworth, Emilie Virgina. Portrait on the Auburn University Web site accessed 28 December 2003: www.auburn.edu/~fitzpjd/ben/images/Emilie.gif (Also in Davis [144, p. 146 (2000)].)

[215] Haynsworth, Emilie V.; Ostrowski, Alexander M. On the inertia of some classes of partitioned matrices. *Basel Mathematical Notes*, #BMN 18, August 1967. (*Basle Mathematical Notes* on cover page.)

[216] Haynsworth, Emilie V.; Ostrowski, Alexander M. On the inertia of some classes of partitioned matrices. *Linear Algebra and its Applications*, 1 (1968), 299–316. (See also Haynsworth & Ostrowski [215].)

[217] Henderson, Charles R. Estimation of genetic parameters. *The Annals of Mathematical Statistics*, 21 (1950), 309–310.

[218] Henderson, C. R.; Kempthorne, Oscar; Searle, S. R.; von Krosigk, C. M. The estimation of environmental and genetic trends from records subject to culling. *Biometrics*, 15 (1959), 192–218. Krosigk, C.M.

[219] Henderson, H. V.; Searle, S. R. On deriving the inverse of a sum of matrices. *SIAM Review*, 23 (1981), 53–60.

[220] Henn, P. *Ahn's Second German Book: Being The Second Division of Ahn's Rudiments of the German Language.* Steiger's German Series, E. Steiger, New York, 1873.

[221] Heyting, A. Die Theorie der linearen Gleichungen in einer Zahlenspezies mit nichtkommutativer Multiplikation. *Mathematische Annalen*, 98 (1927), 465–490.

[222] Hinkelmann, Klaus. Aitken equations. In Kotz, Read & Banks [267, pp. 22–23 (1997)].

[223] Hoaglin, David C.; Welsch, Roy E. The hat matrix in regression and ANOVA. *The American Statistician*, 32 (1978), 17–22; corrigendum: 32 (1978), 146.

[224] Hocquél-Schneider, Sabine. *Alte Nikolaischule Leipzig.* Edition Leipzig, 1994.

[225] Höpfner, Joachim. *On the contribution of the Geodetic Institute Potsdam to the International Latitude Service.* Paper presented at the International Astronomical Union Colloquium 178 "Polar motion: historical and scientific problems", Cagliari, Sardinia, Italy, 23–30 September 1999 on the occasion of the centennial of the first observations of the International Latitude Service [in 1899]. Scientific Technical Report STR99/08 (1999) online accessed 28 December 2003: www.gfz-potsdam.de/bib/pub/str9908/9908.pdf

[226] Horn, Roger A. The Hadamard product. In Johnson [245, pp. 87–169].

[227] Horn, Roger A. Personal communication with Fuzhen Zhang, October 2000.

[228] Horn, Roger A.; Johnson, Charles R. *Matrix Analysis,* Corrected Reprint Edition. Cambridge University Press, 1990. (Original version: 1985.)

[229] Horn, Roger A.; Johnson, Charles R. Hadamard and conventional submultiplicativity for unitarily invariant norms on matrices. *Linear and Multilinear Algebra*, 20 (1987), 91–106.

[230] Horn, Roger A.; Johnson, Charles R. *Topics in Matrix Analysis,* Corrected Reprint Edition. Cambridge University Press, 1994. (Original version: 1991.)

[231] Horn, Roger A.; Mathias, Roy. Cauchy–Schwarz inequalities associated with positive semidefinite matrices. *Linear Algebra and its Applications*, 142 (1990), 63–82.

[232] Horn, Roger A.; Mathias, Roy. An analog of the Cauchy–Schwarz inequality for Hadamard products and unitarily invariant norms. *SIAM Journal on Matrix Analysis and Applications*, 11 (1990), 481–498.

[233] Horn, Roger A.; Mathias, Roy. Block-matrix generalizations of Schur's basic theorems on Hadamard products. *Linear Algebra and its Applications*, 172 (1992), 337–346.

[234] Householder, A. S. Review of Nistor [335, (1975)]. *Mathematical Reviews*, MR0387304 (52 #8147), 1976.

[235] Hu, Shu-An; Smith, Ronald L. The Schur complement interlacing theorem. *SIAM Journal on Matrix Analysis and Applications*, 16 (1995), 1013–1023.

[236] Ikramov, Kh. D. On the inertia law for normal matrices (in Russian). *Doklady Akademii Nauk*, 380 (2001), no. 1, 7–8.

[237] Intellectual Reserve, Inc. *Handwriting Guide: German Gothic Resource Guide.* www.familysearch.org/eng/Search/RG/guide/German_Gothic99-36316.ASP Family History Library, Salt Lake City, Web site accessed 28 December 2003.

[238] James, A. T.; Wilkinson, G. N. Factorization of the residual operator and canonical decomposition of nonorthogonal factors in the analysis of variance. *Biometrika*, 58 (1971), 279–294.

[239] Jbilou, K.; Messaoudi, A. Matrix recursive interpolation algorithm for block linear systems: direct methods. *Linear Algebra and its Applications*, 294 (1999), 137–154.

[240] Jenne, W. Hans Boltz † (in German: Obituary of Hans Boltz.) *Astronomische Nachrichten*, 277 (1949), 47–48.

[241] John, J. A. *Cyclic Designs*. Monographs on Statistics and Applied Probability, Chapman & Hall, London, 1987. (Revised as John & Williams [242, (1995)].)

[242] John, J. A.; Williams, E. R. *Cyclic and Computer Generated Designs,* 2nd edition. Monographs on Statistics and Applied Probability 38, Chapman & Hall, London, 1995. (Revised edition of John [241, (1987)].)

[243] Johnson, Charles R. Partitioned and Hadamard product matrix inequalities. *Journal of Research of the National Bureau of Standards*, 83 (1978), 585–591.

[244] Johnson, Charles R. Inverse M-matrices. *Linear Algebra and its Applications*, 47 (1982), 195–216.

[245] Johnson, Charles R., ed. *Matrix Theory and Applications: Lecture notes prepared for the American Mathematical Society Short Course held in Phoenix, Arizona, January 10–11, 1989.* Proceedings of Symposia in Applied Mathematic, 40: AMS Short Course Lecture Notes. American Mathematical Society, Providence, 1990.

[246] Johnson, Charles R.; Furtado, Susana. A generalization of Sylvester's law of inertia. *Linear Algebra and its Applications*, 338 (2001), 287–290.

[247] Johnson, Charles R.; Zhang, Fuzhen. An operator inequality and matrix normality. *Linear Algebra and its Applications*, 240 (1996), 105–110. (Erratum by R. Bhatia: *Linear Algebra and its Applications*, 260 (1997), 323.)

[248] Jones, Charles W. *Saint Nicholas of Myra, Bari, and Manhattan: Biography of a Legend.* The University of Chicago Press, 1978.

[249] Joseph, Anthony; Melnikov, Anna; Rentschler, Rudolf, eds. *Studies in Memory of Issai Schur.* Papers from the Paris Midterm Workshop of the European Community Training and Mobility of Researchers Network held in Chevaleret, France, May 21–25, 2000, and the Schur Memoriam Workshop held in Rehovot, Israel, December 27–31, 2000. Progress in Mathematics 210, Birkhäuser, Boston, 2003.

[250] Jossa, Franco. Risoluzione progressiva di un sistema di equazioni lineari: analogia con un problema meccanico. *Rendiconto dell'Accademia delle Scienze Fisiche e Matematiche Serie IV (Società Nazionale di Scienze, Lettere e Arti in Napoli)*, 10 (1940), 346–352.

[251] Kadison, Richard V. A generalized Schwarz inequality and algebraic invariants for operator algebras. *Annals of Mathematics: Second Series*, 56 (1952). 494–503.

[252] Kailath, Thomas. A theorem of I. Schur and its impact on modern signal processing. In Gohberg [183, pp. 9–30 (1986)].

[253] Kandić, D. B.; Parlett, B.; Reljin, B. D.; Vasić, P. M. Explicit construction of hyperdominant symmetric matrices with assigned spectrum. *Linear Algebra and its Applications*, 258 (1997), 41–51.

[254] Kanto, Antti J.; Puntanen, Simo. A connection between the partial correlation coefficient and the correlation coefficient of certain residuals. *Communications in Statistics: Simulation and Computation*, 12 (1983), 639–641.

[255] Kedar, Benjamin Z. *The Changing Land between the Jordan and the Sea: Aerial Photographs from 1917 to the Present.* Yad Ben-Zvi Press, Jerusalem & MOD Publishing House, Tel Aviv, 1999.

[256] Kendall, Sir Maurice; Stuart, Alan. *The Advanced Theory of Statistics, Volume 2: Inference and Relationship,* 4th edition. Charles Griffin, London, 1979. (Original version: 1961; 2nd & 3rd editions: 1967 & 1973; 5th edition: 1991; 6th edition by Stuart, Ord & Arnold [430].)

[257] Khatri, C. G. A note on multiple and canonical correlation for a singular covariance matrix. *Psychometrika,* 41 (1976), 465–470.

[258] Khuri, André I. *Advanced Calculus with Applications in Statistics,* 2nd edition. Wiley-Interscience, Hoboken, New Jersey, 2003. (Original version: Wiley, New York, 1993.)

[259] Kiefer, J. Optimum experimental designs (with discussion). *Journal of the Royal Statistical Society, Series B,* 21 (1959), 272–319; reprinted in Brown *et al.* [95, pp. 54–101 (1985)].

[260] Kirkman, Thos. P. Query VI: Fifteen young ladies in a school walk out three abreast for seven days in succession: it is required to arrange them daily, so that no two shall walk twice abreast. *The Lady's and Gentleman's Diary,* 147 (1850), 48. (Solution by the proposer: Kirkman [261, (1851)].)

[261] Kirkman, Thos. P. Answer by the proposer to Query VI. *The Lady's and Gentleman's Diary,* 148 (1851), 48 (*sic*). (Solution to Kirkman [260, (1850)].)

[262] The Kopelman Foundation. Moghilef (Mohilev). JewishGen Belarus SIG, on *The Jewish Encyclopedia* Web site www.jewishgen.org/belarus/je_mogilev.htm conceived, created, and funded by The Kopelman Foundation, accessed 28 December 2003.

[263] Korte, Bernhard. *Modern Applied Mathematics: Optimization and Operations Research.* Papers based on lectures presented at the Summer School on Optimization and Operations Research held at the University of Bonn, Bonn, September 14–22, 1979. North-Holland, Amsterdam, 1982.

[264] Kosaki, Hideki. Remarks on Lebesgue-type decomposition of positive operators. *Journal of Operator Theory,* 11 (1984), 137–143.

[265] Kotz, Samuel; Johnson, Norman L.; Read, Campbell B., eds. *Encyclopedia of Statistical Sciences, Volume 2: Classification–Eye Estimate.* Wiley, New York, 1982.

[266] Kotz, Samuel; Johnson, Norman L.; Read, Campbell B., eds. *Encyclopedia of Statistical Sciences, Volume 3: Faà di Bruno's Formula–Hypothesis Testing.* Wiley, New York, 1983.

[267] Kotz, Samuel; Read, Campbell B.; Banks, David L., eds. *Encyclopedia of Statistical Sciences: Update Volume 1.* Wiley, New York, 1997.

[268] Krämer, Walter. Finite sample efficiency of ordinary least squares in the linear regression model with autocorrelated errors. *Journal of the American Statistical Association,* 75 (1980), 1005–1009.

[269] Kreĭn, M. G. The theory of self-adjoint extensions of semi-bounded Hermitian transformations and its applications, I, II (in Russian). *Matematičeskiĭ Sbornik Novaya Seriya*, 20(62), 431–495; 21(63), 365–404 (1947).

[270] Kunisch. Letter relieving Issai Schur from his duties at the University of Berlin, 28 September 1935. (Reprinted in Soifer [425, p. 9 (1995)], courtesy of Dr. W. Schultze and the Archive of the Humboldt University of Berlin.)

[271] Lakshmikantham, V.; Trigiante, D. *Theory of Difference Equations: Numerical Methods and Applications.* Mathematics in Science and Engineering 181, Academic Press, Boston, 1988.

[272] Lancaster, Peter; Tismenetsky, Miron. *The Theory of Matrices: Second Edition with Applications.* Academic Press, Orlando, 1985. (Original version: *Theory of Matrices* by Peter Lancaster, Academic Press, New York, 1969.)

[273] Laplace, Pierre Simon Marquis de. *Théorie Analytique des Probabilités*, Troisième édition, revue et augmentée par l'auteur (avec trois suppléments, et un quatrième supplément ajouté par l'auteur en 1825). Mme Vve Courcier, Paris, 1820. (Original edition: 1812. Reprinted with four supplements (originally published 1816, 1818, 1820, 1825) in *Œuvres complètes de Laplace, publiées sous les auspices de l'Académie des sciences, Tome septième*, Gauthier-Villars, Paris, 1878–1912.)

[274] Latour, Dominique; Styan, George P. H. Canonical correlations in the two-way layout. In Pukkila & Puntanen [358, pp. 225–243 (1985)].

[275] Le Cam, Lucien M.; Neyman, Jerzy, eds. *Proceedings of the Fifth Berkeley Symposium on Mathematical Statistics and Probability held at the Statistical Laboratory, University of California, June 21–July 18, 1965 and December 27, 1965–January 7, 1966, Volume I: Statistics.* University of California Press, Berkeley, 1967.

[276] Ledermann, W. Issai Schur and his school in Berlin. *Bulletin of the London Mathematical Society*, 15 (1983), 97–106.

[277] Ledermann, Walter; Neumann, Peter M. The life of Issai Schur through letters and other documents. In Joseph, Melnikov & Rentschler [249, pp. xlv–xc (2003)].

[278] Lewis, Michael C.; Styan, George P. H. (1981). Equalities and inequalities for conditional and partial correlation coefficients. In Csörgő *et al.* [133, pp. 57–65 (1981)].

[279] Li, Bishan; Tsatsomeros, M. J. Doubly diagonally dominant matrices. *Linear Algebra and its Applications*, 261 (1997), 221–235.

[280] Li, Chi-Kwong; Mathias, Roy. Extremal characterizations of the Schur complement and resulting inequalities. *SIAM Review*, 42 (2000), 233–246.

[281] Li, Chi-Kwong; Woedermann, Hugo & Zhang, Fuzhen, eds. Special Issue of *Linear Algebra and its Applications* associated with the International Conference on Matrix Analysis and Applications, Nova Southeastern University (Fort Lauderdale, Florida), December 2003; publication expected in early 2005.

[282] Lei, Tian-Gang; Woo, Ching-Wah & Zhang, Fuzhen. Matrix inequalities by means of embedding. *Electronic J. of Linear Algebra*, Volume 11, pp. 66-77, April 2004.

[283] Lieb, E. H. Solution to Problem 76-8 by Anderson & Trapp [14]. *SIAM Review*, 19 (1977), 330.

[284] Lieb, Elliott H.; Ruskai, Mary Beth. Some operator inequalities of the Schwarz type. *Advances in Mathematics*, 12 (1974), 269–273.

[285] Liu, Jianzhou. Some inequalities for singular values and eigenvalues of generalized Schur complements of products of matrices. *Linear Algebra and its Applications*, 286 (1999), 209–221.

[286] Liu, Jianzhou. Some Löwner partial orders of Schur complements and Kronecker products of matrices. *Linear Algebra and its Applications*, 291 (1999), 143–149.

[287] Liu, Jianzhou; Huang, Yunqing; Zhang, Fuzhen. The Schur complements of generalized doubly diagonally dominant matrices. *Linear Algebra and its Applications*, 378 (2004), 231–244.

[288] Liu, Jianzhou; Yuan, Xiu Yu. Interlacing theorems for Schur complements of self-conjugate quaternion matrices. *Natural Science Journal of Xiangtan University: Xiangtan Daxue Ziran Kexue Xuebao*, 20 (1998), no. 4, 1–2.

[289] Liu, Jianzhou; Zhu, Li. A minimum principle and estimates of the eigenvalues for Schur complements of positive semidefinite Hermitian matrices. *Linear Algebra and its Applications*, 265 (1997), 123–145.

[290] Lohan, Ralf. Das Entwicklungsverfahren zum Ausgleichen geodätischer Netze nach Boltz im Matrizenkalkül. *Zeitschrift für angewandte Mathematik und Mechanik*, 13 (1933), 59–60.

[291] Lokki, Olli. Lineaaristen yhtälöryhmien ratkaisemisen perusideoista (in Finnish: On the methods for solving linear equation systems). *Arkhimedes*, 1961 (1961), 18–39.

[292] Lovell, M. C. Seasonal adjustment of economic time series and multiple regression analysis. *Journal of the American Statistical Association*, 58 (1963), 993–1010.

[293] Lyche, Tom. A recurrence relation for Chebyshevian B-splines. *Constructive Approximation*, 1 (1985), 155–173.

[294] Marcus, Marvin. An inequality for linear transformations. *Proceedings of the American Mathematical Society*, 18 (1967), 793–797.

[295] Markham, T. L. Factorizations of completely positive matrices. *Proceedings of the Cambridge Philosophical Society*, 69 (1971), 53–58.

[296] Markham, T. L. Factorizations of nonnegative matrices. *Proceedings of the American Mathematical Society*, 32 (1972), 45–47.

[297] Markham, Thomas L. An application of theorems of Schur and Albert. *Proceedings of the American Mathematical Society*, 59 (1976), 205–210.

[298] Markham, T. L. Oppenheim's inequality for positive definite matrices. *The American Mathematical Monthly*, 93 (1986), 642–644.

[299] Markham, Thomas L.; Smith, Ronald L. A Schur complement inequality for certain P-matrices. *Linear Algebra and its Applications*, 281 (1998), 33–41.

[300] Marsaglia, George; Styan, George P. H. Equalities and inequalities for ranks of matrices. *Linear and Multilinear Algebra*, 2 (1974), 269–292.

[301] Marshall, Albert W.; Olkin, Ingram. *Inequalities: Theory of Majorization and its Applications.* Mathematics in Science and Engineering 143. Academic Press, New York, 1979.

[302] Mashey, Anne B. *A Guide to Olde German Handwriting of the Mid-1800's, No. 2*, Revised edition. Anne B. Mashey, Wexford, Pennsylvania, 1982. (Original version: 1979.)

[303] Mazure, Marie-Laurence; Laurent, Pierre-Jean. Piecewise smooth spaces in duality: application to blossoming. *Journal of Approximation Theory*, 98 (1999), 316–353.

[304] *McGraw-Hill Dictionary of Scientific and Technical Terms,* 6th edition. McGraw-Hill, New York, 2003. (Original version: Daniel N. Lapedes, editor-in-chief, 1974; 3rd edition: Sybil P. Parker, editor-in-chief, 1984.)

[305] Meinardus, Günter; Taylor, G. D. Lower estimates for the error of best uniform approximation. *Journal of Approximation Theory*, 16 (1976), 150–161.

[306] *Merriam-Webster's Geographical Dictionary*, 3rd edition. Merriam-Webster, Springfield, Mass., 1997 & 2001. (Original version: *Webster's New Geographical Dictionary*, 1972 & 1984.)

[307] Messaoudi, A. Matrix recursive projection and interpolation algorithms. *Linear Algebra and its Applications*, 202 (1994), 71–89.

[308] Messaoudi, A. Some properties of the recursive projection and interpolation algorithms. *IMA Journal of Numerical Analysis*, 15 (1995), 307–318.

[309] Messaoudi, A. Recursive interpolation algorithm: a formalism for solving systems of linear equations, I: Direct methods. *Journal of Computational and Applied Mathematics*, 76 (1996), 13–30.

[310] Messaoudi, A. Recursive interpolation algorithm: a formalism for solving systems of linear equations, II: Iterative methods. *Journal of Computational and Applied Mathematics*, 76 (1996), 31–53.

[311] Messaoudi, A. Matrix extrapolation algorithms. *Linear Algebra and its Applications*, 256 (1997), 49–73.

[312] Meurant, G. *Computer Solution of Large Linear Systems.* Studies in Mathematics and its Applications 28. North-Holland, Amsterdam, 1999.

[313] Meyer, Carl D. *Matrix Analysis and Applied Linear Algebra.* SIAM, Philadelphia, 2000.

[314] Mirsky, L. *An Introduction to Linear Algebra.* Oxford University Press, 1955. (Second corrected paperback reprint edition: Dover, New York, 1990.)

[315] Mirsky, L. The spread of a matrix. *Mathematika (London)*, 3 (1956), 127–130.

[316] Mirsky, L. Inequalities for normal and Hermitian matrices. *Duke Mathematical Journal*, 24 (1957), 591–599.

[317] Mitra, Sujit Kumar. Unified least squares approach to linear estimation in a general Gauss–Markov model. *SIAM Journal on Applied Mathematics*, 25 (1973), 671–680.

[318] Mitra, Sujit Kumar; Puntanen, Simo. The shorted operator statistically interpreted. *Calcutta Statistical Association Bulletin*, 40 (1991), 97–102.

[319] Mitra, Sujit Kumar; Puntanen, Simo; Styan, George P. H. Shorted matrices and their applications in linear statistical models: a review. In Tiit, Kollo & Niemi [442, pp. 289–311 (1995)].

[320] Mitra, Sujit Kumar; Puri, Madan L. Shorted operators and generalized inverses of matrices. *Linear Algebra and its Applications*, 25 (1979), 45–56; reprinted in Hall, Hallin & Roussas [201, pp. 555–566 (2003)].)

[321] Mitra, S. Comments on Hamilton [202]. *The American Statistician*, 42 (1988), 89.

[322] Moore, M. H. A convex matrix function. *The American Mathematical Monthly*, 80 (1973), 408–409.

[323] Moore, M. H. Solution to Problem 76-8 by Anderson & Trapp [14]. *SIAM Review*; 19 (1977), 329–330.

[324] Moreland, T.; Gudder, S. Infima of Hilbert space effects. *Linear Algebra and its Applications*, 286 (1999), 1–17.

[325] Morgenstern, Oskar; Woodbury, Max A. The stability of inverses of input-output matrices. *Econometrica*, 18 (1950), 190–192. (Abstract of paper presented at the American Summer Meeting of the Econometric Society, held jointly with the Institute of Mathematical Statistics, at the University of Colorado, Boulder, Colorado, August 29–September 2, 1949; ee also Sherman & Morrison [417, (1949)].)

[326] Mühlbach, G. The general Neville–Aitken algorithm and some applications, *Numerische Mathematik*, 31 (1978/1979), 97–110.

[327] Nakamoto, Ritsuo; Takahasi, Sin-Ei. Generalizations of an inequality of Marcus. *Mathematica Japonica*, 50 (1999), 35-39.

[328] Nashed, M. Zuhair, ed. *Generalized Inverses and Applications: Proceedings of an Advanced Seminar Sponsored by the Mathematics Research Center, The University of Wisconsin–Madison, October 8–10, 1973.* Academic Press, New York, 1976.

[329] Nayak, Tapan K. On best unbiased prediction and its relationship to unbiased estimation. *Journal of Statistical Planning and Inference*, 84 (2000), 171–189.

[330] Nayak, Tapan K. Rao–Cramér type inequalities for mean squared error of prediction. *The American Statistician*, 56 (2002), 102–106.

[331] Ne'eman, Yuval. Issai Schur died here: some background comments, in memoriam. In Joseph, Melnikov & Rentschler [249, pp. xxi–xxx (2003)].

[332] Neumann, M. On the Schur complement and the *LU*-factorization of a matrix. *Linear and Multilinear Algebra*, 9 (1980/1981), 241–254.

[333] Neyman, J.; Scott, Elizabeth L. Consistent estimates based on partially consistent observations. *Econometrica*, 16 (1948), 1–32.

[334] N'Guessan, Assi. Constrained covariance matrix estimation in road accident modelling with Schur complements. *Comptes Rendus Mathématique: Académie des Sciences, Paris*, 337 (2003), 219–222.

[335] Nistor, Gh. Contribuţii la rezolvarea matriceală a compensării pe două grupe după metoda Boltz (in Romanian: Contributions to the matrix solution of two group compensation by the method of Boltz). *Buletinul Institutului Politehnic din Iaşi Serie Nouă, Secţia 1: Matematică, Mecanică Teoretică, Fizică*, 21(25), 79–83 (1975).

[336] O'Connor, J. J.; Robertson, E. F. *MacTutor History of Mathematics Archive* Web site turnbull.mcs.st-and.ac.uk/~history/ School of Mathematics and Statistics, University of St Andrews, St Andrews, Scotland, accessed 28 December 2003.

[337] Odell, Patrick L. Gauss–Markov theorem. In Kotz, Johnson & Read [266, pp. 314–316 (1983)].

[338] Okuguchi, Koji; Irie, Kazuya. The Schur and Samuelson conditions for a cubic equation. *The Manchester School of Economic and Social Studies*, 58 (1990), 414–418.

[339] Olkin, Ingram. Range restrictions for product-moment correlation matrices. *Psychometrika*, 46 (1981), 469–472.

[340] Olkin, Ingram; Siotani, Minoru. *Testing for the equality of correlation coefficients for various multivariate models.* Technical Report 4, Contract 2-10-065, U.S. Office of Education, 1964.

[341] Oppenheim, A. Inequalities connected with definite Hermitian forms. *Journal of the London Mathematical Society*, 5 (1930), 114–119.

[342] Ostrowski, Alexander. A new proof of Haynsworth's quotient formula for Schur complements. *Linear Algebra and its Applications*, 4 (1971), 389–392.

[343] Ostrowski, A. M. On Schur's complement. *Journal of Combinatorial Theory, Series A*, 14 (1973), 319–323.

[344] Ostrowski, A.; Schneider, H. Some theorems on the inertia of general matrices. *Journal of Mathematical Analysis and Applications*, 4 (1962), 72–84.

[345] Ouellette, Diane Valérie. Schur complements and statistics. *Linear Algebra and its Applications*, 36 (1981), 187–295.

[346] Palmgren, Juni. The Fisher information matrix for log linear models arguing conditionally on observed explanatory variables. *Biometrika*, 68 (1981), 563–566.

[347] Pan, Victor Y. *Structured Matrices and Polynomials: Unified Superfast Algorithms.* Birkhäuser Boston & Springer-Verlag, New York, 2001.

[348] Pekarev, È. L. Convolution on an operator domain (in Russian). *Funktsional'nyĭ Analiz i ego Prilozheniya*, 12 (1978), 84–85.

[349] Pekarev, È. L.; Šmul'jan, Ju. L. Parallel addition and parallel subtraction of operators. *Izvestiya Akademii Nauk SSSR: Seriya Matematicheskaya*, 40 (1976), 366–387, 470.

[350] Piegorsch, Walter W.; Casella, George. The early use of matrix diagonal increments in statistical problems. *SIAM Review*, 31 (1989), 428–434; corrected by Piegorsch & Casella [351, (1990)].

[351] Piegorsch, Walter W.; Casella, George. Erratum: Inverting a sum of matrices. *SIAM Review*, 32 (1990), 470. (Includes corrections to Piegorsch & Casella [350, (1989)].)

[352] Pinl, M. Kollegen in einer dunklen Zeit. *Jahresbericht der Deutschen Mathematiker Vereinigung*, 71 (1969), 167–228.

[353] Pinl, Max; Furtmüller, Lux. Mathematicians under Hitler. In Weltsch [457, pp. 129–182 (1973)].

[354] Prasolov, V. V. *Problems and Theorems in Linear Algebra.* Translations of Mathematical Monographs 134, American Mathematical Society, Providence, 1994.

[355] Priest, H. F. Range of correlation coefficients. *Psychological Reports*, 22 (1968) 168–170.

[356] Proskurowski, Wlodzimierz; Widlund, Olof. On the numerical solution of Helmholtz's equation by the capacitance matrix method. *Mathematics of Computation*, 30 (1976), 433–468.

[357] Pukelsheim, Friedrich. *Optimal Design of Experiments.* Wiley, New York, 1993.

[358] Pukkila, Tarmo; Puntanen, Simo, eds. *Proceedings of the First International Tampere Seminar on Linear Statistical Models and their Applications: University of Tampere, Tampere, Finland, August 30th to September 2nd, 1983.* Dept. of Mathematical Sciences/Statistics, University of Tampere, Tampere, Finland, 1985.

[359] Puntanen, Simo. On the relative goodness of ordinary least squares and on some associated canonical correlations. In Puntanen [360, Paper 1 (1987)].

[360] Puntanen, Simo. *On the Relative Goodness of Ordinary Least Squares Estimation in the General Linear Model.* Ph.D. dissertation, Dept. of Mathematical Sciences, University of Tampere, Acta Universitatis Tamperensis Series A 216, University of Tampere, Tampere, Finland, 1987.

[361] Puntanen, Simo; Styan, George P. H. The equality of the ordinary least squares estimator and the best linear unbiased estimator (with comments by Oscar Kempthorne & by Shayle R. Searle and with "Reply" by the authors). *The American Statistician*, 43 (1989), 153–164.

[362] Puntanen, Simo; Styan, George P. H. *Issai Schur and the early development of the Schur complement, with biographical remarks, pictures, applications to statistics, and an extensive bibliography.* Report A 346, Dept. of Mathematics, Statistics and Philosophy, University of Tampere, Tampere, Finland, March 2004.

[363] Puntanen, Simo; Styan, George P. H. Issai Schur and the early development of the Schur complement, with biographical remarks, pictures, and an extensive bibliography. Invited paper in preparation for publication in Li, Woedermann & Zhang [281, (*c.* 2005)].

[364] Puntanen, Simo; Styan, George P. H.; Werner, Hans Joachim. Two matrix-based proofs that the linear estimator Gy is the best linear unbiased estimator. *Journal of Statistical Planning and Inference*, 88 (2000), 173–179.

[365] Quarteroni, Alfio; Valli, Alberto. *Domain Decomposition Methods for Partial Differential Equations.* Oxford University Press, New York, 1999.

[366] Raghavarao, Damaraju. *Constructions and Combinatorial Problems in Design of Experiments.* Wiley, New York, 1971; corrected reprint edition. Dover, New York, 1988.

[367] Rao, C. Radhakrishna. Information and the accuracy attainable in the estimation of statistical parameters. *Bulletin of the Calcutta Mathematical Society*, 37 (1945), 81–91; reprinted in Das Gupta *et al.* [139, vol. 1, pp. 1–13 (1994)].

[368] Rao, C. Radhakrishna. Least squares theory using an estimated dispersion matrix and its application to measurement of signals. In Le Cam & Neyman [275, pp. 355–372 (1967)]; reprinted in Das Gupta *et al.* [140, vol. 3, pp. 92–108 (1995)].

[369] Rao, C. Radhakrishna. Estimation of variance and covariance components: MINQUE theory. *Journal of Multivariate Analysis*, 1 (1971), 257–275; reprinted in Das Gupta *et al.* [140, vol. 3, pp. 263–277 (1995)].

[370] Rao, C. Radhakrishna. Unified theory of linear estimation. *Sankhyā Series A*, 33 (1971), 371–394 & corrigenda: 34 (1972), pp. 194, 477; reprinted in Das Gupta *et al.* [140, vol. 3, pp. 288–311 (1995)].

[371] Rao, C. R. Representations of best linear unbiased estimators in the Gauss–Markoff model with a singular dispersion matrix. *Journal of Multivariate Analysis*, 3 (1973), 276–292; reprinted in Das Gupta *et al.* [140, vol. 3, pp. 383–395 (1995)].

[372] Rao, C. Radhakrishna. *Linear Statistical Inference and its Applications,* 2nd edition. Wiley, New York, 1973. (Original version: 1965. Paperback reprint edition: Wiley Classics Library, 2002.)

[373] Rao, C. Radhakrishna. Projectors, generalized inverses and the BLUE's. *Journal of the Royal Statistical Society Series B*, 36 (1974), 442–448; reprinted in Das Gupta *et al.* [140, vol. 3, pp. 421–428 (1995)].;

[374] Rao, C. Radhakrishna. A lemma on g-inverse of a matrix and computation of correlation coefficients in the singular case. *Communications in Statistics: Theory and Methods*, 10 (1981), 1–10.

[375] Rao, C. Radhakrishna. The inefficiency of least squares: extensions of the Kantorovich inequality. *Linear Algebra and its Applications*, 70 (1985), 249–255.

[376] Rao, C. Radhakrishna. Three score years of research in statistics: 1941–2000 (with bibliography and biography). In Das Gupta *et al.* [140, vol. 5, pp. 393–496 (2001)].

[377] Rao, C. Radhakrishna; Mitra, Sujit Kumar. *Generalized Inverse of Matrices and Its Applications.* Wiley, New York, 1971.

[378] Rao, C. Radhakrishna; Rao, M. Bhaskara. *Matrix Algebra and Its Applications to Statistics and Econometrics.* World Scientific, Singapore, 1998.

[379] Rao, V. R. A note on balanced designs. *The Annals of Mathematical Statistics*, 29 (1958), 290–294.

[380] Rassias, Themistocles M.; Srivastava, Hari M. *Analytic and Geometric Inequalities and Applications.* Mathematics and its Applications 478, Kluwer Academic Publishers, Dordrecht, 1999.

[381] Raveh, Adi. On the use of the inverse of the correlation matrix in multivariate data analysis. *The American Statistician*, 39 (1985), 39–42.

[382] Rees, Rolf S., ed. *Graphs, Matrices, and Designs: Festschrift in Honor of Norman J. Pullman on his Sixtieth Birthday.* Lecture Notes in Pure and Applied Mathematics 139, Marcel Dekker, New York, 1993.

[383] Robinson, G. K. That BLUP is a good thing: the estimation of random effects (with discussion). *Statistical Science*, 6 (1991), 15–51.

[384] Rodriguez, R. N. Correlation. In Kotz, Johnson & Read [265, pp. 193–204 (1982)].

[385] Rohde, Charles A. Generalized inverses of partitioned matrices. *Journal of the Society for Industrial and Applied Mathematics*, 13 (1965), 1033–1035.

[386] Rousseeuw, Peter J.; Molenberghs, Geert. The shape of correlation matrices. *The American Statistician*, 48 (1994), 276–279.

[387] Roy, J. On the efficiency factor of block designs. *Sankhyā*, 19 (1958), 181–188.

[388] Saad, Y. *Iterative Methods for Sparse Linear Systems*, Second edition. SIAM, Philadelphia, PA, 2003. (Original version: PWS, Boston, 1996.)

[389] Salam, A. Noncommutative extrapolation algorithms. *Numerical Algorithms*, 7 (1994), 225–251.

[390] Salam, A. On the vector-valued Pad approximants and the vector ϵ-algorithm. In Cuyt [137, pp. 291–301 (1994)].

[391] Salam, A. An algebraic approach to the vector ϵ-algorithm. *Numerical Algorithms*, 11 (1996), 327–337.

[392] Salam, A. Formal vector orthogonal polynomials. *Advances in Computational Mathematics*, 8 (1998), 267–289.

[393] Salam, A. What is a vector Hankel determinant. *Linear Algebra and its Applications*, 278 (1998) 147-161.

[394] Salam, A. Vector Padé–type approximants and vector Padé approximants. *Journal of Approximation Theory*, 97 (1999), 92–112.

[395] Salam, A. On vector Hankel determinants. *Linear Algebra and its Applications*, 313 (2000) 127-139.

[396] Samuelson, Paul A. Conditions that the roots of a polynomial be less than unity in absolute value. *The Annals of Mathematical Statistics*, 12 (1941), 360–364.

[397] Savage, Leonard J. *The Foundations of Statistics*. Wiley, New York, 1954.

[398] Schall, Robert; Dunne, Timothy T. A unified approach to outliers in the general linear model. *Sankhyā, Series B*, 50 (1988), 157–167.

[399] Scheffé, Henry. *The Analysis of Variance*. Wiley, New York, 1959. (Paperback reprint edition: Wiley Classics Library, 1999.)

[400] Schneider, Claus. Vereinfachte Rekursionen zur Richardson-Extrapolation in Spezialfällen. *Numerische Mathematik*, 24 (1975), 177–184.

[401] Schoeps, Julius H.; Grözinger, Karl E.; Mattenklott, Gert, eds. *Menora: Jahrbuch für deutsch-jüdische Geschichte 1999: im Auftrag des Moses Mendelssohn-Zentrums für europäisch-jüdische Studien*. Philo Verlagsgesellschaft mbH, Berlin & Bodenheim bei Mainz, 1999.

[402] Schur, Issai. Über eine Klasse von Matrizen, die sich einer gegebenen Matrix zuordnen lassen. Doctoral dissertation, Universität Berlin, 1901; reprinted in Brauer & Rohrbach [71, *Band I*, pp. 1–72 (1973)].)

[403] Schur, J. [Schur, Issai.] Bemerkungen zur Theorie der beschränkten Bilinearformen mit unendlich vielen Veränderlichen. *Journal für die reine und angewandte Mathematik*, 140 (1911), 1–28; reprinted in Brauer & Rohrbach [71, *Band I*, pp. 464–491 (1973)].)

[404] Schur, J. [Schur, Issai.] Über Potenzreihen, die im Innern des Einheitskreises beschränkt sind [I]. *Journal für die reine und angewandte Mathematik*, 147 (1917), 205–232; reprinted in Brauer & Rohrbach [71, *Band II*, pp. 137–164 (1973)] & in Fritsche & Kirstein [177, pp. 22–49 (1991)]. (Translated into English in Schur [409, pp. 31–59 (1986)].)

[405] Schur, J. [Schur, Issai.] Über Potenzreihen, die im Innern des Einheitskreises beschränkt sind [II]. *Journal für die reine und angewandte Mathematik*, 148, 122–145 (1918); reprinted in Brauer & Rohrbach [71, *Band II*, pp. 165–188 (1973)] & in Fritsche & Kirstein [177, pp. 22–49 (1991)]. (Translated into English as Schur [409, pp. 61–88 (1986)].)

[406] Schur, Issai. *Biographischen Mitteilungen*, welche die Kaiserliche Leopoldino-Carolina Deutsche Akademie der Naturforscher nach §7 der Statuten von ihren neueintretenden Mitgliedern zur Aufbewahrung im Archiv erbittet (in German: *Biographical communications*, forming the application for membership in the Academia Leopoldino-Carolina Naturae Curiosorum = German Academy of Natural Scientists), 24 June 1919. Reprinted in Fritzsche & Kirstein [177, pp. 214–217 (1991)].

[407] Schur, J. [Schur, Issai.] *Die Algebraischen Grundlagen der Darstellungstheorie der Gruppen*. Vorlesungen über Darstellungstheorie gehalten auf Einladung des Mathematischen Seminars der Eidg. Techn. Hochschule Zürich, bearbeitet und herausgegeben von Dr. E. Stiefel, Graph. Anstalt Gebr. Frey & Kratz, Zürich, 1936. (Mimeographed lecture notes from Schur's 1936 lectures at the Eidgenössische Technische Hochschule (ETH) Zürich, edited and issued by E. Stiefel.)

[408] Schur, Issai. *Vorlesungen über Invariantentheorie*. Bearbeitet und herausgegeben von Helmut Grunsky. Die Grundlehren der mathematischen Wissenschaften in Einzeldarstellungen mit besonder Berücksichtigung der Anwendungsgebiete 143. Springer-Verlag, Berlin, 1968.

[409] Schur, Issai. On power series which are bounded in the interior of the unit circle: I, II. In Gohberg [183, pp. 31–59, 61–88 (1986)]. (English translation of Schur [404, 405, (1917, 1918)].)

[410] Searle, Shayle R. *Linear Models*. Wiley, New York, 1971. (Paperback reprint edition: Wiley Classics Library, 1996.)

[411] Searle, Shayle R. The matrix handling of BLUE and BLUP in the mixed linear model. *Linear Algebra and its Applications*, 264 (1997), 291–311.

[412] Seber, George A. F.; Lee, Alan J. *Linear Regression Analysis*, 2nd edition. Wiley-Interscience, Hoboken, New Jersey, 2003. (Original version by George A. F. Seber, Wiley, New York, 1977.)

[413] Sengupta, Debasis; Jammalamadaka, Sreenivasa Rao. *Linear Models: An Integrated Approach.* World Scientific, Singapore, 2003.

[414] Seshadri, V.; Styan, G. P. H. Canonical correlations, rank additivity and characterizations of multivariate normality. In Gyires [198, pp. 331–344 (1980)].

[415] Sharpe, G. E.; Styan, G. P. H. Circuit duality and the general network inverse. *IEEE Transactions on Circuit Theory*, CT-12 (1965), 22–27.

[416] Sherman, Jack; Morrison, Winifred J. Adjustment of an inverse matrix corresponding to a change in one element of a given matrix. *The Annals of Mathematical Statistics*, 20 (1949), 317. (Abstract of paper presented at the New York Meeting of the Institute of Mathematical Statistics on April 8–9, 1949, and published as [418, (1950)].)

[417] Sherman, Jack; Morrison, Winifred J. Adjustment of an inverse matrix corresponding to changes in the elements of a given column or a given row of the original matrix. *The Annals of Mathematical Statistics*, 20 (1949), 621. (Abstract of paper presented at the Twelfth Summer Meeting of the Institute of Mathematical Statistics, held jointly with the Econometric Society, at the University of Colorado, Boulder, Colorado, on September 1, 1949; see also Morgenstern & Woodbury [325, (1950)].)

[418] Sherman, Jack; Morrison, Winifred J. Adjustment of an inverse matrix corresponding to a change in one element of a given matrix. *The Annals of Mathematical Statistics*, 21 (1950), 124–127. (Abstract: [416, (1949)].)

[419] Shieh, Gwowen. The inequality between the coefficient of determination and the sum of squared simple correlation coefficients. *The American Statistician*, 55 (2001), 121–124.

[420] Simon, Barry. *Trace ideals and their Applications.* London Mathematical Society Lecture Note Series 35, Cambridge University Press, 1979.

[421] Smith, Ronald L. Some interlacing properties of the Schur complement of a Hermitian matrix. *Linear Algebra and its Applications*, 177 (1992), 137–144.

[422] Smith, Steven Bradley. *The Great Mental Calculators: The Psychology, Methods, and Lives of Calculating Prodigies, Past and Present.* With a foreword by Wim Klein & an introduction by Hans Eberstark. Columbia University Press, New York, 1983.

[423] Snyder, Stephen, project coordinator. *A Town Named Libau (Liepaja, Latvia).* JewishGen Web site www.jewishgen.org/yizkor/libau/libau.html accessed 27 December 2003. (Translation of the 36-page booklet: *A Town Named Libau* in English, German and Hebrew and additional material about Libau, Editor and Publisher of booklet unknown, believed to have been published in Israel, 1985.)

[424] Sobel, Marc. Problem 89-7: Let X, Y, and Z be random variables; if the correlations $\rho(X, Y)$ and $\rho(Y, Z)$ are known, what are the sharp lower and upper bounds for $\rho(X, Z)$? *The IMS Bulletin*, 18 (1989), 386. (Solutions: Baksalary [22, (1990)], Elffers [156, (1990)], Good [187, (1990)].)

[425] Soifer, Alexander. Issai Schur: Ramsey theory before Ramsey. *Geombinatorics*, 5 (1995), 6–23.

[426] Stanley, Julian C.; Wang, Marilyn D. Restrictions on the possible values of r_{12}, given r_{13} and r_{23}. *Educational and Psychological Measurement*, 29 (1969), 579–581.

[427] Stewart, G. W. Collinearity and least squares regression (with discussion). *Statistical Science*, 2 (1987), 68–100.

[428] Stigler, Stephen M. *Statistics on the Table: The History of Statistical Concepts and Methods.* Harvard University Press, 1999.

[429] Strøm, Steinar, ed. *Econometrics and Economic Theory in the 20th Century: The Ragnar Frisch Centennial Symposium.* Cambridge University Press, 1998.

[430] Stuart, Alan; Ord, J. Keith; Arnold, Steven F. *Kendall's Advanced Theory of Statistics, Volume 2A: Classical Inference and the Linear Model*, 6th edition. Arnold/Hodder Headline Group, London & Oxford University Press, New York, 1999. (Original version: *The Advanced Theory of Statistics, Volume 2: Inference and Relationship* by Sir Maurice Kendall & Alan Stuart, Charles Griffin, London, 1961; 2nd & 3rd editions: 1967 & 1973; 4th edition: 1979 [256]; 5th edition: 1991.)

[431] Styan, George P. H. Hadamard products and multivariate statistical analysis. *Linear Algebra and its Applications*, 6 (1973), 217–240.

[432] Styan, George P. H. Schur complements and linear statistical models. In Pukkila & Puntanen [358, pp. 37–75 (1985)].

[433] Styan, George P. H.; Subak-Sharpe, Gerald E. Inequalities and equalities associated with the Campbell–Youla generalized inverse of the indefinite admittance matrix of resistive networks. *Linear Algebra and its Applications*, 250 (1997), 349–370.

[434] Sverdrup, Erling. *Laws and Chance Variations: Basic Concepts of Statistical Inference, Volume II: More Advanced Treatment.* Rand McNally, Chicago, 1967.

[435] Swamy, M. N. S.; Thulasiraman, K. *Graphs, Networks, and Algorithms.* Wiley, New York, 1981.

[436] Sylvester, J. J. On the relation between the minor determinants of linearly equivalent quadratic functions. *The London, Edinburgh, and Dublin Philosophical Magazine and Journal of Science, Fourth Series*, 1 (1851), pp. 295–305; reprinted in [21, vol. 1, pp. 241–250 (1973)].

[437] Thibaudeau, Yves ; Styan, George P. H. Bounds for Chakrabarti's measure of imbalance in experimental design. In Pukkila & Puntanen [358, pp. 323–347 (1985)].

[438] Thompson, R. C. On the singular values of matrix products, II. *Scripta Mathematica*, 29 (1973), 111–114.

[439] Thompson, Robert C. Root spreads for polynomials and Hermitian matrix pencils. *Linear Algebra and its Applications*, 220 (1995), 419–433.

[440] Thompson, W. A., Jr. A note on the balanced incomplete block designs. *The Annals of Mathematical Statistics*, 27 (1956), 842–846.

[441] Thompson, W. A., Jr. A note on P.B.I.B. design matrices. *The Annals of Mathematical Statistics*, 29 (1958), 919–922.

[442] Tiit, E.-M.; Kollo, T.; Niemi, H., eds. *Multivariate Statistics and Matrices in Statistics: Proceedings of the Fifth Tartu Conference, Tartu–Pühajärve, Estonia, 23–28 May 1994*. New Trends in Probability and Statistics 3, VSP International Science Publishers, Utrecht & TEV Ltd., Vilnius, 1995.

[443] Uhlig, Frank; Grone, Robert. *Current Trends in Matrix Theory: Proceedings of the Third Auburn Matrix Theory Conference, held March 19–22, 1986, at Auburn University, Auburn, Alabama, U.S.A.* North-Holland, New York, 1987.

[444] *United States Government Printing Office Style Manual*, Revised edition. United States Government Printing Office, Washington, D. C., 1959. (Original version: 1894.)

[445] van der Vorst, Henk A. *Iterative Krylov Methods for Large Linear Systems.* Cambridge Monographs on Applied and Computational Mathematics 13. Cambridge University Press, 2003.

[446] van der Waerden, B. L. Eine Bemerkung zur numerischen Berechnung von Determinanten und Inversen von Matrizen. *Jahresbericht der Deutschen Mathematiker Vereinigung*, 48 (1938), Notizen pp. 29–30.

[447] Van Trees, Harry L. *Detection, Estimation, and Modulation Theory: Part 1, Detection, Estimation, and Linear Modulation Theory*, 2nd edition. Wiley-Interscience, Hoboken, New Jersey, 2001. (Original version: Wiley, New York, 1968.)

[448] Visick, George. A quantitative version of the observation that the Hadamard product is a principal submatrix of the Kronecker product. *Linear Algebra and its Applications*, 304 (2000), 45–68.

[449] Vogt, Annette. Issai Schur: als Wissenschaftler vertrieben. In Schoeps, Grözinger & Mattenklott [401, pp. 217–235 (1999)].

[450] Walz, Guido. *Asymptotics and Extrapolation*. Mathematical Research 88. Akademie Verlag, Berlin, 1996.

[451] Wang, Bo-Ying. *Introduction to Majorization and Matrix Inequalities* (in Chinese). Beijing Normal University Press, 1990.

[452] Wang, Bo Ying; Zhang, Fuzhen. Some inequalities for the eigenvalues of the product of positive semidefinite Hermitian matrices. *Linear Algebra and its Applications*, 160 (1992), 113–118.

[453] Wang, Bo-Ying; Zhang, Fuzhen. Schur complements and matrix inequalities of Hadamard products. *Linear and Multilinear Algebra*, 43 (1997), 315–326.

[454] Watson, Geoffrey S. *Serial Correlation in Regression Analysis*. Ph.D. thesis, Dept. of Experimental Statistics, North Carolina State College, Raleigh, 1951.

[455] Watson, Geoffrey S. Serial correlation in regression analysis, I. *Biometrika*, 42 (1955), 327–341.

[456] Watson, Geoffrey S.; Alpargu, Gülhan; Styan, George P. H. Some comments on six inequalities associated with the inefficiency of ordinary least squares with one regressor. *Linear Algebra and its Applications*, 264 (1997), 13–53.

[457] Weltsch, Robert, ed. *Year Book XVIII*. Publications of the Leo Baeck Institute, Secker & Warburg, London, 1973.

[458] Weniger, Ernst Joachim. Nonlinear sequence transformations for the acceleration of convergence and the summation of divergent series. *Computer Physics Reports*, 10 (1989), 189–371.

[459] Wimp, Jet. *Sequence Transformations and their Applications.* Mathematics in Science and Engineering 154. Academic Press, New York, 1981.

[460] Wolf, Helmut. Die Stellung des Boltzschen Entwicklungs-Verfahrens in der Ausgleichungsrechnung nach bedingten Beobachtungen: Relation between the Boltz developing method and the computation of adjustment with condition equations. In Gigas [181, pp. 11–52 (1949)].

[461] Woodbury, Max A. *Inverting modified matrices.* Memorandum Report 42, Statistical Research Group, Institute for Advanced Study, Princeton, New Jersey, 6 pp., June 14, 1950.

[462] Yanai, Haruo. Vectors and matrices in psychometrics with special emphasis on generalized inverse and projection matrices. In Yanai *et al.* [463, pp. 19–32 (2003)].

[463] Yanai, H.; Okada, A.; Shigemasu, K.; Kano, Y.; Meulman, J. J., eds. *New Developments in Psychometrics: Proceedings of the International Meeting of the Psychometric Society IMPS2001 (Osaka, Japan, 2001).* Springer, Tokyo, 2003.

[464] Yates, Frank. Incomplete randomized blocks. *Annals of Eugenics*, 7 (1936), 121–139; reprinted in Yates [465, pp. 181–199 (1970)].)

[465] Yates, Frank. *Experimental Design: Selected Papers of Frank Yates, C.B.E., F.R.S.* Griffin, London, 1970.

[466] Yatracos, Yannis G. On prediction and mean squared error. *The Canadian Journal of Statistics*, 20 (1992), 187–200.

[467] Zhang, Fuzhen. Another proof of a singular value inequality concerning Hadamard products of matrices. *Linear and Multilinear Algebra*, 22 (1988), 307–311.

[468] Zhang, Fuzhen. *Matrix Theory: Basic Results and Techniques.* Springer-Verlag, New York, 1999.

[469] Zhang, Fuzhen. Schur complements and matrix inequalities in the Löwner ordering. *Linear Algebra and its Applications*, 321 (2000), 399–410.

[470] Zhang, Fuzhen. Matrix inequalities by means of block matrices. *Mathematical Inequalities & Applications*, 4 (2001), 481–490.

[471] Zhang, Fuzhen. A matrix identity on the Schur complement. *Linear and Multilinear Algebra*, in press (received 11 August 2003).

[472] Zielke, Gerhard. *Numerische Berechnung von benachbarten inversen Matrizen und linearen Gleichungssystemen.* Schriften zur Datenverarbeitung 2, Friedrich Vieweg & Sohn, Braunschweig, 1970.

[473] Zyskind, George. On canonical forms, non-negative covariance matrices and best and simple least squares linear estimators in linear models. *The Annals of Mathematical Statistics*, 38 (1967), 1092–1109.

Notation

$\mathbb{R}$	the real numbers
$\mathbb{R}^n$	column vectors of n real components
$\mathbb{C}$	the complex numbers
$\mathbb{C}^n$	column vectors of n complex components
$\mathbb{R}^{m\times n}$	$m \times n$ real matrices
$\mathbb{C}^{m\times n}$	$m \times n$ complex matrices
$\mathbb{H}_n$	$n \times n$ Hermitian matrices
TP	totally positive matrices
$\dim V$	dimension of vector space V
$\langle u, v\rangle$	inner product of vectors u and v
α^c	complement of index set α in $N = \{1, 2, \ldots, n\}$, i.e., $N \setminus \alpha$
$\|\alpha\|$	cardinality of index set α
$\\|x\\|$	norm or length of vector x
$\\|x\\|_p$	l_p norm of vector x, i.e., $\\|x\\|_{l_p} = \left(\sum_i \|x_i\|^p\right)^{\frac{1}{p}}$
I_n	identity matrix of order n
I	identity matrix when the order is implicit in the context
$A = (a_{ij})$	matrix A with entries a_{ij}
$\operatorname{rank}(A)$	rank of matrix A
$\operatorname{tr} A$	trace of matrix A
$\det A$	determinant of matrix A
A^{-1}	inverse of matrix A
A^T	transpose of matrix A
$\bar{A}$	conjugate of matrix A
A^*	conjugate transpose of matrix A, i.e., $A^* = \bar{A}^T$ or adjoint of operator A
$\|A\|$	absolute value of matrix A, that is, $\|A\| = (A^*A)^{1/2}$
$A^{1/2}$	square root of positive semidefinite A
A^-	a generalized inverse of matrix A, i.e., $AA^-A = A$
$A^\dagger$	Moore–Penrose generalized inverse of matrix A
$\operatorname{adj} A$	classical adjoint matrix of A
$A[\alpha, \beta]$	submatrix of matrix A determined by index sets α, β
$A(\alpha, \beta)$	submatrix of matrix A determined by index sets α^c, β^c
$A[\alpha]$	principal submatrix of matrix A indexed by α

$A(\alpha)$	principal submatrix of matrix A indexed by α^c
$A/A[\alpha,\beta]$	Schur complement of $A[\alpha,\beta]$ in A
A/α or $A/A[\alpha]$	Schur complement of $A[\alpha]$ in A
M/A	Schur complement of submatrix A in partitioned matrix M
$\mathcal{R}(A)$	row space of matrix A
$\mathcal{C}(A)$	column space of matrix A
$\mathcal{N}(A)$	null of A, i.e., $\mathcal{N}(A) = \{x : Ax = 0\}$
$\operatorname{Ker}(A)$ or $\ker(A)$	kernel of A, that is, $\{x : Ax = 0\}$
$\operatorname{Im}(A)$ or $\operatorname{ran}(A)$	image or range of A, that is, $\{Ax\}$
$\operatorname{In}(A)$	inertia of matrix A, i.e., $\operatorname{In}(A) = (p(A),\ q(A),\ z(A))$
$\lambda_{\max}(A)$ or $\lambda_1(A)$	largest eigenvalue of matrix A
$\sigma_{\max}(A)$ or $\sigma_1(A)$	largest singular value of matrix A
$\lambda_{\min}(A)$ or $\lambda_n(A)$	smallest eigenvalue of $n \times n$ matrix A
$d(A)$	vector of the main diagonal entries of matrix A
$\lambda(A)$	vector of the eigenvalues of matrix A
$\sigma(A)$	vector of the singular values of matrix A
$s(A)$	spectrum of A, i.e., the set of all eigenvalues of A
$A \geq 0$	A is positive semidefinite
$A > 0$	A is positive definite
$A \succeq 0$	A is nonnegative entrywise; that is, $a_{ij} \geq 0$
$A \geq B$	$A - B$ is positive semidefinite
$\operatorname{diag}(\lambda_1, \lambda_2, \ldots, \lambda_n)$	diagonal matrix with $\lambda_1, \lambda_2, \ldots, \lambda_n$ on the main diagonal
$\operatorname{diag} A$	diagonal matrix formed by the main diagonal entries of A
$A \oplus B$	direct sum of vector spaces or matrices A and B
$A : B$	parallel sum of A and B
$A \circ B$	Hadamard product of matrices A and B
$A \otimes B$	Kronecker product of matrices A and B
$\mathcal{H}$	Hilbert space
$\mathcal{M}^\perp$	orthocomplement of $\mathcal{M}$
$P_{\mathcal{M}}$	orthoprojection to the space $\mathcal{M}$
$A \wedge B$	$\operatorname{ran}(A) \cap \operatorname{ran}(B)$ is the empty set
$A/\mathcal{M}^\perp$	Schur complement
$[\mathcal{M}]A$	shorted operator, Schur complement

Index